Algebraic Cryptanalysis

Gregory V. Bard

Algebraic Cryptanalysis

 Springer

Gregory V. Bard
Department of Mathematics
Fordham University
Bronx, NY 10458
USA
gregory.bard@ieee.org

ISBN 978-1-4899-8450-0   ISBN 978-0-387-88757-9 (eBook)
DOI 10.1007/978-0-387-88757-9
Springer Dordrecht Heidelberg London New York

Printed on acid-free paper

Springer is part of Springer Science+Business Media (www.springer.com)

Preface

Algebraic Cryptanalysis is the process of breaking codes by solving polynomial systems of equations. In some ways this book began when the author began to explore cryptanalysis as a beginning graduate student, and realized with frustration that no book whatsoever existed on the topic. Since that time, some books have been written about Linear Cryptanalysis or Differential Cryptanalysis (e.g. [211] and [214] cover both), but none on Algebraic Cryptanalysis, which is a rich and growing field.

The author had some difficulty entering the field of Algebraic Cryptanalysis. Of course step one is a solid background in Abstract Algebra, and a solid background in cryptography[1]. But after these twin foundations, one is not quite ready to read research papers. This book is intended to be that stepping stone for graduate students wishing to do their dissertation in Algebraic Cryptanalysis, or any other part of cryptanalysis. Furthermore, researchers in other areas of Applied Abstract Algebra or cryptography might benefit from seeing what is going on in cryptanalysis.

The nucleus for the book was my dissertation, under the title "Algorithms for the Solution of Linear and Polynomial Systems of Equations over Finite Fields, with Applications to Cryptanalysis", submitted for the degree Doctor of Philosophy of Applied Mathematics and Scientific Computation, defended in the Summer of 2007, under the guidance of Professor Lawrence C. Washington. The author is *extremely* grateful for Prof. Washington's time, help and assistance at all stages.

In addition to being a text for graduate students, the author hopes the book will be also useful for those currently working in the field as well. The pressures of page counts often require that the internals or variants of algorithms cannot be published in exhaustive detail in the standard scientific literature. Here, we have explained and expanded upon several algorithms previously published by myself and by others. Often the details left out of a published paper are those required to make an algorithm work efficiently.

---

[1] If these subjects are not yet ones that the reader is comfortable with, the author recommends [216] for cryptography, and [119] for undergraduate Abstract Algebra.

The backbone of the theory of polynomial systems of equations, over any field, is algebraic geometry. This topic is exquisitely covered in *Ideals, Varieties, and Algorithms* by Cox, Little and O'Shea [86], also published by Springer-Verlag. The author therefore strongly encourages the reader to read [86] along with this text, but note that [86] is not, by any means, a prerequisite for this book. The topic of Gröbner Bases, in particular, is deferred to [86], because they do an exquisite job there.

Last but not least, a handy desk-reference for finite fields is the encyclopedia [165] by Lidl and Niederreiter. The previous edition of it, [164], was referred to as "the Bible" at several finite field conferences.

Why this Book was Written

Tradition in Applied Mathematics, particularly in the USA, dictates that the research of a doctoral dissertation be divided into journal articles, and published in the years immediately following the defense. However, there is an interesting category of work that is left in limbo. Original research can, of course, be published by normal means. But work that is "dug out of the dust", published but mostly forgotten, cannot be published again. For example, the proofs of the equi-complexity of matrix operations, have been known for a long time, but not published together in one place; the "degree dropper algorithm" (See Section 11.4 on Page 192) must have been known for decades, but the author could not find a proof of it anywhere; the Method of Four Russians for Multiplication was known anecdotally, but the original paper was in Russian [21] and the most recent textbook version found was from 1974 [13, Ch. 6]; much work has been published on SAT-solvers, and how they work, but there is no consolidated elementary introduction for those ignorant of the subject, as the author found himself at the start of this work; many of the algorithms of Nicolas Courtois, including ElimLin, are never fully and explicitly defined anywhere. The author only wishes to see these techniques and algorithms used. While the author is no master of exposition (as the reader is about to discover), he hopes that the space which a book affords has allowed him to render this topic more comprehensible, particularly to a graduate student audience, or even motivated undergraduates.

The author believes SAT-solvers, in particular, are a very underestimated tool. Other communities rely upon them as a computational engine of great power. It is hoped that the chapter on how SAT-solvers work will encourage scholars not to consider them as black-boxes. Furthermore, the author hopes that the two chapters on how to adapt polynomial systems of equations to be solved by SAT-solvers will stimulate research into new and unrelated applications for these techniques, such as solving combinatorial problems outside of cryptography such as graph-coloring. Toward this end we include an appendix introducing this connection.

Advice for Graduate Students

This book is primarily intended for those who are to embark on study in the field of algebraic cryptanalysis, particularly graduate students about to begin a dissertation or a masters thesis in that topic. The author therefore will present the following piece of advice: Read as many papers as possible.

Be sure to include some that are old, as old papers often have excellent ideas. Some of these are not electronically available. Reading a very long paper, in detail, verifying all the small steps with your pencil, is slow but important. There may be some papers which you cannot give this magnitude of time to. If so, it is better to read the first 4 pages of 10 papers than to only read the abstracts of 20 or 30.

With hope for the future,

Gregory V. Bard,
Visiting Assistant Professor,
Department of Mathematics,
Fordham University,
Bronx, New York, USA.

May 4th, 2009

Dedication

With pleasure, I dedicate this dissertation to my family. To my father, who taught me much in mathematics, most especially calculus, years earlier than I would have been permitted to see it. And to my mother and brother, who have encouraged me in every endeavor.

Acknowledgements

I am deeply indebted to my advisor, Professor Lawrence C. Washington, who has read and re-read many versions of this document, and almost every research paper I have written to date, provided me with countless hours of his time, and given me sound advice, on matters technical, professional and personal. He has had many students, and all I have spoken to are in agreement that he is a phenomenal advisor, and it is humbling to know that we can never possibly equal him. He is that rare type of scholar, who is both outstanding in teaching and in research.

I have been very fortunate to work with Dr. Nicolas T. Courtois, currently Senior Lecturer at the University College of London. It suffices to say that the field of algebraic cryptanalysis has been revolutionized by his work. He permitted me to study with him for four months in Paris, and later even to stay at his home in England, after his move there. Much of my work is joint work with him, including everything I have done with Keeloq and ElimLin, as well as my first paper on SAT-solvers. He has always given very freely of his ideas.

The SAGE community has been a true inspiration for me. The fact that so many highly-talented volunteers surrender so much of their free time to work unpaid on a non-for-profit project is heart warming. Knowing that my work is part of SAGE makes me know that my effort was not wasted. This is primarily due to William Stein, who created SAGE and guides the project. Furthermore, he has given me access to certain high-performance computing systems, without which my research would have been essentially impossible. I publicly thank him in particular, and all SAGE volunteers in general.

I would like to particularly thank Michael Abshoff (University of Washington, Seattle) for help with SAGE, and forgiving two unfortunate computational episodes; William Adams (University of Maryland at College Park, Ret.) for introducing me to Gröbner Bases, and for an excellent class in abstract algebra; Martin Albrecht (Royal Holloway College, University of London) for endless support in SAGE as well as finite field linear algebra; Bill Arbaugh (University of Maryland at College Park) for an excellent computer security class; Kostas Arkoudas (Rensselaer Polytechnic Institute) for suggesting I consider SAT-solvers; Shaun Ault (Fordham University) for helping me with analytic combinatorics; Joe Barth (now US Census

Bureau) for help in graduate school, especially with the Algebra Qualifier; Lynn Batten (Deakin University, Australia) for challenging me to prove Theorem 74 on Page 196 during a coffee break on July 10, 2007, at the conference $\mathbb{F}_{q^8}$, in Melbourne; Paul Bello (now Department of Defense) for inspiration, good conversations and career advice; Daniel Bernstein (Univerisity of Illinois at Chicago) for profuse and frequent technical as well as career advice, and numerous useful ideas; Mark Binfield (now Booz-Allen Hamilton) for many long conversations about math and economics; Michael Black (American University) for help with distributed systems, BOINC, SAT-solvers, and teaching advice; Melkana Brakalova (Fordham University) for teaching advice; Selmar Bringsjord (Rensselaer Polytechnic Institute) for teaching two classes in logic, that would turn out to be the foundation of my entire outlook on research and on life; Armand Brumer (Fordham University, Ret.) for technical advice, encouragement, and for reminding me how important computer algebra can be to working pure mathematicians; Ed Casimiro (University of Pheonix) for help with Chapter 6; Carlos Cid (Royal Holloway College, University of London) for career advice, and his excellent book on the AES which inspired this text; Abram Claycomb (now United States Air Force) for much help at RPI and later encouragement as well; Blandine Debraize (Gemalto) for her work in breaking SNOW with SAT-solvers; Theodore Faticoni (Fordham University) for helping me decide to publish this book in the first place; William Randolph Franklin (Rensselaer Polytechnic Institute) for being my undergraduate advisor and supervising my first cryptographic research project—on passwords; William Gasarch (University of Maryland at College Park) for algorithms advice and background on Gray Codes; Theresa Girardi (Fordham University) for much proofreading and guidance in number theory as well as teaching; Alex Golec (Fordham University) for help with programming and compilers; Janusz Golec (Fordham University) for technical advice relating to probability, and guidance in teaching; Virgil D. Gligor (University of Maryland at College Park) for encouraging my switch from engineering to mathematics; Michael Gray (American University) for hiring me, and guidance in matters of teaching; Carmi Gressel (Fortress) for introducing me to the world of patents, and ZK-Crypt; William Hart (University of Warwick) for help with linear algebra and advice related to publishing papers; Michael Headley (American University) for proofreading; Tanja Lange (Technische Universiteit Eindhoven) for practical and professional advice; Bill and Maryam Hastings (Fordham University) for career advice and encouragement; Sebastiaan Indesteege (Katholieke Universiteit Leuven in Belgium) for comments on Keeloq; Peter Jeavons (St Anne's College, Oxford University) for help with SAT problems and polynomials; Chris Jefferson (Oxford University Computing Laboratory) for much help with SAT-solvers; Antoine Joux (Université de Versailles) for help with Gray Codes and Butterfly Transposes; W. David Joyner (United States Naval Academy) for help with SAGE and career advice; Haig Kafafian† for introducing me to cryptography and to matrices; Igor Khalatian (now LiveLOOK, Inc.) for teaching me how to program in C, so many years ago; Jon Katz (University of Maryland at College Park) for two excellent cryptography classes, and guidance in my first cryptographic paper; Kyle Kloster (Fordham University) for proofreading and help with the quadratic sieve;

Susan Lagerstrom-Fife (Springer-Verlag) for patience and guidance in the last stages of the development of this manuscript; Godfrey H. L. Le May (Worcester College, Oxford University, Ret.) for encouragement; David Levermore (University of Maryland at College Park) for career guidance, linear algebra advice, and for admitting me to the PhD program in the first place; Michael Levin (American University) for proofreading and help with Darwinian Gradient Descent; Robert Lewis (Fordham University) for guidance in matters ranging from resultants to teaching, from the role of polynomial systems in chemistry to the role of the college instructor in setting standards; Robert Miller (University of Washington, Seattle) for help with graph theory and linear algebra; Ilya Mironov (Microsoft Research) for discussions about SAT-solvers; Leonard Nissim (Fordham University) for advice in teaching abstract algebra, and on the job hunt; Andrew Novocin (University of Montpellier) for many interesting discussions; Brennan O'Donnell (Fordham University) for vital funding; Sean O'Neil (The VEST Corporation) for help with generating algebraic normal forms, and advice on other topics; Jacques Patarin (Université de Versailles) for encouragement in my studies of both SAT-solvers and the French language; Kenneth Patterson (Royal Holloway College, University of London) for career advice and whose questions at the 2008 Workshop on Mathematical Cryptography in Santander, Spain, that encouraged me to rigorously describe the attack in Section 4.6 on Page 49; Clément Pernet (Université Joseph Fourier, Grenoble) for much useful advice on linear algebra over finite fields, and thinking up the pronunciation "Mary" for M4RI, a mercy to all francophones who will ever mention the library; Carl Pomerance (now Dartmouth University) for encouragement and help with sparse linear algebra; Cris Poor (Fordham University) for many conversations related to NP-Completeness and to teaching; Bill Singer (Fordham University) for advice in matters ranging from teaching to negotiation of the book publishing contract; Simon Schur (now University of Toronto) for letting me use his office; Richard Schwartz (now Brown University) for teaching me abstract algebra, way back when; Yana Shabaev (American University) for tender advice; Harold Snider (National Federation of the Blind, Ret.) for much encouragement, especially related to Oxford; Mate Soos (INRIA Rhone-Alpes) for information about MINISAT internals, and for proofreading Chapter 14; Albert Studdard (University of North Carolina at Pembroke, Ret.) for career advice, and guidance in matters of teaching; Buck Surdu (US Army) for many interesting conversations; Mark Tilmes (University of Maryland at College Park) for assistance, enthusiasm for parallel computing, and forgiving an unfortunate computational episode; Steven Tretter (University of Maryland at College Park) for introducing me to finite fields and their applications; Mak Trifkovic (now University of Victoria) for teaching and career advice; Seena Vali (Fordham University) for proofreading and help with sparse matrices; Christopher Wolf (now University of Bochum) for much advice; Kris Wolff (Fordham University) for help in writing grants; Koon Ho "Kenneth" Wong (University of Queensland Univeristy of Technology) for help with graph theory and partitioning polynomial systems; Angela Wu (American University) for guidance in matters of teaching; and Patrick Studdard, for everything.

The National Security Agency was my first full-time permanent job and I remain very grateful to them not only for the opportunity to serve my country but also for the training and exposure to new ideas that I had there. It was a happy four years for me, and I particularly fondly remember the awe-inspiring dedication its employees had for pushing ahead with agency's mission, even before September 11, 2001. It is a true "puzzle palace" where math nerds can congregate with their own kind and work on deep problems. Both tradition and law forbid me from naming specific individuals who have helped me or guided me, but there were several.

Several texts by Steven D. Krantz were consulted during this project. Principally they are *A Primer on Mathematical Writing* [150], *Mathematical Publishing: A Guidebook* [153], but also *A Mathematician's Survival Guide* [152]. Without his guidance my writing would be even worse than the reader is about to discover it to be. I strongly suggest any graduate student or recent PhD read those books, and also *How to Teach Mathematics* [151]. Note, the vast majority of the contents of these books apply well to Computer Science as well.

Several governmental agencies have contributed financial to my eduction or to my research. They are the National Science Foundation (Division of Mathematical Sciences[2]), the United States Navy (Naval Sea Systems Command), the National Security Agency, and ECRYPT, the European Union's Cryptographic research organization. They have my unending gratitude. I am also happy to thank Fordham University for two faculty research grants and an interdisciplinary seminar grant.

[2] The SAGE grants DMS-0555776 and DMS-0821725, and the University of Maryland at College Park VIGRE grant.

If you Find Any Errors...

Wherfore I trust thei that be learned, and happen to reade this worke, wil beare the moare with me, if thei finde any thyng, that thei doe mislike: Wherein if thei will use this curtesie, either by writynge to admonishe me thereof, either theim selfes to sette forthe a moare perfecter woorke, I will thynke them praise worthie.

Robert Recorde, 1557, quoted from [121, last page]. See also Appendix E on Page 337.

Contents

List of Tables

List of Figures

List of Algorithms

List of Abrreviations

$$f(x) = o(g(x)) \quad \lim_{x \to \infty} \frac{f(x)}{g(x)} = 0$$
$$f(x) = O(g(x)) \quad \exists c, n_0 \qquad \forall n > n_0 \qquad f(x) \leq cg(x)$$
$$f(x) = \Omega(g(x)) \quad \exists c, n_0 \qquad \forall n > n_0 \qquad f(x) \geq cg(x)$$
$$f(x) = \omega(g(x)) \quad \lim_{x \to \infty} \frac{g(x)}{f(x)} = 0$$

$$f(x) = \Theta(g(x)) \quad f(x) = O(g(x)) \text{ while simultaneously } f(x) = \Omega(g(x))$$

$$f(x) \sim g(x) \qquad \lim_{x \to \infty} \frac{f(x)}{g(x)} = 1$$

AES	Advanced Encryption Standard
ANF	Algebraic Normal Form
CPA	Chosen Plaintext Attack
CNF	Conjunctive Normal Form
CNF-SAT	the Conjunctive Normal Form Satisfiability Problem
DES	The Data Encryption Standard
DNF	Disjunctive Normal Form
GCD	Greatest Common Divisor
$\mathbb{GF}(q)$	the Galois (finite) Field of size q
$\mathbb{GL}_n(F)$	the group of $n \times n$ invertible matrices over the field F
HFE	Hidden Field Equations
LUP	Lower triangular-Upper triangular-Permutation (a matrix factorization)
KPA	Known Plaintext Attack
M4RM	The Method of Four Russians for matrix Multiplication
M4RI	The Method of Four Russians for matrix Inversion
$M_n(R)$	the ring of $n \times n$ matrices over the ring R
MC	The Multivariate Cubic problem
MQ	The Multivariate Quadratic problem
OWF	One-way Function
QUAD	The stream cipher defined in [42]
REF	Row Echelon Form
RPA	Random Plaintext Attack
RREF	Reduced Row Echelon Form
SAGE	Software for Algebra and Geometry Experimentation
SAT	the SATisfiability problem
UUTF	Unit Upper Triangular Form
XOR	The exclusive-OR

Chapter 1
Introduction: How to Use this Book

As stated earlier, the purpose of this book is to help graduate students who wish to learn something about algebraic cryptanalysis, or researchers from either other branches of cryptography, computer algebra, or finite fields who may wish to try their hand at code-breaking.

Citations for all the topics described here appear in the respective chapters, and therefore are omitted in the introduction, to preserve readability.

Part One

With that in mind, imagine the typical process of breaking a cipher algebraically. First, one must convert the cipher into a system of polynomial equations. Therefore, this is the topic of Part One of this book. The author has chosen the cipher Keeloq for several reasons. First off, it is a commercially used cipher, from the "real world". Second, it is broken completely, not in some "reduced rounds" version. Third, it is rather simple. Fourth, the direct process (of converting the cipher to equations and then solving it directly) was an absolute failure. Instead, one had to take an indirect approach, and use mathematical properties of the cipher, in order to accomplish anything[1] at all. In any case, Chapter 2 describes how to turn the cipher into polynomials, and Chapter 3 describes the attack from my dissertation, along with another attack. Then, in Chapter 4, we describe the analysis of iterated permutations through analytic combinatorics—a field that uses sequences and series to count objects. That chapter also includes an attack that we believe to be very novel, in which one can target any cipher that has been iterated a large composite number of times, but not one that has been iterated a large prime number of times. Finally, in Chapter 5, we discuss stream ciphers—in particular, QUAD (a family of stream ciphers), as well as Bivium-A, Bivium-B and Trivium.

---

[1] Many researchers who claim that their systems are immune to algebraic attacks may wish to dwell on this point in some detail.

G.V. Bard, *Algebraic Cryptanalysis*, DOI: 10.1007/978-0-387-88757-9_1
© Springer Science + Business Media, LLC 2009

The Keeloq attack in my dissertation was discovered in January of 2007, and much better ones have been published since. Many of those newer attacks are quite interesting, and the author is a coauthor of two papers containing these later attacks. The reader is encouraged to examine those and any other papers on Keeloq as a second stage. They are not included here because once one has seen one algebraic attack, in perhaps excessive detail, then one has a general idea of how the subject functions and one can move on to solving the polynomials themselves. Furthermore, the author has been careful to provide several references for each cipher, to aide the reader in moving from this text directly into the published literature.

Part Two

The second stage, now that a polynomial system of equations describes the cipher, is to solve the system of equations. Many methods of doing this relate to linear systems of polynomial equations. The XL algorithm and the ElimLin algorithm both have linear algebra at their core. Also, several Gröbner Bases finding algorithms, including F4 and F5 by Faugére, also involve linear algebra. Even if one is using SAT-solvers, the ElimLin algorithm is an excellent preprocessor, and so linear algebra is important. Furthermore, working with the linear systems allows us to practice with some of the properties of finite fields, in general.

The theorems of linear algebra are usually true over any field, but others require editing in the case of a finite field, or in the case of a field with non-zero characteristic. In any case, this Chapter 6 discusses a few aspects of finite field linear algebra, principally related to $\mathbb{GF}(2)$, and cryptanalysis. There are many aspects that the reader might find surprising.

Next we propose, in Chapter 7, a model for measuring the complexity of $\mathbb{GF}(2)$-matrix operations. For matrices over $\mathbb{R}, \mathbb{C}$, or $\mathbb{Q}$, usually the number of floating-point operations is counted. However, this makes no sense in our case, as there are no floating point operations. Instead, one could count the number of field operations—but field operations are quite fast (being a single gate). One could count the number of total instructions but then one has to worry about loop increments and other very tedious details of the implementation. Instead, we propose counting the number of matrix memory operations, meaning reads and writes to the original matrix. During that chapter, we also calculate the running times of many basic linear algebra operations.

In Chapter 8 we state the classical theorems that matrix inversion, matrix multiplication, LUP-factorization, matrix squaring, and triangular matrix inversion, are all Big-Θ of each other. For sure, this chapter can be skipped if the reader is not interested, and the material is somewhat disconnected from the rest of the book, but we believe it will be useful to go through them. These facts have been known for a long time, but the author could not find any source that had them all located together. Here, we prove these with maximum generality (i.e. over any field), and all of them in the same notation. This is done for several reasons. First, many times these tricks (used in proving the theorems) are actually useful in using an existing method of solving one problem by switching to another problem—one that the programmer already knows how to solve. Second, the theorems are really elegant and precise explorations of linear algebra. Third, this is excellent practice in the nuances of the complexity of linear algebra.

In particular, the Morgenstern exposition of the Baur-Strassen Theorem (see Theorem 53 on Page 121) is exquisite. By coincidence, as the author was working on the bibliography of this book during Eurocrypt'09, the paper [12] was being presented. In that paper, a proof is given that breaking RSA in a certain model is equivalent to factoring—and the model is remarkably similar to the model in Morgenstern's proof. In fact, several other papers (cited by [12]) have also used this "generic ring" model. Thus there is some connection to cryptography.

The proofs are by no means of theoretical interest only. The so-called "black-box" techniques from sparse linear algebra involve taking a rapid algorithm for multiplying a matrix times a vector, and from there, building all higher matrix operations. Instead, one could imagine some day having some sort of rapid oracle for some particular matrix operation. How would one construct the other matrix operations from this oracle? The proofs are constructive, and so specify exactly how that would be done. We hope that the reader will find our proofs stimulating. Nonetheless, this material is not strictly needed and if the reader finds it boring, the entire chapter can be skipped without harm to the understanding of the other chapters.

Chapter 9, contains two algorithms. The first is the Method of Four Russians for Multiplication, and has been known for a long time, e.g. it appears in [13, Ch. 6]. This runs in time $\Theta(n^3/\log n)$ instead of $\Theta(n^3)$. However, we perform detailed analysis that we are certain has not appeared before, and convert the algorithm from matrices over the boolean semiring into matrices over $\mathbb{GF}(2)$ (see Section 6.2 on Page 81 for a discussion of the distinction). We also introduce the Method of Four Russians for Inversion, the vague outline of which was shown to the author by Nicolas Courtois. He has stated that this algorithm has been known anecdotally in France for a long time, but we perform quite a bit of analysis. Also, this algorithm has been adopted into SAGE, the open source free software competitor to MAGMA, MAPLE, MATLAB, and MATHEMATICA.

In Chapter 10, we step away from algebraic cryptanalysis to discuss factoring. The motivation is that we have now invested many pages in studying linear systems of equations in $\mathbb{GF}(2)$, and now we can apply this to our main problem of breaking stream ciphers and block ciphers. However, linear systems over $\mathbb{GF}(2)$ are also very important in factoring, which in turn is the backbone of part of the cryptanalysis of RSA. Thus, going with the general theme of code-breaking, the author decided to touch on the very basics of the Linear Sieve and Quadratic Sieve (QS). This could have been the topic of an entire other book (e.g. see [54]), and so we only present the tip of the iceberg. Numerous enhancements have been made to the Quadratic Sieve and it also has been surpassed by the Number Field Sieve (NFS). However, the linear algebra step of the NFS is the same as the QS, and the NFS is easier to understand if one is familiar with the QS. We hope the reader will enjoy this side-trip, but again this material is not needed for understanding the other chapters and so can be skipped.

Part Three

In Part Three, we examine how to solve polynomial systems of equations, and this is the very heart of the book. The polynomial systems arise from ciphers in Part One. This is a very hard problem, and if any efficient (i.e. polynomial time) algorithm ever solves this problem, then $P = NP$, which would be a surprise. It is also noteworthy that these polynomials are obviously over finite fields and not the rational, real or complex numbers, and therefore these polynomials might have properties which are alien to those researchers who are perhaps more accustomed to traditional polynomials.

Accordingly, Chapter 11 discusses the properties of polynomials over these finite fields. This includes a discussion on why we should solve polynomials at all. Cryptanalysis is only one of several applications, and we briefly mention some other applications. The concept of a universal map is also introduced. Namely, we prove that any map from any finite set to any finite set can be considered as a polynomial system of equations. After that follows a discussion of the properties of these polynomials. Next, we prove several theorems relating to the fact that any polynomial system of equations can be written with degree 2, via the introduction of new variables. This is true not only over finite fields, but over any ring. We also provide some algorithms for doing this, and show that it is a polynomial time activity for any fixed degree of the input system.

The chapter continues with a discussion of the NP-Completeness of solving polynomial systems of equations. Here we have been excessively detailed because of the fact that any researcher with a strong background in NP-Completeness already knows all the results in that discussion. Therefore, we are compelled to assume that anyone who needs to read that section is one who is new to the NP-Completeness topic. We conclude the chapter with two minor discussions: measures of difficulty of particular cases of polynomial systems, and the role of guessing a few variables prior to computation. This is the "guess-and-determine" paradigm.

After this introduction, we proceed to survey the methods of solving polynomial systems of equations in Chapter 12. There are many methods, and the author hopes he has not omitted any. The methods of Nicolas Courtois have received additional attention, because they work quite well in practice. In fact, the XL and ElimLin algorithms, coupled with the traditional Buchberger algorithm or the newer Faugére algorithms F4 and F5, for finding a Gröbner Basis, or coupled with SAT-Solvers, are sufficient to solve all the systems of polynomial equations over finite fields that the author has ever successfully solved. Furthermore, many researchers see a dichotomy, that one can either take the math-heavy approach, and use Gröbner Bases, or the CS-heavy approach, using SAT-Solvers and XL. The author rejects this dichotomy. For example, ElimLin can be an excellent preprocessor for a Gröbner Bases approach, and so forth.

As mentioned in the preface, the theory of Gröbner Bases is very important to all of Algebraic Cryptanalysis. In fact, the general theory of polynomial systems of equations is far better understood when one becomes familiar with Gröbner Bases methods. The book *Ideals, Varieties and Algorithms* by Cox, Little and O'Shea [86] covers this topic exceptionally well, and is extremely easy to read. We therefore defer all discussion on Gröbner Bases to that text, and strongly encourage the reader to read at least Chapters 2–4 of it.

Also in Chapter 12 we present an application of graph theory, where polynomial systems that are "interconnected" by only a few variables can be greatly simplified. Essentially, we target systems of polynomial equations where there are two sets of equations, and only a few variables are in common between one set and the other. We provide an algorithm where systems with this property can have it discovered. By either guessing the variables that would snap the system in half, or by using resultants to explicitly find values of these variables, the system is then broken into two pieces, which hopefully can be solved more easily. Next, we describe Resultants, a subject which has not been heavily explored by cryptanalysts, but which has proven to be very useful in other areas of applied algebraic geometry. Three additional methods are added in there. The Raddum-Semaev method is very new, and different. The Zhang-zi algorithm is also new, but less has been written about it. Finally, homotopy methods, about which a great deal has been written, are presented but it should be noted that the author is relatively confident they cannot possibly be used in cryptanalysis.

The author is very excited about SAT-Solvers, but, he hopes he has not expended too many pages on this topic. In fact, three chapters are dedicated to this matter. First, we discuss in Chapter 13, how to approach polynomial systems of equations over $GF(2)$, with a SAT-Solver. In the next chapter, we describe how SAT-Solvers actually work inside. We would like to stress that they should not be viewed as black-boxes, though if one wishes to do so they can operate as black-boxes. After that, in Chapter 15, we approach finite fields of characteristic 2, up to size $GF(64)$. In fact, the technique for extension fields can also be applied to Gröbner Bases solutions, not just SAT-solvers, and essentially projects a system of equations from $GF(2^n)$ down to $GF(2)$.

Appendices

There are five appendices. The first is a discussion on block ciphers with very short plaintext inputs, but normal-sized keys. This is more of a philosophical discussion of what is relevant, and what is not, what is common and what is not, and what "faster than brute force" really means. We segregate it from the rest of the text because it is mostly opinion, though backed by evidence, and not a sequence of provable theorems.

The second appendix is the collection of equations used in Chapter 15, to convert the multiply operation over $\mathbb{GF}(2^n)$ into that over $\mathbb{GF}(2)$. We recommend that if needed, they be scanned with Optical Character Recognition, and not typed in manually.

The third appendix is a discussion of how to use the methods of this text to solve four interesting problems: graph coloring, radio channel assignment, and register allocation during compiler optimization, as well as lecture hall scheduling. The latter three are just special cases of the former. We also very briefly discuss interval graphs.

The fourth appendix is a series of references for sparse matrix algorithms, along with a description of Carl Pomerance's "Created Catastrophes" algorithm, which others have called "structured Gaussian Elimination."

The fifth appendix is a small subset of the collection of quotes that were found in my dissertation. It is traditional among older American textbooks to begin the text or even each chapter with a witty quote or quotes. For example, Steve Krantz in [151] and von zur Gathen & Gerhard in [121] continue to do this in the present day. The author likes this habit, but did not have a full enough supply of quotes. But these are really meant to be truly inspirational, and the author hopes that the reader will actually read them.

Suggested Chapter Ordering

Most readers will have to read the book twice. If the reader comes to this topic with foreknowledge of it, then you can read the chapters in any order, of course. But in order to grasp why we wish to do things various ways in the polynomial systems and linear algebra parts of the book, one must know how algebraic cryptanalysis will proceed. Likewise, when reading about algebraic cryptanalysis, all sorts of decisions are made in the process of converting a cipher into a system of equations, in order that the linear and polynomial systems will be efficiently solvable. Therefore, the "math part" and the "crypto part" depend heavily on each other. The best strategy for a beginner is to skim each part (perhaps skipping Chapter 4 and Chapter 8) and then go back to the beginning, and read each chapter in detail.

Theorem Numbering

A brief note is needed, before we continue. The theorems, lemmas, facts, and definitions in this book are numbered sequentially together. That is, if a lemma follows Theorem i, it is Lemma $i + 1$. That means there is no Theorem $i + 1$ or Lemma i. If a theorem follows these, it will be Theorem $i + 2$. This numbering is meant to facilitate cross-referencing, and was recommended by Steven G. Krantz in [150].

Part I
Cryptanalysis

Chapter 2
The Block Cipher Keeloq and Algebraic Attacks

The purpose of this chapter is to supply a (relatively) new, feasible, and economically relevant example of algebraic cryptanalysis. The block cipher "Keeloq"[1] has been used in the remote keyless-entry system of many automobiles. It has a secret key consisting of 64 bits, takes a plaintext of 32 bits, and outputs a ciphertext of 32 bits. The cipher consists of 528 rounds. In this chapter, we define the cipher. We also show some "frontal assaults" that are not effective. In the next chapter, we describe a successful attack from the author's dissertation [31, Ch. 2]. Our attack is faster than brute force by a factor of around $2^{14.77}$ as shown in Section 3.5 on Page 24. A summary of the attack is given in Section 3.6 on Page 24.

Many other attacks on Keeloq are known, discovered since this attack of mine was first written back in January of 2007. In fact, it seems clear from the dates of publication of other attacks, that work on Keeloq was simultaneous among the several research teams involved (see Section 3.8.1 on Page 27). The purpose here is not to describe all or even some of the attacks on Keeloq but to give the reader a non-trivial but straight-forward example of algebraic cryptanalysis succeeding on a real-world cipher.

Notational Convention

For any ℓ-bit sequence, the least significant bit is numbered 0 and the most significant bit is numbered $\ell - 1$.

---

[1] This is to be pronounced "key lock." Some authors typeset it "KeeLoq" but we find the presence of a capital letter in the middle of a word to be offensive to the eye.

G.V. Bard, *Algebraic Cryptanalysis*, DOI: 10.1007/978-0-387-88757-9_2
© Springer Science + Business Media, LLC 2009

2.1 What is Algebraic Cryptanalysis?

Given a particular cipher, algebraic cryptanalysis consists of two steps. First, one must convert the cipher and possibly some supplemental information (e.g. file formats) into a system of polynomial equations, usually over $\mathbb{GF}(2)$, but sometimes over other rings. Second, one must solve the system of equations and obtain from the solution the secret key of the cipher. This chapter deals with the first step only. The systems of equations were solved with SINGULAR [9], MAGMA [2], and with the SAT-solver techniques of Chapter 13, as well as ElimLin, an algorithm by Nicolas Courtois described in Section 12.5 on Page 219.

2.1.1 The CSP Model

In any constraint satisfaction problem, there are several constraints in several variables. A solution must satisfy all constraints, and there might be zero, one, or more than one solution. The constraints are models of a cipher's operation, representing known facts as equations. Most commonly, this includes μ plaintext-ciphertext pairs, $P_1, \ldots, P_\mu$ and $C_1, \ldots, C_\mu$, and the μ facts: $E(P_i) = C_i$ for all $i \in \{1, \ldots, \mu\}$. The key is represented by one unknown for each bit. Almost always there are additional constraints and variables besides these.

If no false assumptions are made, then because these messages were indeed sent, we know there must be some key that was used, and so at least one key satisfies all the constraints. And so it is either the case that there are one, or more than one solution. Generally, algebraic cryptanalysis consists of writing enough constraints to reduce the number of possible keys to one, and few enough constraints that the system is solvable in a reasonable amount of time. In particular, the entire process should be faster than brute force by some margin.

2.2 The Keeloq Specification

In Figure 2.1 on Page 11, the diagram for Keeloq is given. The top rectangle is a 32-bit shift-register. It initially is filled with the plaintext. At each round, it is shifted one bit to the right, and a new bit is introduced. The computation of this new bit is the heart of the cipher.

Five particular bits of the top shift-register are tapped and are interpreted as a 5-bit integer, between 0 and 31. Then a non-linear function is applied, which will be described shortly (denoted NLF).

Meanwhile the key is placed initially in a 64-bit shift-register, which is also shifted one bit to the right at each iteration. The new bit introduced at the left is the bit formerly at the right, and so the key is merely rotating.

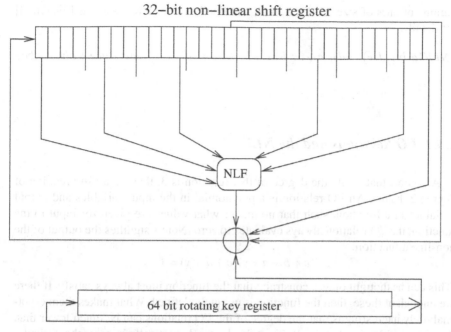

Fig. 2.1 The Keeloq Circuit Diagram

The least significant bit of the key, the output of the non-linear function, and two particular bits of the 32 bit shift-register are XORed together (added in $\mathbb{GF}(2)$). The 32-bit shift-register is shifted right and the sum is now the new bit to be inserted into the leftmost spot in the 32-bit shift-register.

After 528 rounds, the contents of the 32 bit shift-register form the ciphertext. Observe that only one bit of the key is used per round.

2.3 Modeling the Non-linear Function

The non-linear function $NLF(a,b,c,d,e)$ is denoted NLF_{3A5C742E}. This means that if (a,b,c,d,e) is viewed as an integer i between 0 and 31, i.e. as a 5-bit number, then the value of $NLF(a,b,c,d,e)$ is the ith bit of the 32-bit hexadecimal value 3A5C742E. Recall, the least significant bit is numbered as 0.

The following formula is a cubic polynomial and gives equivalent output to the NLF for all input values, and was obtained by a Karnaugh map. In this case, the Karnaugh map is a grid with (for five dimensions) two variables in rows (i.e. 4 rows), and three variables in columns (i.e. 8 columns). The rows and columns are arranged via the Gray Code. This is a simple technique to rapidly arrive at the algebraic normal form (i.e. polynomial), listed below, by first trying to draw boxes around regions of ones of size 32, 16, 8, 4, 2, and finally 1. See a text such as [38, Ch. 3] for details.

$$NLF(a,b,c,d,e) = d + e + ac + ae + bc + be + cd + de + ade + ace + abd + abc$$

2.3.1 I/O Relations and the NLF

Also note that while the degree of this function is 3, there is an I/O relation of degree 2, below. An I/O relation is a polynomial in the input variables and output variables of a function, such that no matter what values are given for input to the function, the I/O relation always evaluates to zero. Note y signifies the output of the non-linear function.

$$(e + b + a + y)(c + d + y) = 0$$

This can be thought of as a constraint that the function must always satisfy. If there are enough of these, then the function is uniquely defined. What makes them cryptanalyticly interesting is that the degree of the I/O relations can be much lower than the degree of the function itself. Since the degree dramatically impacts the difficulty of solving the polynomial system, this is very useful. The I/O degree of a function is the lowest degree of any of its I/O relations, other than the zero polynomial.

Generally, I/O-relations of low degree can be used for generating attacks but that is not the case here, because we have only one relation. Heuristically, relations that are valid with low probability for a random function and random input produce a more rapid narrowing of the keyspace in the sense of a Constraint Satisfaction Problem or CSP. We are unaware of any attack on Keeloq that uses this I/O-relation.

An example of the possibility of using I/O degree to cryptanalytic advantage is the attack from the author's joint paper on DES (the Data Encryption Standard), with Nicolas T. Courtois, where the S-Boxes have I/O degree 2 but their actual closed-form formulas are of higher degree [76].

2.4 Describing the Shift-Registers

The shift-registers that appear in Keeloq are an example of a very common, almost ubiquitous circuit design block in both stream ciphers (see Section 5.1.1.1 on Page 55) and block ciphers.

2.4.1 Disposing of the Secret Key Shift-Register

The 64-bit shift-register containing the secret key rotates by one bit per round. Only one bit per round (the rightmost) is used during the encryption process. Furthermore, the key is not modified as it rotates. Therefore, the key bit being used is the same in round $t, t+64, t+128, t+192, \ldots$

Therefore we can dispose of the key shift-register entirely. Denote $k_{63}, \ldots, k_0$ the original secret key. The key bit used during round t is merely $k_{t-1 \bmod 64}$.

2.4.2 Disposing of the Plaintext Shift-Register

Denote the initial condition of this shift-register as $L_{31}, \ldots, L_0$. This corresponds to the plaintext $P_{31}, \ldots, P_0$. Then in round 1, the values will move one place to the right, and a new value will enter in the first bit. Call this new bit L_{32}. Thus the bit generated in the ith round will be L_{31+i}, and in the 528th and thus last round will be L_{559}. The ciphertext is the final condition of this shift-register, which is $L_{559}, \ldots, L_{528} = C_{31}, \ldots, C_0$, likewise $L_{31}, \ldots, L_0 = P_{31}, \ldots, P_0$.

Change of Indexing

A change of indexing is useful here. The computation of L_i, for $32 \le i \le 559$, occurs during the round numbered $t = i - 31$. Thus the key bit used during the computation of L_i is $k_{i-32 \bmod 64}$.

2.5 The Polynomial System of Equations

This now gives rise to the following system of equations.

$$L_i = P_i \qquad\qquad \forall i \in [0,31]$$

$$L_i = k_{i-32 \bmod 64} + L_{i-32} + L_{i-16} \qquad \forall i \in [32,559]$$
$$ + NLF(L_{i-1}, L_{i-6}, L_{i-12}, L_{i-23}, L_{i-30})$$

$$C_{i-528} = L_i \qquad\qquad \forall i \in [528,559]$$

Note, some descriptions of the cipher omit the $+L_{i-16}$. This should have no impact on the attack at all. Most research papers and the specification given by the company [94] includes the $+L_{i-16}$, as does our diagram; some early papers omitted it.

Since the NLF is actually a cubic function this is a cubic system of equations. Substituting, we obtain

$$L_i = P_i \qquad\qquad\qquad\qquad \forall i \in [0,31]$$

$$\begin{aligned}
L_i = {} & k_{i-32} \bmod 64 + L_{i-32} + L_{i-16} + L_{i-23} + L_{i-30} \\
& + L_{i-1}L_{i-12} + L_{i-1}L_{i-30} + L_{i-6}L_{i-12} + L_{i-6}L_{i-30} \qquad \forall i \in [32,559] \\
& + L_{i-12}L_{i-23} + L_{i-23}L_{i-30} + L_{i-1}L_{i-23}L_{i-30} \\
& + L_{i-1}L_{i-12}L_{i-30} + L_{i-1}L_{i-6}L_{i-23} + L_{i-1}L_{i-6}L_{i-12}
\end{aligned}$$

$$C_{i-528} = L_i \qquad\qquad\qquad\qquad \forall i \in [528,559]$$

In other words, the above equations are to be repeated for each i in the stated intervals, and for each of μ total plaintext-ciphertext message pairs. In practice, $\mu \geq 2$ was both necessary and sufficient.

2.6 Variable and Equation Count

Consider a plaintext-ciphertext pair $\mathbf{P}, \mathbf{C}$. There are 560 equations, one for each L_i, with $i \in [0,559]$, plus another 32 for the C_i, with $i \in [0,32]$. However, the first 32 of these are of the form $L_i = P_i$ for $i \in [0,32]$, and the last 32 of these are of the form $L_i = C_{i-528}$ for $i \in [528,559]$. Thus we can use string substitution and drop down to 528 equations. This is precisely one equation for each round, which defines the new bit introduced into the shift register.

The 64 bits of the key are unknown. Also, of the 560 L_i, the first and last 32 are known, but the inner 496 are not. This yields 560 unknowns. If there are μ plaintext-ciphertext message pairs, then there are 528μ equations. However, there are only $496\mu + 64$ variables, because the key does not change from pair to pair.

2.7 Dropping the Degree to Quadratic

The following is a specific application of the more general technique found in Section 11.4 on Page 192. We will change the system from cubic to quadratic by introducing a few variables. Instead of the previously derived

$$NLF(a,b,c,d,e) = d + e + ac + ae + bc + be + cd + de + ade + ace + abd + abc$$

one can write

$$NLF(a,b,c,d,e) = d + e + ac + \beta + bc + be + cd + de + d\beta + c\beta + \alpha d + \alpha c$$
$$\alpha = ab$$
$$\beta = ae$$

Since the non-linear function was the sole source of non-linear terms, this gives rise to a quadratic rather than cubic system of equations.

This introduces two new variables per original equation, and two new equations as well. Thus m equations and n variables becomes $3m$ equations and $n + 2m$ variables. Therefore with μ plaintext-ciphertext message pairs, we have 1584μ equations and $1552\mu + 64$ variables. Thus, it must be the case that $\mu > 1$ for the system to be expected to have at most one solution. As always with algebraic cryptanalysis, unless we make an assumption that is false, we always know the system of equations has at least one solution, because a message was indeed sent. And thus we very strongly expect to have a unique solution when $\mu > 1$.

$$
\begin{aligned}
L_i &= P_i && \forall i \in [0,31] \\
L_i &= k_{i-32 \bmod 64} + L_{i-32} + L_{i-16} + L_{i-23} + L_{i-30} \\
&\quad + L_{i-1}L_{i-12} + \beta_i + L_{i-6}L_{i-12} + L_{i-6}L_{i-30} + L_{i-12}L_{i-23} \\
&\quad + L_{i-23}L_{i-30} + \beta_i L_{i-23} + \beta_i L_{i-12} + \alpha_i L_{i-23} + \alpha_i L_{i-12} && \forall i \in [32,559] \\
\alpha_i &= L_{i-1}L_{i-6} && \forall i \in [32,559] \\
\beta_i &= L_{i-1}L_{i-30} && \forall i \in [32,559] \\
C_{i-528} &= L_i && \forall i \in [528,559]
\end{aligned}
$$

Even with $\mu = 2$ this comes to 3168 equations and 3168 unknowns, well beyond the threshold of size for feasible polynomial system solving at the time this book was written, in late 2007.

2.8 Fixing or Guessing Bits in Advance

This is a specific application of the guess-and-determine methodology which is very common in many forms of cryptanalysis. For a general discussion, see Section 11.7 on Page 206, Section 12.4.4 on Page 217, or Section 11.7.2 on Page 207.

Sometimes in Gröbner basis algorithms or the XL algorithm, one fixes bits in advance [71, et al]. For example, in $\mathbb{GF}(2)$, there are only two possible values. Thus if one designates g particular variables, there are 2^g possible settings for them, but one needs to try $2^g/2$ on average (if we know exactly one solution exists). Naturally, fewer guesses would be needed if we merely search for one of many solutions. For each guess, one rewrites the system of equations either by substituting the guessed values, or if not, then by adding additional equations of the form: $k_1 = 1, k_2 = 0, \dots$. If the resulting Gröbner Bases method or XL method running time is more than $2^g/2$ times faster as a result of this change, this is a profitable change.

In academic cryptanalysis however, one generates a key, encrypts μ messages, and writes equations based off of the plaintext-ciphertext pairs and various other constraints and facts. Therefore, one knows the key. Instead of guessing all 2^g possible values, we simply guess correctly. However, two additional steps must be required. First, we must adjust the final running time by a factor of 2^g in the worse-case, or $2^g/2$ in the average case. Second, we must ensure that the system identifies a wrong guess as fast, or faster, than solving the system in the event of a correct guess. See Section 11.7.1 on Page 206 for more details on this point.

2.9 The Failure of a Frontal Assault

First we tried a simple CSP. With μ plaintext messages under one key, for various values of μ we encrypted and obtained ciphertexts, and wrote equations as described already, in Section 2.7 on Page 15. We also used fewer rounds than 528, to see the impact of the number of rounds, as is standard. The experiments were an obvious failure, and so we began to look for a more efficient attack, presented in the next chapter. Note, the computer involved was a 1 GHz Intel PC with 1 gigabyte of RAM.

- With 64 rounds, and $\mu = 4$, and 10 key bits guessed, SINGULAR required 70 seconds, and ElimLin in 10 seconds.
- With 64 rounds, and $\mu = 2$ but the two plaintexts differing only in one bit (the least significant), SINGULAR required 5 seconds, and ElimLin 20 seconds. MIN-ISAT[6] [104], using the techniques of Chapter 13, required 0.19 seconds. Note, it is natural that these attacks are faster, because many internal variables during the encryption will be identically-valued for the first and second message.
- With 96 rounds, $\mu = 4$, and 20 key bits guessed, MINISAT and the techniques of Chapter 13, required 0.3 seconds.
- With 128 rounds, and $\mu = 128$, with a random initial plaintext and each other plaintext being an increment of the previous, and 30 key bits guessed, ElimLin required 3 hours.
- With 128 rounds, and $\mu = 2$, with the plaintexts differing only in the least significant bit, and 30 key bits guessed, MINISAT requires 2 hours.

These results on 128 rounds are slower than brute-force. Therefore we did not try any larger number of rounds or finish trying each possible combination of software and trial parameters. Needless to say the 528 round versions did not terminate. Therefore, we needed a new attack, and this is the topic of the next chapter.

Chapter 3
The Fixed-Point Attack

3.1 Overview

Here the author will explain the attack that was the opening of his dissertation. The goal is first, to rewrite the cipher as a function g executed on the output of a function f which is iterated 8 times on the plaintext. Next, we will strip away g, and try to find fixed points of the 8th iterate of f. Some of these are fixed points of f. Using fixed points of f, we can recover the secret key by solving a polynomial system of equations. As you can see, the attack is rather indirect.

3.1.1 Notational Conventions

Evaluating the function f eight times will be denoted $f^{(8)}$.

If $h(x) = x$ for some function h, then x is a fixed point of h. If $h(h(x)) = x$ but $h(x) \neq x$ then x is a "point of order 2" of h. In like manner, if $h^{(i)}(x) = x$ but $h^{(j)}(x) \neq x$ for all $j < i$, then x is a "point of order i" of h. Obviously if x is a point of order i of h, then

$$h^{(j)}(x) = x \text{ if and only if } i | j$$

3.1.2 The Two-Function Representation

Recall that each 64th round uses the same key bit. In other words, the same bit is used in rounds $t, t+64, t+128, \ldots$. Note further, $528 = 8 \times 64 + 16$. Thus the key bits $k_{15}, \ldots, k_0$ are used nine times, and the key bits $k_{63}, \ldots, k_{16}$ are used eight times.

With this in mind, it is clear that the operation of the cipher can be represented as

G.V. Bard, *Algebraic Cryptanalysis*, DOI: 10.1007/978-0-387-88757-9_3
© Springer Science + Business Media, LLC 2009

$$E_k(\mathbf{P}) = g_k(\underbrace{f_k(f_k(\cdots f_k}_{8 \text{ times}}(\mathbf{P})) = g_k(f_k^{(8)}(\mathbf{P})) = \mathbf{C}$$

where the f_k represents 64 rounds, and the g_k the final 16 "extra" rounds.

3.1.3 Acquiring an $f_k^{(8)}$-oracle

Suppose we simply guess the 16 bits of the key denoted $k_{15}, \ldots, k_0$. Of course, we will succeed with probability only 2^{-16}. But at that point, we can evaluate g_k or its inverse g_k^{-1}. Then,

$$g_k^{-1}(E_k(\mathbf{P})) = g_k^{-1}(g_k(f_k^{(8)}(\mathbf{P}))) = f_k^{(8)}(\mathbf{P})$$

and our oracle for E_k now gives rise to an oracle for $f_k^{(8)}$.

3.2 The Consequences of Fixed Points

For the moment, assume we find x and y such that $f_k(x) = x$ and $f_k(y) = y$. How these are obtained will be explained later, but for now assume such points are known. At first, this seems strange to discuss at all. Because $f_k(x) = x$ and therefore $f_k^{(8)}(x) = x$, we know $E_k(x) = g_k(f_k^{(8)}(x)) = g_k(x)$. But, $g_k(x)$ is part of the cipher that we can remove by guessing a quarter (16 bits) of the key. Therefore, if we "know something" about x we know something about multiple internal points, the input, and output of $E_k(x)$. Now we will make this idea more precise.

Intuitively, we now know 64 bits of input and 64 bits of output (32 bits each from each message) of the functions f_k and $f_k^{(8)}$ as well. This forms a very rigid constraint, and it is highly likely that only one key could produce these outputs. This means that if we solve the system of equations for that key, we will get exactly one answer, which is the secret key. The only question is if the system of equations is rapidly solvable or not.

The resulting system must have equations for the 64 rounds of f. For both of x and y, there are equations for $L_0, \ldots, L_{95}$ and 32 additional output equations, but the first 32 of these and last 32 of these (in both cases) are of the forms $L_i = x_i$ and $L_{i-64} = x_i$, and can be eliminated by substituting, as discussed in Section 2.6 on Page 14. Thus there are actually $96 + 32 - 32 - 32 = 64$ equations (again one per round) for both x and y, and thus 128 total equations. We emphasize that this is the same system of equations as Section 2.5 on Page 13 but with only 64 rounds for each message. More precisely $i \in [0, 95]$ not $i \in [0, 559]$, but otherwise the equations are unchanged.

The x_i's and y_i's are known. Thus the unknowns are the 64 bits of the key, and the 32 "intermediate" values of L_i for both x and y. This is 128 total unknowns.

After translating from cubic into quadratic format, it becomes 384 equations and 384 unknowns. This is much smaller than the 3168 equations and 3168 unknowns we had before. In each case, ElimLin, MAGMA [2], SINGULAR[9], and the SAT-solver methods of Chapter 13 solved the system for $k_0, \ldots, k_{63}$ in time too short to measure accurately (i.e. less than 1 minute).

It should be noted that we required two fixed points, not merely one, to make the attack work. One fixed point alone is not enough of a constraint to narrow the keyspace sufficiently. However, two fixed points was sufficient each time it was tried. Therefore, we will assume f has two or more fixed points, and adjust our probabilities of success accordingly. One way to look at this is to say that only those keys which result in two or more fixed points are vulnerable to our attack. However, since the key changes rapidly in most applications (See Section 3.7.1 on Page 25), and since approximately 26.42% of random functions $\mathbb{GF}(2)^{32} \to \mathbb{GF}(2)^{32}$ have this property (see Section 2 on Page 22), we do not believe this to be a major drawback. This is calculated without analytic combinatorics in the author's dissertation [31, Ch. 2.4].

3.3 How to Find Fixed Points

Obviously a fixed point of f_k is a fixed point of $f_k^{(8)}$ as well, but the reverse is not necessarily true. Stated differently, the set of fixed points of $f_k^{(8)}$ will contain the set of all fixed points of f_k.

We will first calculate the set of fixed points of $f_k^{(8)}$, which will be very small. We will try the attack given in the previous subsection, using every pair of fixed points. If it is the case that f_k has two or more fixed points, then one such pair of points of $f^{(8)}$ that we try will indeed be a pair of fixed points of f_k. This will produce the correct secret key. The other pairs will produce spurious keys or inconsistent systems of equations. But this is not a problem because spurious keys can be easily detected and discarded, by a few test encryptions.

The running time required to solve the system of equations is too short to accurately measure, with a valid or invalid pair. Recall, that this is 384 equations and 384 unknowns as compared to 3168, as explained at the end of the previous section.

There are probably very few fixed points of $f_k^{(8)}$, which we will prove below. And thus the running time of the entire attack depends only upon finding the set of fixed points of $f_k^{(8)}$. One approach would be to iterate through all 2^{32} possible plaintexts, using the $f_k^{(8)}$ oracle. This would clearly uncover all possible fixed points of $f_k^{(8)}$ and if f_k has any fixed points, they would be included. However, this is not efficient.

Instead, one can simply try plaintexts in sequence using the $f_k^{(8)}$ oracle. When the ith fixed point x_i is found, one tries the attack with the $i-1$ pairs $\{(x_1, x_i), (x_2, x_i), \ldots, (x_{i-1}, x_i)\}$. If two fixed points of f_k are to be found in $x_1, \ldots, x_i$, the attack will

succeed at this point, and we are done. Otherwise, continue until x_{i+1} is found and
try the pairs $\{(x_1, x_{i+1}), (x_2, x_{i+1}), \ldots, (x_i, x_{i+1})\}$, and so forth.

INPUT: Oracular access to $E_k(x)$ for some unknown k, a randomly generated secret key.
OUTPUT: The secret key k with probability $2^{-17.92}$, or "abort" in all other cases.

1: Guess the 16 bits of the key $k_0, k_1, \ldots, k_{15}$.
2: Define the $f^{(8)}$ oracle to be $f_k^{(8)} \overset{def}{=} g_k^{-1}(E_k(x))$.
3: $P \leftarrow \{\}$
4: For $x = 0 \ldots 2^{32} - 1$ do

    a. If $f_k^{(8)}(x) = x$ then do
        i. For each $y \in P$ do
            A. Write equations assuming $f_k(x) = x$ and $f_k(y) = y$.
            B. Try to solve those equations.
            C. If the equations yield a key k', see if $E_k(x) = g_{k'}(x)$ and $E_k(y) = g_{k'}(y)$.
               • If YES: Halt and report k' is the secret key.
               • If NO: do nothing.
        ii. Insert x into P.

5: Abort.

Algorithm 1: The Fixed Point Attack on Keeloq [G. Bard]

3.4 How far must we search?

The question of how far one must search, in looking for fixed points of $f^{(8)}$,
before one can find 2 fixed points of f, is crucial for determining the running time
of the attack. Here, we present two ways of looking at this.

First, we will scan $\eta 2^{32}$ plaintext-ciphertext pairs, or a fraction η of the entire
codebook. Then, we will only keep fixed points of $f^{(8)}$. A subset of these will be
fixed-points of f. So for each pair of fixed points of $f^{(8)}$ we will "hope" that it
is a pair of fixed-points of f, and then try to solve our system of equations. Our
"hope" will succeed eventually if f has 2 fixed points inside the subset of the $\eta 2^{32}$
plaintext-ciphertext pairs which we searched, and if not, it will not. We use analytic
combinatorics to calculate precisely how long we should anticipate that this will
take. This is what was done in [75].

The other way of looking at this is essentially to set $\eta = 0.6$, and calculate the
probability of success. This is what was done in the author's dissertation [31, Ch.
2].

3.4.1 *With Analytic Combinatorics*

As stated before, we require two fixed points of f to write the system of polynomial equations that will yield a key for us. The question becomes how to obtain those pairs.

First, we will query $\eta 2^{32}$ random plaintext-ciphertext pairs, or a fraction η of the entire codebook. After guessing 16 bits of the secret key, g_k^{-1} can be used and applied to the codebook. That allows for (p,c), the plaintext-ciphertext pairs to be replaced by $(p, g_k^{-1}(c))$ which are now actually $(p, f_k^{(8)}(p))$.

Among these, if any have the same thing on both sides of the comma, then these are points fixed by $f_k^{(8)}$ and so by Corollary 13, they are points of order $\{1,2,4,8\}$ for f_k. Thus, the fixed points of f_k, which are useable for the cryptanalysis, are a subset of those for $f_k^{(8)}$, which we can find.

Theorem 1. *Let π be a random permutation from S_n. The probability that π has c_1 fixed points and c_2 cycles of lengths 2, 4, or 8, is given by*

$$\frac{1}{c_1! c_2!} \left(\frac{7}{8}\right)^{c_2} e^{-15/8}$$

in the limit as $n \to \infty$,

Proof. Note that the set of permutations on n elements, with c_1 fixed points, and c_2 cycles of length 2, 4, or 8, can be thought of as a triple labelled product. The first item in the product is from $\mathscr{P}^{(\{1\}, c_1)}$, the second item from $\mathscr{P}^{(\{2,4,8\}, c_2)}$, and the third item from $\mathscr{P}^{\overline{\{1,2,4,8\}}}$. We must now calculate the EGF.

The first item has $\alpha(z) = z$, and $\beta(z) = z^{c_1}/c_1!$, for an EGF of $\beta(\alpha(z)) = z^{c_1}/c_1!$. The second item has $\alpha(z) = z^2/2 + z^4/4 + z^8/8$, and $\beta(z) = z^{c_2}/c_2!$, therefore an EGF of $\beta(\alpha(z)) = \frac{1}{c_2!} \left[z^2/2 + z^4/4 + z^8/8\right]^{c_2}$. Finally, the third item has EGF given by Lemma 9,

$$exp\left(\log\left(\frac{1}{1-z} - \sum_{i|8} z^i/i\right)\right) = \frac{1}{1-z} exp\left(-\sum_{i|8} z^i/i\right)$$

giving a final, total EGF of

$$\frac{z^{c_1}}{(1-z)c_1! c_2!} \left[\frac{z^2}{2} + \frac{z^4}{4} + \frac{z^8}{8}\right]^{c_2} exp\left(-\sum_{i|8} z^i/i\right)$$

Multiplying by $1-z$ and taking the limit as $z \to 1^-$, via Theorem 3 we obtain

$$\frac{1}{c_1! c_2!} \left[\frac{1}{2} + \frac{1}{4} + \frac{1}{8}\right]^{c_2} exp(-\sigma(8)/8) = \frac{1}{c_1! c_2!}(7/8)^{c_2} e^{-15/8}$$

$\square$

Corollary 2. *Let π be a random permutation from S_n. The probability that π has c_1 fixed points is given by $\frac{1}{e(c_1!)}$ in the limit as $n \to \infty$.*

Proof. Since this is essentially the previous theorem, but allowing c_2 to be any possible value, we can just sum the formula we just derived over $c_2 = 0, 1, \ldots,$ and obtain

$$\sum_{c_2=0}^{\infty} \frac{1}{c_1! c_2!} (7/8)^{c_2} e^{-15/8} = \frac{e^{-15/8}}{c_1!} \sum_{c_2=0}^{\infty} \frac{1}{c_2!} \left(\frac{7}{8}\right)^{c_2}$$

$$= \frac{e^{-15/8}}{c_1!} \left(\sum_{c_2=0}^{\infty} \frac{x^{c_2}}{c_2!}\right)\Bigg|_{x=7/8}$$

$$= \frac{e^{-15/8}}{c_1!} e^{7/8} = \frac{1}{e(c_1!)}$$

as desired.

Alternatively, we could have just used Theorem 12 on Page 44. $\square$

The method requires $c_1 \geq 2$, otherwise the attack fails. This can be easily calculated as $1 - \Pr\{c_1 = 0\} - \Pr\{c_1 = 1\} = 1 - 2/e = \approx 0.2642$ probability of success.

Second, suppose that η is the fraction of the code-book available. Then any given fixed point is found with probability η in the known part of the code-book, and so at least two will be found with probability

$$1 - \binom{c_1}{0} \eta^0 (1-\eta)^{c_1} - \binom{c_1}{1} \eta^1 (1-\eta)^{c_1-1} = 1 - (1-\eta)^{c_1-1} [1 - (c_1 + 1)\eta]$$

and so the following η and success probabilities can be found, generated by Theorem 12 and listed in Table 3.4.1. Note, these are absolute probabilities, not probabilities given $c_1 \geq 2$.

Table 3.1 Success Probabilities of Bard's Dissertation Attack

η	10%	20%	30%	40%	50%
Success	0.47%	1.75%	3.69%	6.16%	9.02%

η	60%	70%	80%	90%	100%
Success	12.19%	15.58%	19.12%	22.75%	26.42%

Using MAPLE [3], one can also calculate exactly when the probability of having the two fixed points in the η fraction of the code-book is one-half. This is at $\eta = 63.2\%$ remarkably close to the empirical calculation in [31, Ch. 2].

Note that while finding two fixed points of f_k is enough to break the cipher, using SAT-solvers as noted above, the fixed points of $f_k^{(8)}$ are still an annoyance. Our post-processed code-book will have all the fixed points of $f_k^{(8)}$ in it, and at worst we must try all pairs.

If π has c_1 fixed points, and c_2 cycles of length 2, 4, or 8, then π^8 has at most $c_1 + 8c_2$ fixed points, as each cycle of length 2 produces 2, of length 4 produces 4, and of length 8 produces 8. Thus of the c_2 cycles of length 2, or 4, or 8, at most $8c_2$ fixed points are produced. This means in the code-book we have at most $c_1 + 8c_2$ fixed points, or $(c_1 + 8c_2)(c_1 + 8c_2 - 1)/2$ pairs of them. At absolute worst, we have to check all of them. The expected value of the number of pairs, given $c_1 \geq 2$ can be calculated with MAPLE [3], and is $113/2 - 105/e \approx 17.87$. As each pair takes less than a minute, this is not the rate-determining step.

The post-processing of the code-book will take much more time, $\eta 2^{32}$ Keeloq encryptions, but this is still much smaller than brute-forcing the 2^{64} keys.

3.4.2 Without Analytic Combinatorics

These few paragraphs were written during the author's dissertation, before he learned about analytic combinatorics. They might therefore be useful to a reader who is not interested in learning analytic combinatorics, but is interested in Keeloq.

One could generate a probability distribution on the possible values of n_1 and n_8, the number of fixed points of f_k and $f_k^{(8)}$. However, if all we need to know is how many plaintexts must be tried until two fixed points of f are discovered, then this can be computed by an experiment.

We generated 10,000 random permutations of size $2^{12}, 2^{13}, 2^{14}, 2^{15}$ and $100,000$ of 2^{16}. Then we checked to see if they had two or more fixed points, and aborted if this were not the case. If two or more fixed points were indeed present, we tabulated the number of fixed points of the eigth power of that permutation on composition. Finally, we examined at which value the second fixed point of f was found, when iterating through the values of $f^{(8)}$ and searching for its fixed points. The data is given in Table 3.2 on Page 24. It shows that we must check around 60% of the possible plaintexts (60% of the entire code-book). It also confirms the values of $n_1 = 2.39$ (calculated in the author's dissertation [31, Ch. 2.4.8], or in Corollary 2 on Page 22 here) and $n_8 = 5.39$ (also calculated in the author's dissertation [31, Ch. 2.4.8]). Note that n_8 is the expected number of points of order 1,2,4, or 8 for f, or equivalently the expected number of fixed points of $f^{(8)}$, both assuming that f has at least two fixed points.

3.5 Comparison to Brute Force

Here we compare the attack with $\eta = 1$ (using the entire codebook) with a brute force attack. Of course, only $\eta = 0.6$ is necessary on average, and the attack can succeed with smaller η as explained in Section 3.4 on Page 20, but it is good to be conservative, and so we pessimistically calculate with $\eta = 1$ here.

Table 3.2 Fixed points of random permutations and their 8th powers

Size	2^{12}	2^{12}	2^{13}	2^{14}	2^{15}	2^{16}
Experiments	1000	10,000	10,000	10,000	10,000	100,000
Aborts ($n_1 < 2$)	780	7781	7628	7731	7727	76,824
Good Examples ($n_1 \geq 2$)	220	2219	2372	2269	2273	23,176
Average n_1	2.445	2.447	2.436	2.422	2.425	2.440
Average n_8	4.964	5.684	5.739	5.612	5.695	5.746
Average Location	2482	2483	4918	9752	19,829	39,707
Percentage (η)	60.60%	60.62%	60.11%	59.59%	60.51%	60.59%

Recall, that f has two or more fixed points with probability $1 - 2/e$, and that we require f to have two or more. Our success probability is $2^{-16}(1 - 2/e) \approx 2^{-17.92}$. A brute force attack which would itself have probability $2^{-17.92}$ of success would consist of guessing $2^{46.08}$ possible keys and then aborting, because $46.08 + 17.92 = 64$, the length of the key. Therefore, our attack must be faster than $2^{46.08}$ encryptions of guesses, or $528 \times 2^{46.08} \approx 2^{55.124}$ rounds.

We require, for our attack, $g_k^{-1}(E_k(\mathbf{P}))$, which will need an additional 16 rounds. Even if we use the whole dictionary of 2^{32} possible plaintexts, this comes to $(528 + 16)2^{32} \approx 2^{41.087}$ rounds, which is about $2^{14.04}$ times faster than brute force.

3.6 Summary

The attack in this chapter is a constraint satisfaction problem (CSP), like all algebraic attacks. Normally a CSP has zero, one, or more than one solution. In the case of algebraic cryptanalysis, unless a false assumption is made, there is always a solution because a message was sent. Therefore, we have only to ensure that the constraints are sufficient to narrow down the keyspace to a single key, which is our objective. A secondary, but crucial, objective is that the attack must finish within a reasonable amount of time, namely faster than brute force by a wide margin.

If one has μ plaintext-ciphertext pairs encrypted with the same key, then one has a set of constraints. Here, with Keeloq, we have one (cubic) equation for each round that we take under consideration (See the equations at the start of Section 2.5 on Page 13, where NLF is defined as a cubic polynomial in Section 2.3 on Page 11). Thus there are 528μ constraints. This becomes 3 equations for each round, when we convert into quadratic degree (see Section 2.7 on Page 14) by introducing a new variable similar to the degree-dropper algorithm (see Section 11.4 on Page 192).

One approach is to therefore generate a key, generate μ plaintexts, encrypt them all, write down the system of equations, and solve it. Because this might take too long, we may elect to "guess" g bits of the key to the system of equations and adjust the final running time by 2^g, or equivalently the final probability of success by 2^{-8}, as described in Section 2.9 on Page 16. This is an example of a guess-and-determine attack (see Section 11.7 on Page 206). Upon doing that, we in fact do solve the systems (See bulleted list in Section 2.9 on Page 16), but discover that the attack is far worse than brute force.

Instead, a fixed point is a very attractive target, in place of a plaintext-ciphertext pair. The entire description of a fixed point of f is concerned only with the first 64 rounds. Therefore, only 64 equations are needed. However, the first objective, namely narrowing the key down to one possibility, is not accomplished here. Instead, two fixed points are needed. This is still a very limited number of equations, roughly a factor of $3168/384 = 8.25$ times smaller than the attack in Section 2.9 on Page 16, both in terms of number of equations and in terms of number of variables.

If the degree were a linear system, this would be faster by a factor of $8.25^3 \approx 561.5$ or $8.25^{2.807} \approx 373.7$ depending on the algorithm used. Of course, solving a polynomial system of equations is far harder than solving a linear system, so the speed-up is expected to be very much larger than that. And so, our second objective, which is speed, is accomplished. This leaves us with the following attack:

- Search the code-book for fixed points of $f^{(8)}$.
- Find two fixed points of f by trying pairs of fixed points of $f^{(8)}$.
- Write down the equations that describe the Constraint Satisfaction Problem (CSP) of f having those two fixed points.
- Solve the equations.

A more detailed version of this is given in Algorithm 1 on Page 20.

3.7 Other Notes

3.7.1 A Note about Keeloq's Utilization

An interesting note is Keeloq's utilization in at least some automobiles. Specifically, it encrypts the plaintext 0 and then increments the key by arithmetically adding one to the integer represented by the binary string $k_{63}, k_{62}, \ldots, k_1, k_0$. This way the same key is never used twice. This is rather odd, of course, but if one defines the dual of a cipher as interchanging the plainspace with the keyspace, then the dual of Keeloq has a 64-bit plaintext, and a 32-bit key. The cipher is operating in precisely counter-mode in that case, with a random initial counter, and fixed key of all zeroes. However, not all automobiles use this method. See [234] [106] and [94].

For those interested, this is an example of a dual cipher. A cipher E' is dual to E if and only if

$$E'_{\mathbf{P}}(k) = E_k(\mathbf{P})$$

3.7.2 RPA vs KPA vs CPA

Because this attack will use a large portion of the codebook (from 10%–60% of all possible plaintext-ciphertext pairs), it will be easy to see that random plaintexts would permit the attack to proceed identically. The random plaintext attack model (RPA) is between the more commonly discussed known plaintext attack (KPA) and chosen plaintext attack (CPA) models. In the known plaintext case, the adversary is given plaintext-ciphertext pairs, and must find the secret key. In the random plaintext model, the adversary can further be assured that those plaintexts were generated randomly—much like the distinction between average-case and worse-case running time in quicksort. In the chosen plaintext case, the attacker actually choses the plaintexts according to properties that he/she wishes to exploit. This distinction is not important here, but could be elsewhere.

3.8 Wagner's Attack

Again, in this attack (first published in [84]), we will iterate over some portion of the code-book. One property of the cipher Keeloq, is that only one bit is changed per round. Thus the last sixteen rounds, represented by $g_k(x)$, only affect sixteen bits of the ciphertext. Thus, if x is a fixed point of $f_k^{(8)}$, then 16 out of the 32 bits will match, compared between the plaintext and the ciphertext. To be precise, 16 bits will be different, and 16 will match though be shifted by 16 positions. One can easily scan a code-book for this property.

This matching property will always occur for a fixed point of $f_k^{(8)}$, but it also happens by coincidence with probability 2^{-16}, between two random words. Therefore, the number of code-book entries with this property will be the number of fixed points of $f_k^{(8)}$, plus an expected $2^{-16}2^{32} = 2^{16}$ "red herrings". As you can see, the "red herrings" out number the true fixed points of $f_k^{(8)}$ by a large margin.

If the effect is not a "red herring" for a particular code-book entry (in other words, the plaintext really is a fixed point of $f_k^{(8)}$) then there is an easy algorithm for finding the 16 key bits. Simply repeatedly apply the following formula, which can be calculated because ciphertext is the last few L's, and the plaintext itself, being a fixed point, is known to be what is in the linear shift register at the end of round 512. More precisely, the ciphertext $C_0, \ldots, C_{31}$ is $L_{528}, \ldots, L_{559}$ and the plaintext $P_0, \ldots, P_{31}$ (because[1] we have a fixed point) is $L_{512}, \ldots, L_{543}$. The formula to apply is

$$k_{i-32} = L_i + L_{i-32} + L_{i-16} + NLF(L_{i-1}, L_{i-6}, L_{i-12}, L_{i-23}, L_{i-30})$$

---

[1] Here the sharing of 16 bits among the plaintext and the ciphertext of a fixed-point of $f^{(8)}$ is rendered explicitly obvious because $L_{528}, \ldots, L_{543}$ appear in both the range given by the P's and the range given by the C's.

where NLF was defined in Section 2.3 on Page 11. Therefore, each code-book entry with the matching property can be tagged with a 16-bit potential sub-key. Of course, for the 2^{16} red-herrings, the sub-key will be unpredictable. But, for each of the genuine fixed-points, it will be correct!

As it turns out, the 16-bit sub-key, as well as any single plaintext-ciphertext pair that is a fixed point of f_k, not merely of $f_k^{(8)}$, is enough to mount an algebraic attack. Thus we have the following steps. Let c_3 denote the number of fixed points of $f^{(8)}$.

1. Check all 2^{32} code-book entries for the matching property.
2. Of these (roughly $2^{16} + c_3$) plaintext-ciphertext pairs, compute the sub-key that they imply. Recall, c_3 is the number of fixed points of $f^{(8)}$.
3. For each plaintext-ciphertext pair with the property, set up an algebraic cryptanalysis problem with the one pair, assuming it is a fixed point of f, and assuming the sub-key is correct.
4. If an answer is obtained, verify assumptions. If assumptions turned out to be false, or if the problem is "unsatisfiable", go to Step 3.

Sorting upon this sub-key between Step 2 and Step 3 would reveal which are the likely pairs, as the same sub-key will tag all the fixed points of f_k and $f_k^{(8)}$. We expect each of the 2^{16} "red-herrings" to be tagged with uniformly randomly distributed potential sub-keys. Therefore, in the first few Step 3 & 4 executions, we would obtain the key.

What is needed for success? First, that f_k have at least one genuine fixed point. This occurs with probability $1 - 1/e$, as proven in Corollary 11, and is roughly 0.6321. Second, the expected amount of work in Step 1 is at most 2^{32} Keeloq Encryptions, and a more precise estimate is found in [84]. Third, Step 2 is negligible. Fourth, for Step 3 and Step 4, we must execute these stages for each potential sub-key. Given the model of the previous attack, and using Theorem 1, we can obtain a bound on the expected number of repetitions of Steps 3 and 4. This is upper-bounded by the expected value of $c_1 + 8c_2$ given that $c_1 > 0$. Using MAPLE [3], this comes to $113/2 - 46/e \approx 39.58$, the difference being that we now allow $c_1 = 1$, which was previously forbidden. Of course, without the sorting explained in the previous paragraph, the expected number of Step 3 and Step 4 executions would be around 2^{15}. For certain, the longest step is Step 1, iterating over the code-book.

3.8.1 Later Work on Keeloq

The author's attack was done in January of 2007, and between that time and the time of the writting of this book, several other attacks on Keeloq have been published, including some by the author of this book. They are found in the bibliography under the following entries, which we list in no particular order:

Three things are noteworthy about this list. First, it is remarkably long considering that it covers papers about a cipher which was first broken in 2007. Second, this shows the remarkable importance of Keeloq in the evolution of cryptanalysis as an academic discipline. Third, the fact that some extremely prestigious names are in the author list, such as Eli Biham, Bart Preneel, and Nicolas Courtois, signify that cryptanalysis is indeed a mainstream and legitimate branch of academic cryptography—a statement that few would have made 10–15 years ago. Last, the author has included some relatively non-traditional sources, including blogs. The reason for this is that blogs are often *extremely* expository, and therefore can be a good starting place for someone first learning about a new topic.

- "Cryptanalysis of the Keeloq block cipher" by Andrey Bogdanov [50], published as an e-print.
- "Algebraic and Slide Attacks on Keeloq" by Nicolas Courtois, Gregory Bard, and David Wagner [84], published in the proceedings of FSE'08.
- "Attacks on the Keeloq Block Cipher and Authentication Systems" by Andrey Bogdanov [49], published in the proceedings of RFIDSec'07.
- "Code Hopping Decoder using a PIC16C56" by S. Dawson, a manufacturer's technical report about keeloq [94], with a URL given in the bibliography.
- "Algebraic and Slide Attacks on Keeloq" by Nicolas Courtois and Gregory Bard [74], published as an e-print.
- "Linear Slide Attacks on the Keeloq Block Cipher" by Andrey Bogdanov [51], published in the proceedings of INSCRYPT'07.
- "How to Steal Cars" by Eli Biham, Orr Dunkelman, Sebastiaan Indesteege, Nathan Keller, and Bart Preneel [47], a rump-sesssion talk at CRYPTO'08.
- "Periodic Ciphers with Small Blocks and Cryptanalysis of Keeloq" by Nicolas Courtois, Gregory Bard, and Andrey Bogdanov [83], published in Tatra Mountains Mathematical Publications, the mathematics journal of the Slovak Academy of Sciences.
- "A Practical Attack on Keeloq" by Sebastiaan Indesteege, Nathan Keller, Orr Dunkelman, Eli Biham, and Bart Preneel [142], published in EUROCRYPT'08.
- "Remote keyless entry system for cars and buildings is hacked; RUB security experts discover major vulnerability; Access from a distance of 300 feet without traces" by Christof Paar [184], published as a press-release, but URL given in the bibliography.
- Chapter 2 of "Algorithms for the Solution of Linear and Polynomial Systems of Equations over Finite Fields, with Applications to Cryptanalysis", the author's dissertation [31].
- "Researchers Crack Keeloq Code for Car Keys", by Kim Zetter [234], published in WIRED magazine, but a URL is given in the bibliography.
- "On the Power of Power Analysis in the Real World: A Complete Break of the Keeloq Code-Hopping Scheme" by Thomas Eisenbarth, Timo Kasper, Amir Moradi, Christof Paar, Mahmoud Salmasizadeh, and Mohammad T. Manzuri Shalmani [106], published in the proceedings of CRYPTO'08.

Chapter 4
Iterated Permutations

The purpose of this chapter is to allow us to calculate what fraction of permutations in S_n have a particular property ϕ, in the limit as n tends to infinity. This will be accomplished via analytic combinatorics.

Combinatorics is the branch of mathematics concerned with counting objects. The technique of using a function of a variable to count objects of various sizes, using the properties of multiplication and addition of series as an aid, is accredited to Pierre-Simon Laplace [116, Ch. "Invit."].

Consider the following problem. We have boxes that can be filled with 1, 2, 3, or 4 items, among an identical set of 5000 indistinguishable items. How many different ways are there to dispose of the items? Note that we can use as many as 5000 boxes, or as few as 1250 boxes. Using the methods of this chapter, we will address precisely this question in Section 4.2.4.3 on Page 35.

An ordinary generating series associated with a set of objects assigns as the coefficient of the z^i'th term, the number of objects of size i. An exponential generating series is merely this, with each term divided by $i!$. In particular, this can be used to describe permutations drawn at random from S_n, which is what concerns us here. In combinatorial arguments, OGFs and EGFs abound [116, Ch. "Invit."] [200] and are especially useful in counting partitions of sets. Here, we will use this family of techniques, now called "analytic combinatorics" to count permutations of particular types.

The content of this chapter is joint work with Prof Shaun Ault of Fordham University.

4.1 Applications to Cryptography

As we saw in Section 3.1.2 on Page 17, the cipher Keeloq can be written as the eighth iterate of a permutation followed by one more permutation [31, Ch. 2]. This eighth power naturally affects the cycle structure; for example, we will prove that the fixed points of the eighth power are those of order $\{1, 2, 4, 8\}$ under the

G.V. Bard, *Algebraic Cryptanalysis*, DOI: 10.1007/978-0-387-88757-9_4
© Springer Science + Business Media, LLC 2009

original. There are many other properties of these repeated permutations that follow from the factorization of the number of iterations, and we will show cryptanalytic consequences of several.

In particular, in the event a user is so unwise as to iterate a cipher a very large number times, then we show, depending on the factorization of that number of times, the number of fixed points of the cipher will change. In fact, this leads to a distinguisher attack if the number of iterations has a large number of divisors. Finally, we show how to turn this into a key-recovery attack in the extremely unlikely scenario that cipher is used as part of a super-encipherment.

The famous cryptanalyst Adi Shamir, the "S" in RSA, has stated that these properties of random permutations have been known primarily from the theory of random graphs, and for a long time, but this is possibly their first application to cryptanalysis [11].

4.2 Background

4.2.1 Combinatorial Classes

A combinatorial class $\mathscr{C}$ is a set of objects C together with a function $\ell_C : C \to \mathbb{Z}^{\geq 0}$, which assigns to each element a non-negative integer "size". For example, if $\mathscr{P}$ is the set of permutation groups S_n for all positive integers n, then we may use the size function $\ell_P(\pi) = n$, for any $\pi \in S_n$, to make $\mathscr{P}$ into a combinatorial class. Sometimes the size function is a matter of context, for example the size function for matrices might be the dimension or the number of elements, depending on what is being counted. An element $c \in \mathscr{C}$ is said to be maximal if there is no element $d \in \mathscr{C}$ such that $\ell_C(d) > \ell_C(c)$. Note, there might well be many maximal elements, all of the same size.

Let C_i be the cardinality of the set of elements in $\mathscr{C}$ with size i. Thus in our example, $P_i = i!$ for $i \geq 0$. It will be useful to represent C_i by either an exponential or generating function an ordinary generating function (OGF or EGF). First, a brief discussion of generating functions is in order.

4.2.2 Ordinary and Exponential Generating Functions

Given a set of constants indexed by $\mathbb{Z}^{\geq 0}$, say $c_0, c_1, c_2, \ldots$, the *ordinary generating function* (or OGF) is defined as the formal power series:

$$\mathscr{C}(z) \overset{def}{=} \sum_{i=0}^{\infty} c_i z^i = c_0 + c_1 z + c_2 z^2 + c_3 z^3 + \cdots.$$

The EGF is defined as the formal power series:

$$\mathscr{C}_e(z) \overset{def}{=} \sum_{i=0}^{\infty} \frac{c_i}{i!} z^i = c_0 + \frac{c_1}{1!} z + \frac{c_2}{2!} z^2 + \frac{c_3}{3!} z^3 + \cdots.$$

and as you can see, the EGF is just the OGF term-wise divided by $i!$.

Sometimes the infinite sum that is presented is actually the Taylor Series of a well-known elementary function. In that case, we will use that function as an abbreviation for the sequence. Also, it is clear that the OGF or EGF is a polynomial if and only if the combinatorial class has a maximal element.

4.2.3 Operations on OGFs

In order to understand why OGFs and EGFs are useful, we must first see how simple combinatorial operations on classes affect the OGFs. This, in a way, motivated the original concept of an OGF.

4.2.3.1 Simple Sum

The description of the sum of two combinatorial classes can be broken into two cases. The first case, which is very simple, occurs if the set of objects in the class have empty intersection. The second case will be handled in Section 4.2.3.3 on Page 32.

If this is the case, let $\mathscr{A} = \mathscr{B} + \mathscr{C}$ be defined as follows. The set of objects in $\mathscr{A}$ is the union of the set of objects in $\mathscr{B}$ and $\mathscr{C}$. The size of an object in $a \in \mathscr{A}$ will be determined as follows: if it is from $\mathscr{B}$ then let it be $\ell_B(a)$ and if it is from $\mathscr{C}$ then let it be $\ell_C(a)$.

The number of objects of size n in $\mathscr{A}$, or A_n, therefore is $B_n + C_n$. This means that the OGF would be

$$(B_0 + C_0) + (B_1 + C_1)z + (B_2 + C_2)z^2 + (B_3 + C_3)z^3 + \cdots = \mathscr{B}(z) + \mathscr{C}(z)$$

4.2.3.2 Cartesian Product

Suppose you take two combinatorial classes $\mathscr{B}$ and $\mathscr{C}$, and build a combinatorial class $\mathscr{A}$ as follows. Each element of $\mathscr{A}$ shall be an ordered pair, with the first entry coming from $\mathscr{B}$ and the second from $\mathscr{C}$. The size of an element (b, c) will be

$$\ell_{\mathscr{A}}((b,c)) = \ell_{\mathscr{B}}(b) + \ell_{\mathscr{C}}(c)$$

This is called the Cartesian Product, because the set of objects of $\mathscr{A}$ will be the Cartesian Product of the objects in $\mathscr{B}$ and those in $\mathscr{C}$. Let B_i indicate the number of objects in $\mathscr{B}$ that have size equal to i, and likewise C_i and A_i. At first it looks like computing the OGF of $\mathscr{A}$ might be challenging, since, for example,

$$A_4 = B_0 C_4 + B_1 C_3 + B_2 C_2 + B_3 C_1 + B_4 C_0$$

but observe the product of the OGFs produces the required result

$$(B_0 + B_1 z + B_2 z^2 + B_3 z^3 + \cdots)(C_0 + C_1 z + C_2 z^2 + C_3 z^3 + \cdots)$$
$$= (B_0 C_0) + (B_1 C_0 + B_0 C_1)z + (B_2 C_0 + B_1 C_1 + B_0 C_2)z^2 +$$
$$(B_3 C_0 + B_2 C_1 + B_1 C_2 + B_0 C_3)z^3 + \cdots$$

We will write this operation as $\mathscr{B} \times \mathscr{C} = \mathscr{A}$. Furthermore, we can abbreviate $\mathscr{B} \times \mathscr{B} = \mathscr{B}^2$, along with higher powers. As noted above, the OGF of a Cartesian Product is the product of the OGFs. There is a similar interpretation for EGFs and products of EGFs, called the labelled product, see Section 4.2.5.1 on Page 36. See Section 4.4.2 or Theorem 15 as an example.

4.2.3.3 Sum with Non-Empty Intersection

When explaining the set theoretic operation of "disjoint union" to students, the author once used the analogy of painting all the elements of $\mathscr{B}$ with red, and all the elements of $\mathscr{C}$ with blue. Thus if the set of objects of $\mathscr{B}$ and $\mathscr{C}$ are not of empty intersection, then each element of the intersection will appear twice, once in each color. If we are asked the size of an element of $\mathscr{B} + \mathscr{C}$ then the color indicates whether we should use ℓ_B or ℓ_C.

The combinatorial way to achieve this painting is to make two combinatorial classes both consisting of one object each, namely $\mathscr{I}_1$ with "∘" and $\mathscr{I}_2$ with "·". Both of these objects will have size 0, and thus the OGFs of $\mathscr{I}_1$ and $\mathscr{I}_2$ are both $1 + 0z + 0z^2 + 0z^3 \cdots = 1$. Then consider $(\mathscr{B} \times \mathscr{I}_1) + (\mathscr{C} \times \mathscr{I}_2)$.

This will consist of ordered pairs, where either the first element is from $\mathscr{B}$ and the second element is "∘" or the first element is from $\mathscr{C}$ and the second element is "·". Furthermore, all the lengths are unchanged. We have now used set theory to paint the elements of $\mathscr{B}$ and $\mathscr{C}$ different colors.

This is what we mean by $\mathscr{B} + \mathscr{C}$ when the set of objects in $\mathscr{B}$ and $\mathscr{C}$ are not disjoint (i.e. they do not have empty intersection).

4.2.3.4 Semiring of Combinatorial Classes

With two trivial examples, we will be able to show ring-like structure for combinatorial classes. Let the empty class $\mathscr{E}$ be one with no objects, and so the number of objects in $\mathscr{E}$ with size n is always 0. This means that the OGF is $0 + 0z + 0z^2 + 0z^3 + \cdots = 0$.

Let the singleton class $\mathscr{I}$ be one with only one object, and let its size be 0. Thus the number of objects in $\mathscr{E}$ with size n is always 0, except if $n = 0$, in which case it is 1. This means that the OGF is $1 + 0z + 0z^2 + 0z^3 + \cdots = 1$. By convention, $\mathscr{B}^0 = \mathscr{I}$, for all combinatorial classes $\mathscr{B}$.

We shall state, without proof, the following facts.

Sums are Closed: For any two combinatorial classes $\mathscr{C}$ and $\mathscr{D}$ we have that $\mathscr{C} + \mathscr{D}$ is a combinatorial class.

Sums are Associative: For any three combinatorial classes $\mathscr{B}$, $\mathscr{C}$, and $\mathscr{D}$, it is the case that $\mathscr{B} + (\mathscr{C} + \mathscr{D}) = (\mathscr{B} + \mathscr{C}) + \mathscr{D}$.

Sums have Identity: For any combinatorial class $\mathscr{C}$ it is the case that $\mathscr{C} = \mathscr{C} + \mathscr{E} = \mathscr{E} + \mathscr{C}$.

Sums are Commutative: For any two combinatorial classes $\mathscr{C}$ and $\mathscr{D}$ we have that $\mathscr{C} + \mathscr{D} = \mathscr{D} + \mathscr{C}$.

Products are Closed: For any two combinatorial classes $\mathscr{C}$ and $\mathscr{D}$ we have that $\mathscr{C} \times \mathscr{D}$ is a combinatorial class.

Products are Associative: For any three combinatorial classes $\mathscr{B}$, $\mathscr{C}$, and $\mathscr{D}$, it is the case that $\mathscr{B} \times (\mathscr{C} \times \mathscr{D}) = (\mathscr{B} \times \mathscr{C}) \times \mathscr{D}$.

Products have Identity: For any combinatorial class $\mathscr{C}$ it is the case that $\mathscr{C} = \mathscr{C} \times \mathscr{I} = \mathscr{I} \times \mathscr{C}$.

Products are Commutative: For any two combinatorial classes $\mathscr{C}$ and $\mathscr{D}$ we have that $\mathscr{C} \times \mathscr{D} = \mathscr{D} \times \mathscr{C}$.

The Distributive Law: For any three combinatorial classes $\mathscr{B}$, $\mathscr{C}$, and $\mathscr{D}$, it is the case that $\mathscr{B} \times (\mathscr{C} + \mathscr{D}) = (\mathscr{B} \times \mathscr{C}) + (\mathscr{B} \times \mathscr{D})$

Of course, we have neglected to say what $\mathscr{B} = \mathscr{C}$ means. These symbols indicate an isomorphism of combinatorial classes. This means first that there is some bijection between the objects in $\mathscr{B}$ and $\mathscr{C}$ (as sets). Call this ϕ. Second, we require that for any $b \in \mathscr{B}$ that $\ell_B(b) = \ell_C(\phi(b))$. That is to say, any pair of objects identified as matched by ϕ to each other must have equal sizes. Clearly, this notion is an equivalence relation.

Under the above equivalence relation, the only thing separating combinatorial classes from being a commutative ring is the absence of additive inverses. This structure is called a commutative semiring, a structure used elsewhere in this book (see Section 6.2 on Page 81). The proofs are not interesting and so are omitted. But it is pleasing that these laws hold, and we will make implicit use of them at times.

4.2.3.5 Sequences of Objects

An object in $SEQ(\mathscr{B})$ is a finite sequence of objects from $\mathscr{B}$, with the additional artificial member that is a sequence of length zero (the empty sequence). The size of an object in $SEQ(\mathscr{B})$ is the sum of the sizes of the elements of $\mathscr{B}$ which compose it. For example, $L = SEQ(\mathbb{Z}^{\geq 0})$ would be finite length sequences of positive integers. The size of one of those would be the sum of those integers. Thus L_n is the number of ways to write n as a sum of positive integers, but including ordering. We must require that $\mathscr{B}$ have no element of size 0 for this construction to work. Furthermore, it is clear that

$$SEQ(\mathscr{B}) = \mathscr{I} + \mathscr{B} + \mathscr{B}^2 + \mathscr{B}^3 + \mathscr{B}^4 + \cdots$$

and therefore if $\mathscr{B}$ has an OGF of $\mathscr{B}(z)$ then $SEQ(\mathscr{B})$ will have an OGF of

$$1 + \mathscr{B} + \mathscr{B}^2 + \mathscr{B}^3 + \mathscr{B}^4 + \cdots = \frac{1}{1 - \mathscr{B}(z)}$$

4.2.3.6 Other Operations

Flajolet and Sedgewick [116, Ch 1.2] have described several other operations, which we will not require here. These include cycles (sequences that are considered identical if a rotation-like shift will transform one into another), the power set, and multisets (sequences where order does not matter).

4.2.4 Examples

Some examples might be useful for practice at this point.

4.2.4.1 Permutations in General

For our example combinatorial class, the permutations $\mathscr{P}$, its OGF is $\mathscr{P}(z) = z + 2z^2 + 6z^3 + 24z^4 + 120z^5 + \cdots$, and its EGF is $\mathscr{P}_e(z) = z + z^2 + z^3 + z^4 + z^5 + \cdots = z/(1 - z)$. At times, we may wish to add S_0, the unique permutation of the empty set, which we state has size equal to zero.

4.2.4.2 The Non-Negative Integers

The series $1 + z + z^2 + z^3 + z^4 + z^5 + \cdots = 1/(1 - z)$ represents the OGF of the non-negative integers, $\mathbb{Z}^{\geq 0}$ with "size" function being the identity: $\ell(n) = n$. The EGF is therefore $1 + z + z^2/2! + z^3/3! + z^4/4! + z^5/5! + \cdots = e^z$.

The even integers would have OGF of $1 + 0z + z^2 + 0z^3 + z^4 + 0z^5 + \cdots$. The EGF is therefore $1 + z^2/2! + z^4/4! + z^6/6! + \cdots = \cosh z$. Likewise the odd integers would have OGF of $z + z^3/3! + z^5/5! + \cdots = \sinh z$

And one can see that the sum law works, because the even non-negative integers and odd non-negative integers are disjoint, and therefore can be summed simply, and produce all the non-negative integers. Also, recall

$$\sinh z + \cosh z = e^z$$

4.2.4.3 Partitions into Boxes

Now we return to our motivating example. Our boxes could contain 1, 2, 3, or 4 objects. The size of such a box should be 1, 2, 3, or 4 accordingly, and there is one such box of each size. Therefore, the OGF of the combinatorial class of boxes $\mathscr{B}$ would be

$$\mathscr{B}(z) = 0 + z + z^2 + z^3 + z^4 + 0z^5 + 0z^6 + \cdots$$

and we can use the sequence operation to represent an arbitrarily long sequence of boxes from $\mathscr{B}$. Then we have the OGF

$$SEQ(\mathscr{B})(z) = \frac{1}{1 - z - z^2 - z^3 - z^4}$$

and the 5000th degree term of the Taylor Series of that function would be the number of ways to produce a sequence of boxes that collectively contain a total of exactly 5000 objects.

This simple problem can be generalized tremendously. For example, let A_1, A_2, $\ldots$, A_k be sets of whole numbers. The number of all distinct ways that n identical objects can be placed into k containers, where container j must have some number of objects that occurs in the set A_j will be the coefficient of z^n in the OGF:

$$\left(\sum_{i \in A_1} z^i \right) \left(\sum_{i \in A_2} z^i \right) \cdots \left(\sum_{i \in A_k} z^i \right),$$

a function that we will use in the proof of Lemma 6. Notice that the jth factor in the entire product is the OGF that represents the set A_j.

4.2.4.4 Cycles

A less trivial example of a combinatorial class is the class $\mathscr{O}$ of n-cycles of S_n, for all $n > 0$, with size function $\ell(\pi) = n$ if $\pi \in S_n$. In other words, size n members of $\mathscr{O}$ comprise the subset of permutations of S_n where the permutation has exactly one orbit. For any $n > 0$ there are $n!/n$ or $(n-1)!$ of these. Thus the OGF is $z + z^2 + 2z^3 + 6z^4 + 24z^5 + 120z^6 + \cdots$, and the EGF is $z + z^2/2 + z^3/3 + z^4/4 + z^5/5 + z^6/6 + \cdots$. Thus the probability that a random permutation from S_n has only one cycle is given by the coefficients of the z^n terms in the EGF. Namely, $(n-1)!/n! = 1/n$.

The EGF for $\mathscr{O}$ also converges:

$$z + \frac{z^2}{2} + \frac{z^3}{3} + \frac{z^4}{4} + \frac{z^5}{5} + \frac{z^6}{6} + \cdots = \log\left(\frac{1}{1-z} \right)$$

as can be verified by term-by-term integration of the power series for $\frac{1}{1-z}$.

4.2.4.5 Morse Code

The alphabet of Morse Code has two symbols: "dot" and "dash". Let the size of "dot" be 1 and the size of "dash" be 2. Then the OGF is $M(z) = z + z^2$. The set of Morse Code words is the set of all sequences of dots and dashes. This will have OGF of $1/(1 - M(z))$. That function has Taylor Series

$$\frac{1}{1 - z - z^2} = 1 + z + 2z^2 + 3z^3 + 5z^4 + 8z^5 + 13z^6 + 21z^7 + 34z^8 + 55z^9 + 89z^{10} + 144z^{11} + \cdots$$

which is clearly Fibonacci's series.

4.2.4.6 Zig-Zag Arrangement

Consider an arrangement of $\{1, 2, 3, \ldots, n\}$ that is written as one list, such that the first element gets mapped to one higher than itself, which gets mapped to one lower than itself, which gets mapped to one higher than itself, *et cetera*. For example, $(3, 4, 1, 5, 2)$ would qualify, where as $(2, 3, 4, 1, 5)$ would not. Furthermore, let us require n to be always odd, and let the size of a zig-zag arrangement be n.

The number of arrangements of this type can be counted by computer and has an OGF which begins

$$0 + z + 0z^2 + 2z^3 + 0z^4 + 16z^5 + 0z^6 + 272z^7 + 0z^8 + 7936z^9 + 0z^{10} + 353792z^{11} + \cdots$$

which does not appear to be familiar at first glance, at least to the author.

However, the corresponding EGF is actually $\tan z$. Using the definition of a zigzag arrangement, Flajolet and Sedgewick [116, Ch. "Invit"] prove this result, by using an integro-differential equation! However, they accredit the result to Désiré André, published in 1881.

4.2.5 Operations on EGFs

Now that we have some examples, and have seen how the OGF reacts to certain operations, we (perhaps) understand the motivation for the definition of an OGF. However, the motivation for the definition of an EGF might seem less clear. The "labelled product" operation will shed more light, as will the connection with probability.

4.2.5.1 The Labelled Product

Every permutation can be written as a product of disjoint cycles [101, Ch. 1.3 and Ch. 4.1]. We later will make use of this form, including the number of the cycles, which we call the cycle count. We have already examined, in the form of $\mathcal{O}$, the set of permutations of cycle length one. We call such a permutation a "uni-cycle".

What about cycle count of two? We call these "bi-cycles", or $\mathcal{B}$. Naturally, a permutation which is the product of two cycles is fully described by the two cycles which comprise it. But, the Cartesian Product as described above is not sufficient. In other words, we will now show that $\mathcal{O} \times \mathcal{O} \neq \mathcal{B}$.

Consider how we combine perhaps $(1,3,2)$ and $(1,3,4,2)$ to build a permutation on seven letters. We begin with seven symbols, which to avoid confusion we will call $\{a,b,\ldots,g\}$. Some subset of these three will be assigned to be governed by the first cycle. The remaining four will be assigned to be governed by the second cycle.

There is still further murkiness. If we get, for example, $\{c,d,e\}$ to be governed by $(1,3,2)$, then the assignment $\{(c=1,d=3,e=2)\}$ is absolutely indistinguishable from $\{(c=3,d=2,e=1)\}$ because these result in (c,d,e) and (e,c,d) which are equal as members of S_3. Yet, the assignment of $\{(c=2,d=3,e=1)\}$ results in (e,d,c), which is not equal to the above. But on the other hand, it is equal to $\{(c=3,d=2,e=1)\}$ being assigned to the very distinct cycle $(1,2,3)$ from $\mathcal{O}$. It takes some work to see that once we know the subset of 3 out of the 7 letters assigned to the first cycle, that we need not count how to assign them. This is covered in more detail in [116, Ch 2.2].

Thus making a bicycle on c letters requires first a cycle σ_x and a cycle σ_y both from $\mathcal{O}$ such that $\ell_{\mathcal{O}}(\sigma_x) = x$, $\ell_{\mathcal{O}}(\sigma_y) = y$, and $x+y = c$. Then we choose either x letters out of the c to be governed by σ_x, or equivalently y letters out of the c to be governed by σ_y. Luckily these are the same because

$$\binom{c}{x} = \binom{c}{c-x} = \binom{c}{y}$$

Now suppose the coefficient of the z^x term in the OGF of $\mathcal{O}$ is O_x and likewise of the z^y term is O_y. Then, for fixed values of x and y, we desire the coefficient of the $z^{x+y} = z^c$ term of the OGF of $\mathcal{B}$ to be

$$\binom{c}{x} O_x O_y = \frac{c!}{x!y!} O_x O_y$$

and in general the cth term should be

$$\sum_{x=0}^{x=c} \frac{c!}{x!(c-x)!} O_x O_{c-x} = \sum_{x=0}^{x=c} \binom{c}{x} O_x O_{c-x}$$

which we will now construct.

Multiply the EGF of $\mathcal{O}$ with itself. We claim that this is the EGF of $\mathcal{B}$. Note the z^x term will be $O_x/x!$ and the z^{c-x} term will be $O_{c-x}/(c-x)!$, by definition. Then, since the length of a two-cycle permutation is the sum of the lengths of its two cycles, these two will contribute to the z^c term of the product of the EGFs. We get that this term will be

$$\sum_{x=0}^{x=c} \frac{O_{c-x}}{(c-x)!} \frac{O_x}{x!} = \frac{1}{c!} \sum_{x=0}^{x=c} \binom{c}{x} O_x O_{c-x}$$

and so the corresponding term of the OGF of $\mathscr{B}$ will be that times $c!$ by definition, giving the desired result.

This long and convoluted operation is called "the labelled product" and the EGF of the labelled product is always the product of the EGFs, as described in [116, Ch. 2].

4.2.5.2 Random Permutations

In cryptography and other disciplines, we are often concerned with determining whether or not a random permutation has some given property ϕ. We can calculate then the OGF of the combinatorial class $\mathscr{F}$ of permutations with that property, and divide term-wise with the same term from the OGF of $\mathscr{P}$, the combinatorial class of all permutations. But this is the same as the coefficients of the EGF of $\mathscr{F}$.

Thus, if we calculate an EGF, and are interested in permutations from S_8 for example, then we simply read off the coefficient of the 8th degree term, and that will be exactly the probability. This renders superfluous any experiments that one might perform on random permutations to verify any assumptions on the probability of a particular property, which is a relief to those concerned with even S_{16}, where it would be hard to generate enough random trials to cover a significant portion of of the 16! possibilities. For further details, see the "note" on Page 43.

4.2.5.3 Asymptotic Probabilities

The above works for any specific size n, but first, it might be difficult to calculate, and second we might want to know the limit of this probability as the size goes to infinity. The following is much easier, and is found in [116, Ch. 1].

Theorem 3. *Let $\mathscr{F} \subset \mathscr{P}$ be the combinatorial class of permutations with property ϕ. Suppose further $\mathscr{F}$ has EGF equal to $f(z)$. Then the limit (as n goes to infinity) of the probability that a random permutation of size n has property ϕ is given by*

$$p = \lim_{z \to 1^-} (1-z)f(z)$$

provided that $(1-z)f(z)$ is continuous from the left at $z = 1$.

Proof. Let the OGF of $\mathscr{F}$ be given by $A_0 + A_1 z + A_2 z^2 + A_3 z^3 + A_4 z^4 + A_5 z^5 + \cdots$. Consider the following function

$$g_n(z) = \frac{A_0}{0!} + \sum_{1 \le i \le n} \left(\frac{A_i}{i!} - \frac{A_{i-1}}{(i-1)!} \right) z^i,$$

which when evaluated at $z = 1$, the sum telescopes,

$$= \frac{A_0}{0!} + \left(\frac{A_1}{1!} - \frac{A_0}{0!}\right)(1) + \left(\frac{A_2}{2!} - \frac{A_1}{1!}\right)(1)^2 + \cdots + \left(\frac{A_n}{n!} - \frac{A_{n-1}}{(n-1)!}\right)(1)^n = \frac{A_n}{n!}.$$

Thus $g_n(1)$ is the desired probability, for size n.

The limit $g(z) = \lim_{n\to\infty} g_n(z) = \frac{A_0}{0!} + \sum_{i\geq 1}\left(\frac{A_i}{i!} - \frac{A_{i-1}}{(i-1)!}\right)z^i$ does not necessarily exist for all such series, but when it does, we have

$$\begin{aligned}
g(z) = \lim_{n\to\infty} g_n(z) &= \lim_{n\to\infty} \frac{A_0}{0!} + \left(\sum_{i=1}^{n} \frac{A_i}{i!}z^i\right) - \left(\sum_{i=1}^{n} \frac{A_{i-1}}{(i-1)!}z^i\right) \\
&= \lim_{n\to\infty} \left(\sum_{i=0}^{n} \frac{A_i}{i!}z^i\right) - z\left(\sum_{j=0}^{n} \frac{A_j}{j!}z^j\right) \\
&= (1-z)\lim_{n\to\infty} \left(\sum_{i=0}^{n} \frac{A_i}{i!}z^i\right) = (1-z)f(z)
\end{aligned}$$

Thus we have the following relation among limits

$$p = \lim_{n\to\infty} g_n(1) = \lim_{n\to\infty} \lim_{z\to 1^-} g_n(z) = \lim_{z\to 1^-} \lim_{n\to\infty} g_n(z) = \lim_{z\to 1^-}(1-z)f(z)$$

Note, we implicitly assumed that $g(z)$ is continuous (from the left) near $z = 1$ in order to reverse the order of the limits in the last step, but this will be the case in all of our examples. □

Theorem 3 is exploited extensively in a paper by Marko R. Riedel dedicated to random permutation statistics, but in a different context (see [200]).

4.2.6 Notation and Definitions

The somewhat unusual notation of $exp(C)$ where C is a series, means precisely substituting the entire series C for z into the Taylor expansion for $e^z = \sum_{i\geq 0} z^i/i!$, similar to matrix exponentiation.

It is well-known that any permutation may be written uniquely as a product of disjoint cycles, up to reordering of the cycles and cyclic reordering within each cycle; indeed, for any given permutation π consisting of k disjoint cycles, having cycle lengths $c_1, c_2, c_3, \ldots, c_k$, there are exactly $k!c_1c_2c_3\cdots c_k$ ways to reorder to obtain an equivalent expression for π. Any counts we make of symmetric group elements must take this fact into account. Note, we use the convention that if π has a fixed-point, a, then the 1-cycle (a) is part of the expression for π as disjoint cycles. In particular, the identity of S_n is written $(1)(2)(3)\cdots(n)$. We use the term *cycle-count* for the number of disjoint cycles (including all 1-cycles) in the expression of a permutation. It shall be convenient to include in our analysis the unique permutation of no letters, which has by convention cycle-count 0. We may view this element as the sole member of S_0, the unique permutation of the empty set.

4.3 Strong and Weak Cycle Structure Theorems

Let A be a subset of the positive integers. We consider the class of permutations that consist entirely of disjoint cycles of lengths in A, and denote this by $\mathscr{P}^{(A,\mathbb{Z}^{\geq 0})}$. Furthermore, if $B \subseteq \mathbb{Z}^{\geq 0}$, we may consider the subclass $\mathscr{P}^{(A,B)} \subseteq \mathscr{P}^{(A,\mathbb{Z}^{\geq 0})}$ consisting of only those permutations whose cycle count is found in B. That is, any permutation of cycle count not in B, or containing a cycle length not in A, are prohibited. Note, that having a cycle-count of zero means that the permutation is S_0, the unique permutation of the empty set.

The following theorems were first proven (presumably) long ago but can be derived from [116, Ch. 2.2] and also [200], and it is commonly noted that the technique in general was used by Laplace in the late 18th century. The nomenclature is the author's, and the proofs are due to Shaun Ault of Fordham University, on topics suggested by Nicolas Courtois.

Theorem 4. *The Strong Cycle Structure Theorem:*

The combinatorial class $\mathscr{P}^{(A,B)}$ *has associated EGF,* $\mathscr{P}_e^{(A,B)}(z) = \beta(\alpha(z))$, *where* $\beta(z)$ *is the EGF associated to B and* $\alpha(z) = \sum_{i \in A} \dfrac{z^i}{i}$.

However, we only need a weaker form in all but one case in this paper:

Theorem 5. *The Weak Cycle Structure Theorem:*

The combinatorial class $\mathscr{P}^{(A,\mathbb{Z}^{\geq 0})}$ *has associated EGF,* $\mathscr{P}_e^{(A,\mathbb{Z}^{\geq 0})}(z) = \exp(\alpha(z))$, *where* $\alpha(z)$ *is as above:* $\alpha(z) = \sum_{i \in A} \dfrac{z^i}{i}$

This is clearly a special case of the Strong Cycle Structure Theorem with $\beta(z) = 1 + z + z^2/2! + z^3/3! + z^4/4! + \cdots = e^z$ (the EGF of $\mathbb{Z}^{\geq 0}$). Interestingly, if $A = \mathbb{Z}^+$, then $\alpha(z) = z + z^2/2 + z^3/3 + z^4/4 + z^5/5 + \cdots = \log\left(\frac{1}{1-z}\right)$, which provides a verification of the theorem in this special case:

$$\exp\left(\log\left(\frac{1}{1-z}\right)\right) = \frac{1}{1-z} = 1 + z + z^2 + z^3 + z^4 + \cdots,$$

which is the EGF for the combinatorial class $\mathscr{P}$ of all permutations (together with the unique permutation on 0 letters), as expected.

Since the proof of the strong version is not fundamentally more difficult than the weak version, we shall provide a proof of Theorem 4. While this has been proven already in [116, Ch. 2.2], we feel that a more expository proof is appropriate in this context. First, a lemma which proves the case $B = \{k\}$.

Lemma 6. *The combinatorial class* $\mathscr{P}^{(A,\{k\})}$ *has associated EGF,*

$$\mathscr{P}_e^{(A,\{k\})}(z) = \frac{1}{k!} \left(\sum_{i \in A} \frac{z^i}{i} \right)^k.$$

Proof. Let $A \subseteq \mathbb{Z}^+$. For a given cycle-count, k, we must only include cycles of lengths found in A. Begin with an OGF. If $\pi \in S_n$ has k cycles, then its cycle structure defines a partition of n identical objects into k containers, where each container cannot have any number of objects that does not occur as a member of A. The OGF that generates this is $\left(\sum_{i \in A} z^i \right)^k$, as stated in Section 4.2.4.3. Now, we must remember that those objects in the containers are *not* identical! Think of each cycle-structure as being a template onto which we attach the labels $1, 2, 3, 4, \ldots, n$ in some order. A priori, this provides a factor of $n!$ for each partition of n, and so the coefficient of z^n in the above OGF should be multiplied by $n!$. The best way to accomplish this is to simply consider our OGF as an EGF: In our OGF, if C_n is the coeffiecient of z^n, then as EGF, $n!C_n$ is the coefficient of $z^n/n!$. Now, for each disjoint cycle of length i, there are i ways of cyclically permuting the labels, each giving rise to an equivalent representaion of the same i-cycle. Thus, we have over-counted unless we divide each term z^i by i. Finally, each rearrangement of the k cycles among themselves gives rise to an equivalent expression for the permuation, so we must divide by $k!$, and our EGF for permutations of cycle-count k with cycle-lengths in A now has the required form, $\mathscr{P}_e^{(A,\{k\})}(z) = \frac{1}{k!} \left(\sum_{i \in A} z^i/i \right)^k.$ $\square$

The proof of Theorem 4 then follows easily:

Proof. Let $A \subseteq \mathbb{Z}^+, B \subseteq \mathbb{Z}^{\geq 0}$. Categorize all permutations in $\mathscr{P}$ by cycle-count. Only permutations with cycle-counts $k \in B$ will contribute to our total, so by Lemma 6,

$$\mathscr{P}_e^{(A,B)}(z) = \sum_{k \in B} \mathscr{P}_e^{(A,\{k\})}(z) = \sum_{k \in B} \frac{1}{k!} \left(\sum_{i \in A} \frac{z^i}{i} \right)^k = \sum_{k \in B} \frac{\alpha(z)^k}{k!} = \beta(\alpha(z)),$$

since $\sum_{k \in B} z^k/k!$ is the EGF associated to B. The Weak Cycle Structure Theorem then follows as an immediate corollary. $\square$

4.3.1 Expected Values

While OGFs and EGFs are very useful for the study of a one-parameter family of constants, $A_0, A_1, A_2, A_3, \ldots$, we often wish to work with a two-parameter family, $\{A_{s,t}\}_{s,t \geq 0}$. This is accomplished using *double* generating functions. The double OGF, $A(y, z)$ of a two-parameter family of constants, $\{A_{s,t}\}$ is defined to be the formal sum:

$$A(y,z) = \sum_{s=0}^{\infty}\sum_{t=0}^{\infty} A_{s,t} y^s z^t,$$

and the EGF $A_e(y,z)$ is defined to be the formal sum:

$$A_e(y,z) = \sum_{s=0}^{\infty}\sum_{t=0}^{\infty} \frac{A_{s,t}}{(s+t)!} y^s z^t.$$

For our purposes in proving Theorem 7, we will be interested in a combinatorial class of permutations categorized not only by the order of the symmetric group S_n in which the permutation lies, but also by the number of fixed points that the permutation possesses.

Theorem 7. *Let $\mathscr{F} \subset \mathscr{P}$ be a combinatorial class of permutations with double EGF $a(y,z)$, where the coefficient of $y^s z^t /(s+t)!$ is the number of permutations π with property ϕ_s such that $\pi \in S_{s+t}$. Then the limit (as $n = s+t$ goes to infinity) of the expected value of s such that a random permutation of size n satisfies ϕ_s is given by:*

$$\lim_{z\to 1^-}(1-z)a_y(z,z)$$

provided $(1-z)a_y(z,z)$ is convergent and continuous from the left at $z=1$.

Proof. Let $a(y,z) = \sum_{s\geq 0}\sum_{t\geq 0} y^s z^t A_{s,t}/(s+t)!$. The coefficient of $y^s z^t$ is the probability that a random permutation of S_{s+t} has property ϕ_s, by construction. Consider the partial derivative with respect to y:

$$a_y(y,z) = \sum_{s\geq 0}\sum_{t\geq 0} \frac{sA_{s,t}}{(s+t)!} y^{s-1} z^t.$$

The probabilities are now multiplied by the corresponding value of s. Now, letting $y = z$ produces:

$$a_y(z,z) = \sum_{s\geq 0}\sum_{t\geq 0} \frac{sA_{s,t}}{(s+t)!} z^{s+t-1} = \sum_{n\geq 0}\left(\sum_{s+t=n}\frac{sA_{s,t}}{n!}\right)z^{n-1}.$$

Thus, $a_y(z,z)$ is the OGF that computes the expected value of s such that a random permutation of size n satisfies ϕ_s (shifted by one degree). Using the same technique as in the proof of Thm 3, we find that

$$\lim_{z\to 1^-}(1-z)a_y(z,z) = \lim_{n\to\infty}\left(\sum_{s+t=n}\frac{sA_{s,t}}{n!}\right).$$

$\square$

4.4 Corollaries

Corollary 8. *The probability that a random permutation (in the limit as the size grows to infinity) does not contain cycles of length k is given by $e^{-1/k}$.*

Proof. The set A of allowable cycle lengths is $\mathbb{Z}^+ - \{k\}$, and so has EGF given by artificially removing the term for k from the EGF of $\mathscr{O}$:

$$z + \frac{z^2}{2} + \frac{z^3}{3} + \cdots + \frac{z^{k-1}}{k-1} + 0 + \frac{z^{k+1}}{k+1} + \frac{z^{k+2}}{k+2} + \cdots = \log\left(\frac{1}{1-z}\right) - \frac{z^k}{k},$$

and thus by the Weak Cycle Structure Theorem, the combinatorial class in question has EGF equal to

$$a(z) = exp\left(\log\left(\frac{1}{1-z}\right) - \frac{z^k}{k}\right) = \frac{1}{1-z}e^{-z^k/k}$$

Thus the probability of a random permutation (as the size tends toward infinity) not having any cycles of length k is given by $\lim_{z \to 1^-}(1-z)a(z) = e^{-1/k}$ □

Note: On the Precision of these estimations:

This result means that $p \to e^{-\frac{1}{k}}$ when $N \to \infty$. What about when $N = 2^{32}$? We can answer this question easily by observing that the Taylor expansion of the function $a(z)$ is the EGF and therefore gives all the *exact* values of $A_n/n!$. For example when $k = 4$ we computed the Taylor expansion of $g(z)$ at order 201, where each coefficient is a computed as a ratio of two large integers. This takes less than a second with the computer algebra software MAPLE [3]. The results are surprisingly precise: the difference between the $A_{200}/200!$ and the limit is less than 2^{-321}. Thus convergence is very fast and even for very small permutations (on 200 elements). See also, Section 4.2.5.2 on Page 38.

Returning to the proving of corollaries, let us define $\mathscr{P}^{\overline{A}} = \mathscr{P}^{(\mathbb{Z}^+ - A, \mathbb{Z}^{\geq 0})}$ and find its EGF.

Lemma 9. *The EGF of $\mathscr{P}^{\overline{A}}$ is given by $exp(f(z))$, where*

$$f(z) = \sum_{i \notin A} z^i/i = \log\left(\frac{1}{1-z}\right) - \sum_{i \in A} z^i/i$$

Proof. Because $\mathscr{P}^{\overline{A}} = \mathscr{P}^{(\mathbb{Z}^+ - A, \mathbb{Z}^{\geq 0})}$ we can use the Weak Cycle Structure Theorem. The EGF of the combinatorial class of cycles with size from the set $\mathbb{Z}^+ - A$ is given by that of $\mathscr{O}$ (the class of all cycles) with the "forbidden lengths" artificially set to zero, namely

$$\sum_{i\in(\mathbb{Z}^+ - A)} z^i/i = \sum_{0<i\notin A} z^i/i = \left(\sum_{i>0}\frac{z^i}{i}\right) - \left(\sum_{i\in A}\frac{z^i}{i}\right) = \log\left(\frac{1}{1-z}\right) - \sum_{i\in A} z^i/i$$

The correct answer follows.   □

Corollary 10. *Let A be a subset of the positive integers. The probability that a random permutation (in the limit as the size grows to infinity) does not contain cycles of length in A is:*

$$\prod_{i\in A} e^{-1/i} = e^{-\sum_{i\in A} 1/i}$$

Proof. Using Lemma 9 we obtain an EGF of

$$exp\left(\log\left(\frac{1}{1-z}\right) - \sum_{i\in A} z^i/i\right) = \frac{1}{1-z}\prod_{i\in A} e^{-z^i/i}$$

then multiplying by $(1-z)$ and taking the limit as $z \to 1$ gives the desired result.   □

This offers confirmation of Corollary 8 when substituting $A = \{k\}$. A permutation with no fixed points is called a derangement. Using a similar strategy, we can calculate the probability of a derangement.

Corollary 11. *Let π be a permutation taken at random from S_n. The probability that π is a derangement is $1/e$ in the limit as $n \to \infty$.*

Proof. Just apply Corollary 10 to the case of $\mathscr{P}^{\overline{\{1\}}}$.   □

Suppose we wish to consider if a permutation has exactly t cycles of length from a set $C \subset \mathbb{Z}^+$, in other words, all the other cycles are of length not found in C. In that case, we can consider such a permutation π as a product of π_A and π_B such that π_A has only t cycles of length found in A, and nothing else, and π_B has only cycles of length not found in A. This is an example of a "labelled product" [116, Ch 2.2], as discussed in Section 4.2.5.1 on Page 36. Recall, the EGF of a labelled product is merely the product of the EGFs.

Theorem 12. *Let π be a permutation taken at random from S_n. The probability that π has c fixed points is $1/(c!e)$.*

Proof. Consider $\pi = \pi_A\pi_B$, where π_A consists of exactly c fixed points, and nothing else, while π_B is a derangement of the remaining $n-c$ points. We must compute the labelled product

$$f(z) = \mathscr{P}_e^{(\{1\},\{c\})} \times \mathscr{P}_e^{\overline{\{1\}}}$$

Thus, by the Strong and Weak Cycle Structure Theorems,

$$f(z) = \frac{z^c}{c!}exp\left(\log\left(\frac{1}{1-z}\right) - z\right) = \frac{z^c}{(1-z)c!}e^{-z}$$

An application of Thm 3 provides the result:

$$\lim_{z \to 1^-} (1-z)f(z) = \lim_{z \to 1^-} \frac{z^c}{c!}e^{-1} = \frac{1}{c!e}$$

□

4.4.1 On Cycles in Iterated Permutations

Theorem 13. *Let π be a permutation in S_n. A point x is a fixed point for π^k if and only if x is a member of a cycle of length i in π, for some positive integer i dividing k.*

Proof. Write π in disjoint cycle notation, and then x appears in only one cycle (hence the name "disjoint.") Call this cycle ψ. Since all other cycles do not contain x, then $\pi^m(x) = \psi^m(x)$ for all integers m. Of course, ψ is of order i in S_n, thus $\psi^i = id$, the identity element of S_n.

If x is in a cycle of length i then that means that i is the smallest positive integer such that $\psi^i(x) = x$. Write $k = qi + r$ with $0 \le r < i$. Then

$$x = \psi^k(x) = \psi^r(\psi^{iq}(x)) = \psi^r((\psi^i)^q(x)) = \psi^r(id^q(x)) = \psi^r(id(x)) = \psi^r(x)$$

so $\psi^r(x) = x$ but we said that i is the least positive integer such that $\psi^i(x) = x$ and $r < i$. The only way this is possible is if r is not positive, i.e. it is zero. Thus $k = qi$ or i divides k.

The reverse assumes that i divides k so write $iq = k$ then

$$\psi^k(x) = \psi^{iq}(x) = (\psi^i)^q(x) = (id)^q(x) = id(x) = x$$

□

4.4.1.1 An Example

Before we continue, observe what happens to a cycle of π when evaluating π^2. First, if the cycle is of odd length,

$$(x_1, x_2, \ldots, x_{2c+1}) \mapsto (x_1, x_3, x_5, \ldots, x_{2c+1}, x_2, x_4, x_6, \ldots, x_{2c})$$

but if the cycle is of even length,

$$(x_1, x_2, \ldots, x_{2c}) \mapsto (x_1, x_3, x_5, \ldots, x_{2c-1})(x_2, x_4, x_6, x_8, \ldots, x_{2c})$$

One can rephrase Theorem 13 as follows:

Corollary 14. *Let π be a permutation from S_n. Let k be a positive integer, and let the set of positive integer divisors of k be D. Then the set of fixed points of π^k is precisely the set of points under π in cycles of length found in D.*

4.4.2 Limited Cycle Counts

Theorem 15. *Let k be a positive integer, and π a permutation from S_n. The expected number of fixed points of π^k is $\tau(k)$, taken in the limit as $n \to \infty$. Note, $\tau(k)$ is the number of positive integers dividing k.*

Proof. We shall construct a double EGF, $a(y,z)$, where the coefficient of $y^s z^t$ is the probability that the kth power of a random permutation of S_{s+t} has s fixed points. Let π be a permutation taken at random from S_n. A point x is a fixed point under π^k if and only if x is a member of a cycle of order dividing k under π, via Corollary 13. Note also π^k has exactly t fixed points if and only if $\pi = \pi_A \pi_B$, where $\pi_A \in S_t$ consists only of cycles of length dividing k, and $\pi_B \in S_{n-t}$ consists only of cycles of length *not* dividing k. Let D_k be the set of all positive divisors of k. The double EGF that counts the number of such permutations $\pi_A \pi_B$ will be given by the labelled product $\mathscr{P}_e^{(D_k, \mathbb{Z}^{\geq 0})}(y) \cdot \mathscr{P}_e^{\overline{D_k}}(z)$. By the Weak Cycle Structure Theorem and Lemma 9, we obtain:

$$a(y,z) = exp\left(\sum_{i|k} \frac{y^i}{i}\right) exp\left(\log\left(\frac{1}{1-z}\right) - \sum_{i|k} \frac{z^i}{i}\right)$$

$$= exp\left(\log\left(\frac{1}{1-z}\right)\right) exp\left(\sum_{i|k} \frac{y^i}{i} - \sum_{i|k} \frac{z^i}{i}\right)$$

$$= \frac{1}{1-z} exp\left(\sum_{i|k} \frac{y^i - z^i}{i}\right).$$

Theorem 7 provides the correct expected value. First observe that

$$a_y(y,z) = \frac{1}{1-z} exp\left(\sum_{i|k} \frac{y^i - z^i}{i}\right) \sum_{i|k} y^{i-1}.$$

Then $a_y(z,z) = \frac{1}{1-z} exp(0) \sum_{i|k} z^{i-1}$. Finally,

$$\lim_{z \to 1^-} (1-z)a_y(z,z) = \lim_{z \to 1^-} \sum_{i|k} z^{i-1} = \sum_{i|k} 1 \overset{def}{=} \tau(k).$$

$\square$

4.4.3 Monomial Counting

For the case of $\mathbb{GF}(2)$, it is easy to count how many monomials are possible, of degree d among n variables. This is equal to $\binom{n}{d}$, because for each subset of the variables, there is exactly 1 possible monomial.

However, for $\mathbb{GF}(8)$ just as an example, we know from Fermat's "little" theorem that $x^8 = x$. (In general, in $\mathbb{GF}(q)$ we will have $x^q = x$). Therefore, there is no reason to distinguish between $x^9 y$ and $x^2 y$, as these are equal for all field elements that could be assigned to x and y. See Section 11.3.2 on Page 191 for details on this point.

How do we count the number of possible monomials of degree d, among n variables, over the field $\mathbb{GF}(q)$? The answer is actually strikingly similar to the 5000 items placed in boxes that could contain 1, 2, 3, or 4 objects, that we studied in Section 4.2.4.3 on Page 35.

Each monomial can have degree $0, 1, 2, \ldots$, up to $q - 1$ on any particular variable. Thus, each variable is a box that can contain $0, 1, \ldots, q - 1$ "degree points". We have precisely n such boxes, which will represent one variable each. Note, while the items going into the boxes are indistinguishable, the boxes are not. This is because in $x_i x_i x_j$, the first x_i is indistinguishable from the second, but not from the x_j. Then, all we need to do is multiply out the OGFs, and take the dth degree term.

Thus each box has OGF $\mathscr{B}(z) = 1 + z + z^2 + z^3 + \cdots + z^q$ and a sequence of length n of them (a monomial) has OGF $\mathscr{M}(z) = \left(1 + z + z^2 + z^3 + \cdots + z^q\right)^n$ and finally, we have

Theorem 16. *Counting equivalent monomials as identical, over the field* $\mathbb{GF}(q)$, *we have k monomials of degree d among n variables where k is the coefficient of the* z^d *term in*

$$\left(1 + z + z^2 + z^3 + \cdots + z^q\right)^n$$

Of course, for ease of calculation, one can truncate the sum at z^d instead of at z^q if $d < q$. This is because there is no monomial of degree 11 that contains x_i^{12}, just as an example. Truncating the series makes it easier to calculate.

Note, this formula appeared in [232], as Proposition 3.1, but without proof or derivation, and using different notation.

4.5 Of Pure Mathematical Interest

The authors encountered the following interesting connections with some concepts in number theory, but they turned out to be not needed in the body of the book. We present them here for purely scholarly interest.

4.5.1 The Sigma Divisor Function

Lemma 17. *The sum* $\sum_{i|k} 1/i = \frac{1}{k}\sigma(k)$ *where both i and k are positive integers, and where* $\sigma(k)$ *is the divisor function (i.e. the sum of the positive integers which divide k).*

Proof.

$$\sum_{i|k} 1/i = \frac{k}{k}\sum_{i|k} 1/i = \frac{1}{k}\sum_{i|k} k/i = \frac{1}{k}\sum_{i|k} i \overset{def}{=} \frac{1}{k}\sigma(k)$$

□

Corollary 18. *Let π be a permutation taken at random from S_n. The probability that π^k is a derangement is $e^{-\sigma(k)/k}$, in the limit as $n \to \infty$.*

Proof. Let D be the set of positive integers dividing k. From Corollary 14, we know that x is a fixed point of π^k if and only if x is in a cycle of length found in D for π. We will use Corollary 9, with $\overline{A} = \overline{D}$. We obtain the probability is $e^{-\sum_{i \in D} 1/i}$, and Lemma 17 gives the desired result.   □

Note that substituting $A = \{1\}$ into the above yields the same result as Corollary 11.

4.5.2 The Zeta Function and Apéry's Constant

Corollary 10 provides an amusing connection with Riemann's zeta function. Recall, for complex s, the infinite series,

$$\sum_{n \geq 1} 1/n^s \overset{def}{=} \zeta(s)$$

defines the "zeta function" $\zeta(s)$, provided the series converges.

Corollary 19. *The probability that a random permutation (in the limit as the size grows to infinity) does not contain cycles of square length is:*

$$e^{-\sum_{i \geq 1} 1/i^2} = e^{-\zeta(2)} = e^{-\pi^2/6} \approx 0.19302529,$$

or roughly $1/5$.

Corollary 20. *The probability that a random permutation (in the limit as the size grows to infinity) does not contain cycles of cube length is:* $e^{-\zeta(3)} \approx 0.30057532$

Note, $\zeta(3)$ is known as Apéry's Constant [223] [18], and occurs in certain quantum electrodynamical calculations, but is better known to mathematicians as being the probability that any three integers chosen at random will have no common factor dividing them all [227].

4.5.3 Greatest Common Divisors and Cycle Length

Theorem 21. *Let π be a permutation from S_n. If x is in a cycle of length ℓ under π, then x will be in a cycle of length $\ell/gcd(k,\ell)$ under π^k.*

Proof. Consider a cycle $(x_1, x_2, \ldots, x_\ell)$ of a permutation π. In the group of integers modulo ℓ under addition (denoted $\mathbb{Z}_\ell$) the element k is a generator if and only if k is coprime to ℓ. Repeated evaluations of π send x_1 to $x_2, x_3, x_4, \ldots, x_\ell, x_1, x_2, \ldots$, and repeated evaluations of π^k send x_1 to $x_{1+k}, x_{1+2k}, x_{1+3k}, x_{1+4k}, \ldots$, where the subscripts are taken as addition modulo ℓ. We can see that the subscripts are the coset containing 1, of the subgroup generated by k in $\mathbb{Z}_\ell$

If k is coprime to ℓ, then this subgroup is the whole group, and thus there is only one coset, the entire group of ℓ elements. Thus x_i is in a cycle of length ℓ.

If k is not comprime to ℓ, then let the greatest common divisor of k and ℓ be g. Surely k generates a subgroup of order ℓ/g, and the coset containing 1 must be the same size as this subgroup, because all cosets are the same size. Thus, x_1 is in a cycle of length $\ell/\gcd(\ell,k)$.

Of course, we can rewrite this cycle ℓ ways, so that any of its elements is x_1, by shifting it one place each time we write it. Thus all x_i in the cycle of π have the same cycle length as x_1 under π^k.   □

Corollary 22. *Let π be a permutation from S_n, and let k, ℓ be positive integers with $\gcd(k,\ell) = g$. If ψ is a cycle in the disjoint cycle decomposition of π, and has length ℓ, then in the disjoint cycle decomposition of π^k, it will be replaced by g cycles of length ℓ/g.*

Proof. Let x be an element not fixed by ψ. Obviously, there are ℓ such elements. Using Theorem 21 we know that under π^k, and thus ψ^k, that x will be in a cycle of length ℓ/g. Since all the ℓ elements of the cycle have to have the same outcome, then all ℓ of them are in cycles of length ℓ/g, see Theorem 13 on Page 45. Therefore, there must be g such cycles in the decomposition of ψ.

Since the cycles making up the decomposition of π are disjoint, and since disjoint cycles commute, then the decomposition of ψ as a permutation, and its decomposition as a "factor" of π are identical, since no cancellation could possibly occur among disjoint cycles.   □

4.6 Highly Iterated Ciphers

Here we present two attacks, which while no where near practical feasibility, present surprising results that the author did not anticipate.

4.6.1 Distinguishing Iterated Ciphers

Suppose there were three naïve cryptography students, who choose to use 3-DES iterated[1] roughly one million times, because they are told that this will slow down a brute force attacker by a factor of one million. Alice will choose 1,000,000 iterations, Bob will choose 1,081,079 iterations and Charlie will choose 1,081,080 iterations. Intuitively, one would not expect these three choices to have significantly different security consequences.

However, assuming that the 3-DES cipher for a random key behaves like a randomly chosen permutation from $S_{2^{64}}$, these permutations will have

$$\tau(1,000,000) = 49 \qquad \tau(1,081,079) = 2 \qquad \tau(1,081,080) = 256$$

fixed points which allows for the following distinguisher attack. It is noteworthy that Charlie's number is the lowest positive integer x to have $\tau(x) = 256$, while Bob's number (only one less) is prime, and thus has $\tau(x-1) = 2$. This enables the dramatic difference in vulnerability to the attack.

In a distinguishing attack, the attacker is presented either with a cipher under a random key, or with a random permutation from the set of those with the correct domain. The attack will proceed as follows: Randomly iterate through $1/64$ of the plain-space. If a fixed point is found, guess that one is being given a user cipher. If no fixed point is found, guess random.

In the case of Alice's implementation, there will be an expected value of ≈ 0.766 fixed points. In the case of Bob's, $1/32$ expected fixed points. In the case of Charlie's, 4 expected fixed points. A random permutation would have $1/64$ expected fixed points. Thus, we can see that Charlie's would be easily distinguishable from a random permutation, but Bob's much less so. Against Alice, the attack could definitely still be mounted but with an intermediate probability of success. To make this notion precise, we require the probability distribution of the number of fixed points of π^k.

Theorem 23. *Let $\pi \in S_n$ be a permutation chosen at random, then the cth term of the following EGF*

$$exp\left(\sum_{i|k} \frac{y^i - 1}{i} \right)$$

is the probability that π^k has exactly c fixed points.

Proof. Consider the double EGF of Theorem 15, $a(y,z) = \frac{1}{1-z}exp(\sum_{i|k} \frac{y^i - z^i}{i})$. Recall, the coefficient of $y^s z^t$ is the probability that $\pi^k \in S_{s+t}$ has s fixed points. Now, for any given s, we can find the probability that $\pi^k \in S_n$ has s fixed points (in the limit

---

[1] Since the brute force attack is the optimal attack known at this time, it is perhaps not completely unreasonable. The classic UNIX implementations encrypt with a variant of DES 25 times, for example [120, Ch. 8].

as $n \to \infty$), by evaluating $\lim_{z \to 1^-} (1-z)a(y,z)$. The result is the EGF $exp(\sum_{i|k} \frac{y^i-1}{i})$.

$\square$

However, the above requires us to have 256 terms inside of the exponentiation, for there are 256 positive integers dividing 1,081,080, and we will need to know the coefficient of the cth term for at least 1000 terms. Therefore, we are compelled to leave this expansion as a challenge for the computer algebra community.

Meanwhile, we performed the following experiment. We generated 10,000 random permutations π from $S_{10,000}$ and raised π to the kth power for the values of k listed. Then we calculated c, the number of fixed points of π^k, and determined if a search of the first 1/64th of the domain would reveal no fixed points. That probability is given by

$$(1 - c/n)^{n/64} \approx e^{-c/64}$$

and taking the arithmetic mean over all experiments, one obtains

No fixed points One or more

	No fixed points	One or more	
$k = 1$	0.985041	0.014959	Random
$k = 1000000$	0.797284	0.202716	Alice
$k = 1081079$	0.984409	0.015591	Bob
$k = 1081080$	0.418335	0.581665	Charlie

Perhaps this is unsurprising, as in the case of Charlie, we expect 256 fixed points, and so it would be surprising if all of those were missing from a part of the domain equal to 1/64th of the total domain in size. On the other hand, for Bob we expect only 2 fixed points, and it is exceptional that we find one by accident.

Finally, we observe that if there is an equal probability of an adversary being presented with a random cipher from $S_{2^{64}}$ or 3-DES in the key of one of our three users, iterated to their exponent, then the success probability of the attacker would be for Alice 59.39%, for Bob 50.03%, and for Charlie 78.34%. For example $(0.9850\cdots)/2 + (0.5817\cdots)/2 = 0.7834\cdots$ for Charlie. Note in each case, we check only $2^{64}/64 = 2^{58}$ plaintexts, and so this attack is $2^{112}/2^{58} = 2^{54}$ times faster than brute-force.

4.6.1.1 Repeating the Attack

Suppose we iterate the distinguishing attack an odd number of times, and take the "majority vote" as a "best guess". This is done in many human situations, perhaps it will have an effect here?

In asymptotic cryptography there is a well-known result that if against a cipher (parameterized by a security parameter k), there is some attack A which succeeds with probability "non-negligibly different from one-half" then iterating A polynomially-many times will raise that probability to any particular desired value between 1/2 and 1 (e.g. 2/3 or 9/10). Here, the cryptic phrase "$p(k)$ is non-negligibly different from one-half" means that there is some polynomial $f(k)$ such that $1/(p(k) - 1/2) < f(k)$ for k sufficiently large. An example of a $p(k)$ that is acceptable is $1/2 + 1/k$ and an example of one that is not acceptable is $1/2 + 2^{-k}$. The concept of negligible functions is also used in Section 5.2.2.1 on Page 68.

And so, in the world of asymptotic cryptography, it is merely necessary to calculate the success probability at the end of a distinguishing attack, and show that it differs from 1/2 non-negligibly. Finally note "polynomially-many" times refers to k as well, in that the number of repetitions is upper bounded by some polynomial in terms of k.

In concrete security, it is very difficult to try to understand what the translation of the above means. However, if we look at the success probability against Bob, for the naïve definition of "negligible", we would say that the attack against his cipher (50.03%) has negligibly different probability of success from 1/2; meanwhile, for Charlie's cipher, we would say that the success probability of the attack (78.34%) indeed is negligibly different from 1/2. The following table shows how we can very easily amplify the success probability against Charlie, but that this fails miserably against Bob. As always, Alice's cipher performs in between.

Executions	1 of 1	2 of 3	3 of 5	4 of 7	5 of 9
Alice	$0.5939\cdots$	$0.6392\cdots$	$0.6719\cdots$	$0.6983\cdots$	$0.7205\cdots$
Bob	$0.5003\cdots$	$0.5005\cdots$	$0.5006\cdots$	$0.5007\cdots$	$0.5008\cdots$
Charlie	$0.7834\cdots$	$0.8795\cdots$	$0.9285\cdots$	$0.9562\cdots$	$0.9727\cdots$
Plainspace Used	1/64	3/64	5/64	7/64	9/64

4.6.1.2 A General Maxim:

The grand conclusion is that if a permutation must be iterated for some reason, then it should be iterated a prime number of times, to avoid fixed points.

4.6.2 A Key Recovery Attack

Consider the cipher given by

$$F_{k_1,k_2}(p) = E_{k_1}(E_{k_2}^{(n)}(E_{k_1}(p))) = c$$

where k_1 and k_2 are keys, and $E_k(p) = c$ is encryption with a block cipher (let $D_k(c) = p$ denote decryption). If E is DES and $n = 1$, then this is the "triple DES" construction. Here, we consider that E is AES-256 as an example, and n is Charlie's number, 1081080. Then F is a block cipher with 512-bit key and 128-bit plaintext block. We will refer to k_1 as the outer key, and k_2 as the inner key.

Suppose an attacker had an oracle for F that correctly encrypts with the correct k_1 and k_2 that the target is using. Call this oracle $\phi(p)$. Observe that $G_{k_3}(x) = D_{k_3}(\phi(D_{k_3}(x)))$ will have $G_{k_3}(x) = E_{k_2}^{(n)}(x)$ if and only if $k_3 = k_1$. Thus if we can correctly guess the outer key, we have an oracle for the nth iteration of encryption under the inner key. If $k_3 \neq k_1$, then provided that E_{k_1} is computationally indistinguishable from a random permutation from $S_{2^{128}}$ when k_1 is chosen uniformly at random (a standard assumption) then $G_{k_3}(x)$ also behaves as a random permutation.

Thus, for $k_1 = k_3$, we can expect $G_{k_3}(x)$ to behave like Charlie's cipher in the previous section, and for $k_1 \neq k_3$, we can expect $G_{k_3}(x)$ to behave like a random permutation in the previous section.

Let one execution of the distinguishing attack signify guessing all possible k_3 values, and executing the previous section's attack for each key. If "random" is indicated (i.e. no fixed point found), then we reject the k_3 but if "real" is indicated (i.e. at least one fixed point found), then we add k_3 to a "candidate list."

After one run of this distinguishing attack, we would have a candidate list of outer keys of expected size

$$(0.014959)(2^{256} - 1) + (0.581665)(1)$$

where the success probabilities are given in the previous section, for the attack on Charlie. Note, we are using no majority voting scheme, because the majority voting schemes of the previous section have too many false positives (see Section 4.6.1.1 on Page 51).

If we repeat the distinguisher attack on these candidate keys, taking care to use a distinct set of plaintexts in our search, the success probabilities will be the same. This non-overlapping property of the plaintext search could be enforced by selecting the six highest-order bits of the plaintext to be the value of n. After n runs, we would expect the list to contain

$$(0.014959)^n(2^{256} - 1) + (0.581665)^n(1)$$

candidate keys.

Of course, the true $k_3 = k_1$ key will be present with probability 0.581665^n. Next, for each key k_c on the candidate list, we will check all possible 2^{256} values of k_2 (denoted k_x), via checking if

$$p = \phi(D_{k_c}(D_{k_x}^{(n)}(D_{k_c}(p))))$$

which will be true if $k_x = k_2$ and $k_c = k_1$. This check should be made for roughly 4–6 plaintexts, to ensure that the match is not a coincidence. This necessity arises from the fact that the cipher has a 512-bit key and 128-bit plaintext. We will be very conservative, and select 6.

The number of encryptions required for the n runs is

$$(1081080 + 4)\left(\frac{2^{128}}{64}\right)\left(2^{256} + (0.014959)(2^{256}) + (0.014959)^2(2^{256})\right)$$

$$+(0.014959)^3(2^{256}) + \cdots + (0.014959)^n(2^{256})\big) =$$

$$= (1081080 + 2)(2^{378}) \frac{1 - (0.014959)^{n+1}}{1 - 0.014959}$$

$$= 2^{398.06579\cdots}(1 - 0.014959^{n+1})$$

and for the second stage

$$(6)(2)(2 + 1081080)(2^{256})(0.014959^n)(2^{256}) = (2^{535.6290\cdots})(0.014959^n)$$

$$= 2^{535.6290 - 6.062842n}$$

for a success probability of $(0.581665)^n$.

Using MAPLE [3], we find that $n = 23$ is optimal, leaving a candidate list of $2^{116.555\cdots}$ possible keys, and requiring $2^{398.41207\cdots}$ encryptions, but with success probability $(0.581665)^{23} \approx 2^{-17.98001\cdots}$. A brute-force search of the 2^{512} possible keys would have $(6)(2)(1081082)2^{512}$ encryptions to perform, or $2^{535.629007\cdots}$. Naturally, if a success probability of $2^{-17.98001\cdots}$ were desired, then only $2^{517.649\cdots}$ encryptions would be needed for that brute-force search.

Therefore this attack is $2^{119.237}$ times faster than brute-force search.

Chapter 5
Stream Ciphers

While Keeloq is an important cipher, used in industry and with an interesting set of methods for its cryptanalysis, the author believes that some more examples might be useful. Here, we present the ciphers Trivium and Bivium, as well as QUAD.

The purpose of this chapter is not only to exposit on how the ciphers presented here can be or cannot be attacked. Instead, the main purpose is to share some interesting ciphers and exposit on how those ciphers are converted into a system of equations. This relationship between the cipher and the equations is not trivial. The task of converting a cipher to equations, and doing so efficiently, is a major task in algebraic cryptanalysis. Also, because QUAD is based on random systems of equations, it is an endless source of cryptanalytic examples. The Bivium and Trivium equations are an excellent source for testing new techniques.

While great care has been taken to cite the work of others carefully, and note who has done what, the author wishes to be rather clear that *nothing* contained in this chapter *whatsoever* is his own idea, but rather taken from cited published papers, and presented in a more pedagogical style.

5.1 The Stream Ciphers Bivium and Trivium

5.1.1 Background

In order to understand Bivium, what it is, what it is not, and why we should be interested in it, it is necessary to present a large quantity of background.

5.1.1.1 What is a Stream Cipher?

A stream cipher has some kind of internal state that is being continually updated. Suppose this is n bits, then we can write a function $f : \mathbb{GF}(2)^n \to \mathbb{GF}(2)^n$ that rep-

G.V. Bard, *Algebraic Cryptanalysis*, DOI: 10.1007/978-0-387-88757-9_5
© Springer Science + Business Media, LLC 2009

resents this "state-update function." Alternatively, it could be over some finite field
$\mathbb{GF}(q)$. At each cycle, some "filter function" takes this state, and outputs usually one
field element. QUAD, which is discussed below, is an exception, and will output several, but this is usually not the case. Thus g is usually written $\mathbb{GF}(q)^n \rightarrow \mathbb{GF}(q)$. So
for stream ciphers using $\mathbb{GF}(2)$, which is most of them, the output of g is a single
bit.

These field elements, outputted by the stream cipher as a "keystream" are added
to the plaintext, one symbol at a time. At the other end, they are subtracted. Of
course, over $\mathbb{GF}(2)$ these are the same operation. The internal state is initially some
setup-function of the secret key (shared between transmitter and receiver) and an
initialization vector. There will be more said on this in Section 5.1.1.4 on Page 59.

In summary, the entire operation can be represented

- $\mathbf{x_i} \leftarrow f(\mathbf{x_{i-1}})$
- $k_i \leftarrow g(\mathbf{x_i})$
- $c_i = p_i + k_i$
- Transmit c_i

which is repeated for all p_i in $p_1, p_2, \ldots, p_\ell$.

Historically, f represented a linear-feedback shift-register or some kind of non-linear version of one. Today, essentially all stream ciphers are much more complicated than that. In fact, they may be composed of several linear-feedback or
non-linear-feedback shift-registers, and it is crucial not to mistake their own state-update functions for the state-update function of the entire cipher. Also historically,
g merely took the most-significant or least-significant bit of the internal state. This
was found to be unwise due to several examples of cryptanalysis.

Stream ciphers are often extremely fast, on the order of $2\times-100\times$ as fast as a
block cipher. They are also often designed to be implemented in hardware, with
remarkably small and efficient circuits. That makes them ideal for use in mobile
devices where large circuits drain the battery too fast, and where memory is limited.
For a historical text on stream ciphers (up-to-date when written) see [125].

5.1.1.2 What was eSTREAM?

Some stream ciphers of the 20th century had been successfully broken in its last
decade. Most notable among these was AS5/1, and in particular the paper "Real
Time Cryptanalysis of AS5/1 on a PC" by Alex Biryukov, Adi Shamir and David
Wagner [48]. The cipher AS5/1, at that time, was in use by approximately 130 million European cell-phone users with GSM phones, according to that paper. Another
triumph was AS5/2, and in particular the paper "Instant Ciphertext-Only Cryptanalysis of GSM Encrypted Communication" by Elad Barkan, Eli Biham, and Nathan
Keller [36], on the cipher AS5/2, also used with GSM phones.

Recognizing this state of affairs at that time, and that there was no stream cipher
which held the same magnitude of trust and confidence as, for example, the block

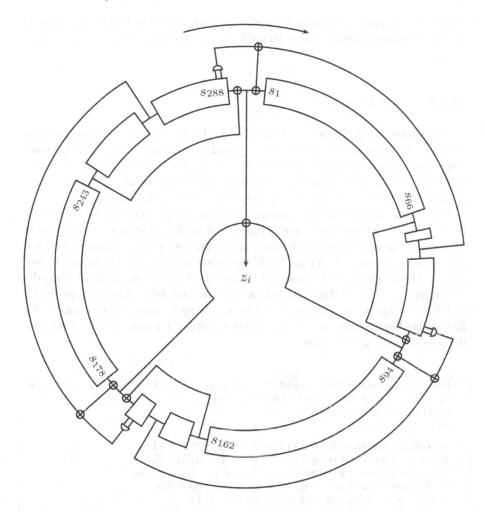

Fig. 5.1 The Stream Cipher Trivium

ciphers AES or 3-DES, the European Union cryptographic organization ECRYPT began the eSTREAM project.

A previous effort, the NESSIE project (New European Schemes for Signatures, Integrity and Encryption), had a stream cipher search which failed at the same task. Six stream-ciphers had been selected for NESSIE, and each one had a successful cryptanalytic attack published against it [194, Ch II.B.3], and so none were adopted.

And so a call for stream ciphers was announced in November of 2004, and the eS-TREAM competition began. In total, 34 stream ciphers were submitted. These were divided into software-based and hardware-based. Only 7 ciphers were placed in the eSTREAM "portfolio," of which 4 are software-based and 3 are hardware-based. One of those three is Trivium. It is an excellent example for algebraic cryptanalysis, particularly because it has yet to be broken (at least at the time this was written). If the reader breaks Trivium, surely the method and approach would be highly interesting.

5.1.1.3 What is Trivium?

The stream cipher Trivium was designed by Christophe De Cannière and Bart Preneel [56]. Because only 7 of the 34 proposed stream ciphers in eSTREAM were placed in the final "portfolio", and of these 7, only 3 are hardware-based, Trivium is worthy of our attention. Moreover, the design of Trivium is remarkably simple. The algorithm for Trivium encryption is given as Algorithm 2 on Page 58.

Christophe De Cannière has graciously made the image[1] in Figure 5.1 on Page 57 freely available for any one to use, reproduce or modify. It shows the relationship between the three shift registers which are at the core of the stream cipher. My own drawing, Figure 5.2 on Page 59, is somewhat less readable but includes more details. The idea is that Bivium-A and Bivium itself (sometimes called Bivium-B) are simpler versions of Trivium, and one can see that by looking at Figures 5.3 and 5.4 on Page 62, which are formed by removing elements of Trivium. So far Bivium has been broken, as well as Bivium-A, but not Trivium (at least at the time this was written, April of 2009).

INPUT: An initial condition of 93 bits $X_0, \dots, X_{92}$, as well as 84 bits $Y_0, \dots, Y_{83}$ and 111 bits $Z_0, \dots, Z_{110}$, for the NLFSRs, and N bits of plaintext $p_1, \dots, p_N$.
OUTPUT: Ciphertext of N bits $c_1, \dots, c_N$.
1: for $i = 1, 2, \dots, N$ do

    1: (Generate new bit for register X): $b_y \leftarrow X_{65} + X_{92} + X_{90}X_{91} + Y_{77}$
    2: (Generate new bit for register Y): $b_z \leftarrow Y_{68} + Y_{83} + Y_{81}Y_{82} + Z_{86}$
    3: (Generate new bit for register Z): $b_x \leftarrow Z_{65} + Z_{110} + Z_{108}Z_{109} + X_{68}$
    4: (Generate the key bit): $k \leftarrow X_{65} + X_{92} + Y_{68} + Y_{83} + Z_{65} + Z_{110}$
    5: (Encrypt the plaintext bit with the key bit): $c_i \leftarrow p_i + k_i$
    6: (Shift the register X by 1): for $j \in 92, 91, \dots, 1$ do $X_j \leftarrow X_{j-1}$
    7: (Put the new bit into X): $X_0 \leftarrow b_x$
    8: (Shift the register Y by 1): for $j \in 83, 82, \dots, 1$ do $Y_j \leftarrow Y_{j-1}$
    9: (Put the new bit into Y): $Y_0 \leftarrow b_y$
    10: (Shift the register Z by 1): for $j \in 110, 82, \dots, 1$ do $Z_j \leftarrow Z_{j-1}$
    11: (Put the new bit into Z): $Z_0 \leftarrow b_z$

Algorithm 2: The Stream Cipher Trivium [De Cannière and Preneel]

---

[1] http://www.ecrypt.eu.org/stream/ciphers/trivium/trivium.pdf

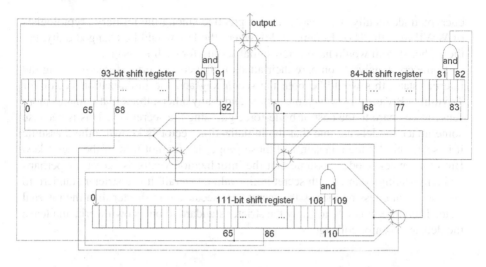

Fig. 5.2 The Stream Cipher Trivium

5.1.1.4 Secret Key versus Initial State

Trivium has, as do all eSTREAM ciphers, an 80-bit secret key, and an 80-bit "initialization vector", for a total of 160-bits of initial information. However, the state of the cipher is 288 bits. How does 160-bits of input "inflate" into a 288-bit initial state?

This situation is remarkably common and not discussed in many books on cryptography, so we will explain in detail here. One option could have been to simply give the system an initial key of 288 bits, since the internal state is 288 bits. After all, secret keys are to be kept entirely secret, be generated by means mathematically equivalent to fair coins, and so on. Surely that would result in a secure initial condition.

However, actually generating random numbers is expensive and slow. Generating them badly is extremely unwise (e.g. the Venona Code (see [220])). Also, in hardware situations, you do not want to rekey very often, because that is an expensive operation. Often, sharing the secret key (for example between a cell phone and the tower) involves RSA (see Section 10.1.1 on Page 160), or some other public-key system. These are multi-step number-theoretic operations and certainly involve much more computation than running the stream cipher itself.

Therefore, one takes a secret key of size decided to be sufficient for the security of the data involved (80 bits). For other security objectives, an initialization vector is provided (in this case it is also 80 bits) and then the 160-bit input is inflated into the initial state. One use of initialization vectors is that two consecutive messages should probably not be encrypted identically, but at the same time, rekeying after each message might be extremely inconvenient. Thus, having the initialization vector available prevents rekeying too often, but also prevents two distinct messages from being encrypted identically. This was common practice among the military cryptologists of WWII (e.g. the Nazi Enigma code), where the key would be changed daily, but the cipher system would have a new initial setting for each message.

Some cryptanalysts compare their attack to the cost of brute-force guessing the internal state (this would be both the secret key and the initialization vector for Trivium or Bivium). However, the author, and many others, think that it makes more sense to compare to the cost of brute-force guessing the secret key. This is because some attack models assume that the initialization vector is known, others assume it is selectable by the attacker, and those properties are not true of the secret key. But other attack models assume that the initialization vector is secret, or perhaps an incrementing counter with secret initial value. Overall, it is a stricter standard to be faster than guessing merely the secret key, because it is shorter than the internal state. Therefore, it is safe to apply the stricter standard to one's own work, and leave the debate to senior colleagues.

5.1.1.5 Initialization Stage in Trivium

The way that this is done in Trivium is that the shift-register X is preset to the 80-bit secret key, padded with 13 bits of zero to make 93 total bits. The shift-register Y is preset to the 80-bit initialization vector, padded with 4 bits of zero to make 84 bits. The padding is done so that X_0 is the first bit of the key and Y_0 is the first bit of the initialization vector, and X_{80} as well as Y_{80} are the first bits of the padding. And the shift-register Z is preset to all zeroes, except the 3 most significant bits Z_{108}, Z_{109}, and Z_{110} in our notation, are set to one. Then, after this preset, the system is clocked $4 \times 288 = 1152$ times.

The last step, of clocking the system more than a thousand times, is very important, and would be present even if there was a 288-bit secret key re-chosen every message. The reason is that transmitting the "all-zero plaintext" or other types of simple message might leak bits of the key into a form readily read by the attacker. The idea is that k_{t+1152} is essentially uncorrelated to k_t. This process of allowing the cipher to run for a while before using its output is called an "initialization phase", "key setup phase", or for historical reasons a "key schedule."

5.1.1.6 Two Types of Attack

A key-recovery attack on Trivium/Bivium would try to discover the secret key, based on plaintext-ciphertext pairs encrypted under that key. This means one must have equations for the setup phase as well as for the encryption of those messages. Consequently, there will be many equations, and one must solve them in time faster than required to check all 2^{80} possible keys in a brute-force key search.

Another type of attack—state recovery—is as follows. The cryptanalyst simply takes the initial 288-bit state to play the role of the unknown that he/she is trying to solve for. One ignores the setup phase. Nonethless, the attack must be faster than 2^{80}

test encryptions, not 2^{288} encryptions, in order to be considered "faster-than-brute-force." If the attacker can recover the initial state at any time, then the message can be read from that point forward.

To recover the secret key and initialization vector from the initial state might be very difficult or impossible—a process called rewinding. This is in stark contrast to Keeloq, which could be rewound. Of course, since Keeloq is a block cipher, there is no key setup phase either.

Part of the irreversibility of the Trivium/Bivium family comes from the AND-gate. Given a 1 at the output of an AND-gate, one knows that both inputs were a 1. But given a 0 at the output of an AND-gate, it could be that the inputs were either 00, 01, or 10. This ambiguity is part of the irreversibility.

However, if the secret key is desired for special reasons (see Appendix A.3 on Page 303), then it can be found. Once the internal state is solved for, in many cases including Trivium, the operation of the cipher can be executed in reverse, and the state at time 1152 rewound to $t = 1151, t = 1150, \ldots, t = 1$. Then, the initial key and IV will be known.

5.1.1.7 What is Bivium?

In the cryptanalysis of block ciphers, sometimes one has an attack that is not faster than brute force, but can work successfully on "reduced" versions of the cipher. For example, the author has a paper with Nicolas Courtois on performing an algebraic cryptanalysis on 6 rounds of the Data Encryption Standard—a block cipher with 16 rounds [76]. Attacks against fewer rounds than the total are an important step toward advancing the techniques of cryptanalysis.

For stream ciphers, however, there is no obvious equivalent to simply reducing the number of rounds. The stream cipher Bivium was created, by Hårvard Raddum [196], to play that role for Trivium. Bivium has a design which is extremely similar to Trivium. In fact, Bivium has two non-linear feedback shift-registers (NLFSRs), while Trivium has three NLFSRs. In Bivium, they are of length 93 and 84, or 177 bits in total length. In Trivium, they are of length 93, 84, and 111, or 288 bits in total length.

5.1.2 Bivium as Equations

Bivium can be specified in terms of the circuit diagram given in Figure 5.4 on Page 62, but instead it can also be specified in terms of an algorithm, which is given in Algorithm 3. Note that the key-setup phase was stated to be $4 \times 177 = 708$ clocks instead of 1152 in Trivium.

The algorithm for Bivium encryption is given as Algorithm 3 on Page 63.

While Keeloq had only really one shift-register (as the key itself was simply rotating), you can see from the diagram or the algorithm that Bivium has two shift

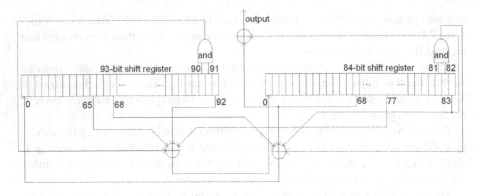

Fig. 5.3 The Stream Cipher Bivium-A

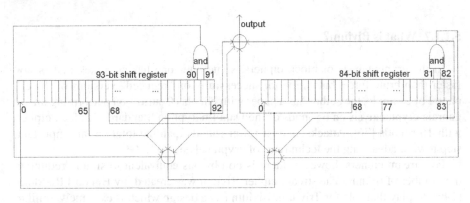

Fig. 5.4 The Stream Cipher Bivium, sometimes called Bivium-B

registers. We will call these X and Y respectively, and they are of length 93 bits and 84 bits. As before, the bits are numbered X_0 for the leftmost, and X_{92} for the rightmost, and Y_0 through Y_{83}, likewise.

The first cycle will be $t = 1$. At $t = 200$, suppose X_{40} is s. Then at $t = 201$ we know $X_{41} = s$; at $t = 202$ we know $X_{42} = s$, and so forth. Therefore at time t we know $s = X_{t-160}$, which is valid for $252 \geq t \geq 160$. At $t = 252$, then $s = X_{92}$ and the bit "falls off" the left side of the shift register. The bit was inserted as the first bit of the shift register X_0 in time $t = 160$, which means it was calculated from the circuit in $t = 159$.

INPUT: An initial condition of 93 bits $X_0, \ldots, X_{92}$, as well as 84 bits $Y_0, \ldots, Y_{83}$, for the NLFSRs, and N bits of plaintext $p_1, \ldots, p_N$.
OUTPUT: Ciphertext of N bits $c_1, \ldots, c_N$.
1: for $i = 1, 2, \ldots, N$ do

    1: (Generate new bit for register X): $b_x \leftarrow X_{65} + X_{92} + X_{90}X_{91} + Y_{77}$
    2: (Generate new bit for register Y): $b_y \leftarrow X_{68} + Y_{83} + Y_{81}Y_{82} + Y_{68}$
    3: (Generate the key bit): $k \leftarrow X_{65} + X_{92} + Y_{68} + Y_{83}$
    4: (Encrypt the plaintext bit with the key bit): $c_i \leftarrow p_i + k_i$
    5: (Shift the register X by 1): for $j \in 92, 91, \ldots, 1$ do $X_j \leftarrow X_{j-1}$
    6: (Put the new bit into X): $X_0 \leftarrow b_x$
    7: (Shift the register Y by 1): for $j \in 83, 82, \ldots, 1$ do $Y_j \leftarrow Y_{j-1}$
    8: (Put the new bit into Y): $Y_0 \leftarrow b_y$

Algorithm 3: The Stream Cipher Bivium [Hårvard Raddum]

We will renumber the subscripts for convenience. Let X_i at time t be represented by x_{92+t-i}. Thus, in the previous paragraph s is actually x_{252}. We can calculate that x_n will "fall off" the left side of the shift register at time $t = n$, because $x_n = X_{92}$ at that moment. Likewise, $x_n = X_0$ when $t = n - 92$, and so if $n > 93$ then x_n was calculated from the circuit at time $t = n - 93$. Of course, x_1 was bit X_{92} when $t = 1$ and x_{93} was bit X_0 at that time.

Likewise, let Y_i at time t be represented as y_{83+t-i}. If $n > 84$, then y_n was calculated from the circuit at time $t = n - 84$. Lastly, y_1 was bit Y_{83} when $t = 1$ and y_{84} was bit Y_0 at that time.

Therefore, the initial state of X, which is $X_0, X_1, \ldots, X_{92}$ at $t = 1$, is recorded as $x_{93}, x_{92}, \ldots, x_1$ (note the reversal of ordering). Similarly, the initial state of Y, which is $Y_0, Y_1, \ldots, Y_{83}$ at $t = 1$, is recorded as $y_{84}, y_{83}, \ldots, y_1$.

All x_n for $n > 93$ are created by the circuit, given by the following formula:

$$b_x \leftarrow X_{68} + Y_{68} + Y_{83} + Y_{81}Y_{82}$$

so at time t this becomes

$$x_{t+93} = x_{24+t} + y_{15+t} + y_t + y_{t+2}y_{t+1}$$

On the other hand,

$$b_y \leftarrow Y_{77} + X_{65} + X_{92} + X_{90}X_{91}$$

so at time t this becomes

$$y_{t+84} = y_{15+t} + x_{27+t} + x_t + x_{t+1}x_{t+2}$$

As you can see, we introduce two new variables per time-cycle. The key bit at time t is given by $X_{92} + X_{65} + Y_{68} + Y_{83}$. Since, for all stream ciphers, $k_t + p_t = c_t$, we have

$$p_t + c_t = x_t + x_{27+t} + y_{15+t} + y_t$$

and note that p_t as well as c_t are both known.

Thus for t cycles where we have both plaintext and ciphertext, we will write 3 equations, and add 2 new variables. The initial setting for X and Y represent 177 additional variables as well. It turns out [176] that 177 clock cycles is necessary and sufficient. This yields 531 variables and 531 equations.

5.1.2.1 Features of these Equations

These equations have several features which make them noteworthy:

- They are of degree two.
- They have a very short length, consisting of only five terms for all three types. This makes them ideal for the Raddum-Semaev method (see Section 12.10 on Page 238).
- They are very sparse. With 531 variables, there are 141246 possible monomials, and so the sparsity is $\beta = 3.54 \times 10^{-5}$. This makes them very susceptible to SAT-Solvers [176], see Chapter 13.
- There are very few quadratic terms. In absolute terms, only the b_x and b_y equations have quadratic terms at all, and they have only one. The $p_i + c_i$ equations have none. In relative terms, one anticipates roughly 266 times as many quadratic terms as linear, but in reality, there are roughly 0.154 quadratic terms per linear term.
- The pattern of the equations is highly structured and predictable. Therefore, they differ greatly from random.

Note that the equations $c_i + p_i = x_t + x_{27+t} + y_{15+t} + y_t$ are what will use our knowledge of the plaintext-ciphertext pairs. Suppose the last one is $c_f + p_f$. Then x_{27+f} is the last value of f and y_{15+f} is the last value of y that we need. If we take the equations as specified here, we would continue to x_{f+93} and y_{f+84}. This means that the least 66 equations of x and the last 69 equations of y will calculate information that is never used. Thus we can delete these 135 equations, at considerable savings.

5.1.3 An Excellent Trick

The following example is given in Section 5.2 of [176]. The term $s_{162} + s_{177}$ appeared in three of their equations. They replaced it with a dummy variable a, and added an equation $a = s_{162} + s_{177}$. This shortens the lengths of the sums, and so dramatically shortens the clauses (see Section 13.4.1.1 on Page 249) and though it introduces a new variable, it also drops the density of the system.

5.1.4 Bivium-A

The two wires which are designated in the diagram with dotted lines are removed to convert Bivium-B into Bivium-A, which is far less secure. It is surprising that merely "cutting" two wires could have this effect. See Figure 5.3 on Page 62. The change to the algorithm is extremely simple, in that the line

$$k \leftarrow X_{65} + X_{92} + Y_{68} + Y_{83}$$

should become instead

$$k \leftarrow Y_{68} + Y_{83}$$

and the change to our equations only affects the following:

$$p_t + c_t = y_{15+t} + y_t$$

but recall that p_t and c_t are known. The effect is rather dramatic.

Bivium-A was broken in 21 seconds by [176] and about a day by [196]. Meanwhile, Bivium-B is estimated to take 2^{56} and 2^{52} bit operations respectively, by those authors.

5.1.5 A Notational Issue

Beware that in the papers under discussion, McDonald, Charnes and Pieprzyk as well as Raddum use a different bit numbering for the internal state. They denote $X_0, X_1, \ldots, X_{92}$ as $s_1, s_2, \ldots, s_{93}$, as well as $Y_0, Y_1, \ldots, Y_{83}$ by by $s_{94}, s_{95}, \ldots, s_{177}$. Likewise, for Trivium, they denote $Z_0, Z_1, \ldots, Z_{110}$ as $s_{178}, s_{179}, \ldots, s_{288}$. But the two or three shift-registers are distinct, and so it is not clear why this was done. In the interest of pedagogy, we renumbered.

5.1.6 For Further Reading

For further reading, the following may be useful

- "Trivium: A Stream Cipher Construction Inspired by Block Cipher Design Principles" by Christophe De Cannière [56], published at ISC'06.
- "Cube Attacks on Tweakable Black Box Polynomials", by Itai Dinur and Adi Shamir [98], published at EUROCRYPT'09.
- "Breaking One.Fivium by AIDA an Algebraic IV Differential attack" by M. Vielhaber [218], published as an e-print.
- "Differential Fault Analysis of Trivium" by Michal Hojsík and Bohuslav Rudolf [139], published in the proceedings of FSE'08.

- "An Algebraic Analysis of Trivium Ciphers based on the Boolean Satisfiability Problem", by Cameron McDonald, Chris Charnes, and Josef Pieprzyk [176].
- "Floating Fault Analysis of Trivium" by Michal Hojsík and Bohuslav Rudolf [140], published in the proceedings of INDOCRYPT'08.
- "Two Trivial Attacks on Trivium", by Alexander Maximov and Alex Biryukov [175], published at *Selected Areas of Cryptography* in 2007.
- "Slid Pairs in Salsa20 and Trivium", by Deike Priemuth-Schmid and Alex Biryukov [195], and a response to it "Response to Slid Pairs in Salsa20 and Trivium" by Daniel Bernstein [44].
- "Cryptanalytic Results on Trivium", by Hårvard Raddum [196], published as an eSTREAM techincal report.
- "Attacking Bivium with SAT-Solvers", by Tobias Eibach, Enrico Pilz and Gunnar Völkel [105], published in the proceedings of the 2008 SAT-Solver conference.

5.2 The Stream Cipher QUAD

QUAD is not a stream cipher *per se* but a family of stream ciphers, parameterized by a positive integers m, n, and a finite field $\mathbb{GF}(q)$. The designers of QUAD, Côme Berbain, Henri Gilbert, and Jacques Patarin, stated designing a relatively efficient stream cipher with provable security as their goal [42]. The heart of QUAD is a random system of m polynomials, of degree 2, in $n < m$ unknowns. It is a stream cipher with internal state being a sequence of n field elements, and outputting $m - n$ keystream field elements per cycle. This is unusual, in that most stream ciphers output one keystream bit per cycle for $\mathbb{GF}(2)$, or one field element if over larger finite fields.

The stream cipher is unique in that there is a proof of security. What this means is that asymptotically, an adversary who can break a version of QUAD at certain settings (in the sense of distinguishing it from a random generator) can solve the NP-Complete problem MQ for other settings. Since MQ is NP-Complete (see Section 11.5 on Page 199), this is not believed to be possible for "sufficiently large" problems. However, we will see that this has little practical impact for the recommended parameter choices.

5.2.1 How QUAD Works

Let the internal state then be $\mathbf{x} = (x_1, x_2, \ldots, x_n)$, where each x_i is a field element from the finite field in question $\mathbb{GF}(q)$. Each of the m polynomials can be evaluated for a particular value of $\mathbf{x}$, and so we can call the polynomials $f_1, f_2, \ldots, f_m$, and think of them as functions. Then

$$\mathbf{x} \leftarrow (x_1, x_2, \ldots, x_n) = (f_1(\mathbf{x}), f_2(\mathbf{x}), \ldots, f_n(\mathbf{x}))$$

$$\mathbf{k} \leftarrow (k_1, x_2, \ldots, k_{m-n}) = (f_{n+1}(\mathbf{x}), f_{n+2}(\mathbf{x}), \ldots, f_m(\mathbf{x}))$$

represents one cycle. In other words, each of the m polynomials is evaluated at $\mathbf{x}$, giving m field elements. The first n of these become the new $\mathbf{x}$, and the remaining $m - n$ are outputted as $m - n$ keystream field elements.

Another way to look at it is that $f_1, f_2, \ldots, f_n$ taken together make a map $GF(q)^n \rightarrow GF(q)^n$, and this is the state update function. The filter function is $GFq^n \rightarrow GF(q)^{m-n}$ given by the remaining $m - n$ polynomials $f_{n+1}, f_{n+2}, \ldots, f_m$.

The creators of QUAD suggest, for example, $n = 160$ and $m = 320$, over $GF(2)$ [42]. They recommend that the polynomials be dense, but leave open the option for future research into sparse polynomials to make faster versions of the cipher, provided that they can also be proven secure. So far, only the $GF(2)$-and-dense case has been proven.

As in Bivium and Trivium, there is also a "key setup" or initialization process (see Section 5.1.1.4 on Page 59). The cipher has a key and an initialization vector, which are combined in a complex process to produce the initial internal state. After that point, the cipher operates as above. We do not attack the initialization process here, and so we omit its description.

5.2.2 Proof of Security

5.2.2.1 Computationally Indistinguishable

First, we must define what it means for a set A to be computationally indistinguishable from a set B. The notion of "computationally indistinguishable" is somewhat confusing. Essentially it means that no polynomial-time algorithm can distinguish between an element drawn at random from set A from an element drawn at random from set B, both of a size k, with sufficiently high success. It is the notion of "sufficiently high" that will cause us some difficulty. If there was a fixed probability of an item from A versus B appearing, then we could talk of the probability of correctness. However, there is a more general notion.

Let us say that the algorithm will output either "A" or "B", and we will measure the absolute value of the difference of the probability that the algorithm will output "A" given that the object in question really is from A versus the probability that the algorithm will output "A" given that the object in question really is from B. This difference is called the "advantage" of the algorithm.

While it is not normal to do so, the language of combinatorial classes (see Section 4.2.1 on Page 30) is useful here. The advantage of algorithm D shall be calculated as follows. Let $Adv(D, k) = |p_{A,k} - p_{B,k}|$, where $p_{A,k}$ is the probability that D outputs "A" given that it is presented with an item drawn uniformly at random from the set of objects in the combinatorial class $\mathscr{A}$ that have size equal to k. Likewise, $p_{B,k}$ is the same, but the object is drawn from $\mathscr{B}$. Now, either $Adv(D, k)$ is negligible or non-negligible compared to k.

Definition 24. We write that $f(k)$ is *negligible compared to* k if there does not exist
a polynomial $p(k)$ such that $1/f(k) \le p(k)$ for all k sufficiently large.

An example of an $f(k)$ that is negligible is 2^{-k} and one that is not negligible is
k^{-2}. The concept of negligible functions is also used in Section 4.6.1.1 on Page 52. If
for all polynomial time algorithms D we have that $Adv(D,k)$ is negligible compared
to k then the sets A and B are computationally indistinguishable.

5.2.2.2 The Objective

A pseudorandom generator g is a family of functions, indexed by k, such that
$g_k : \mathbb{F}^k \rightarrow \mathbb{F}^{m_k}$, with $m_k > k$, with the following further property. The set given by
$g_k(x)$ for all $x \in \mathbb{F}^k$ should be computationally indistinguishable from the set $\mathbb{F}^{m_k}$.
In other words, the image of g, for any k, lives inside of $\mathbb{F}^{m_k}$. This subset, formed
by applying g to every possible input from the strictly smaller set $\mathbb{F}^k$, should be
computationally indistinguishable from the whole of $\mathbb{F}^{m_k}$.

A pseudorandom generator should not be considered a pseudorandom function,
but all pseudorandom functions (with range larger than their domain) are pseudorandom generators. The distinction is that the input to a pseudorandom function can
be chosen by the distinguishing algorithm, but the distinguishing algorithm against
a pseudorandom generator must consider only the case of the inputs to the function
being generated uniformly at random over the entire set of possible inputs.

As it turns out, QUAD is not the first stream cipher family for which an attempt
has been made to prove pseudorandom generator status. In fact, [42] lists several
previous attempts at the end of Section 1. But, QUAD is special for several reasons.
First, the proof is based on a universally agreed upon hard problem, that of solving
a quadratic system of equations over a finite field (particularly $\mathbb{GF}(2)$). Second, the
stream cipher is fairly efficient for certain parameter settings.

Unfortunately, the statement of status as a pseudorandom generator essentially
only promises that security exists for security parameter choices above a certain
threshold, and it is by no means easy to figure out what those settings should be.
Furthermore, [232] shows that the recommended settings are not secure. Moreover,
this then requires the user to select larger settings, and this adverse affects the performance of the stream cipher. Even with the recommended settings, the authors of
QUAD [42] note that QUAD is very significantly slower than the AES. Finally, a
proof over fields other than $\mathbb{GF}(2)$ was not attempted (though it might be proven
between the time of writing and the moment that the reader reads this).

5.2.2.3 The Underlying Hard Problem: A Pre-Image Finder

The underlying hard problems are MQD and MQ. The MQ problem is as fol-
lows. Given a quadratic system of equations over $\mathbb{GF}(2)$, can one find a solution (a
setting for each variable that makes each polynomial equation true). The decision
problem MQD is related, where one must merely state if a solution exists or not. We

will prove the MQD problem NP-Complete in Section 11.5 on Page 199 and also the MQ problem NP-hard in the same section. Note the connection with the satisfiability problems, which merely substitute logical sentences for quadratic equations, as discussed in Chapter 13.

The problem can also be seen as similar to a numerical analysis problem. Given an m-dimensional vector $\mathbf{b}$, and an $m \times n$-dimensional matrix A, can one find an n-dimensional vector $\mathbf{x}$ such that $A\mathbf{x} = \mathbf{b}$, or certify that none exists. This is solving a linear system of equations. Instead, if we replace the rows of A with quadratic, rather than linear polynomials, we get a polynomial system of equations. Either can be seen as a map from $\mathbb{F}^n \to \mathbb{F}^m$.

In the linear case, it is a problem solvable in time $\Theta(nm \min(m, n))$ as shown in Section 7.5.3 on Page 97. In the quadratic case, it is NP-hard, as shown in Section 11.5 on Page 199 for $\mathbb{GF}(2)$.

Restricting to $\mathbb{GF}(2)$, for now, we can see that there is the interesting question of the number of expected solutions. If $m = n$ then we can expect probably 1 or 2 solutions. The number of solutions is either 2^{n-r+1}, where r is the rank of matrix A, or there are no solutions (see Corollary 28 on Page 88). The quantity $n - r$ is known as the "nullity", and Table 9.3 on Page 145 lists common nullities for $\mathbb{GF}(2)$-matrices, along with their probabilities. If $m > n$ by a significant margin, we expect 0 or 1 solutions.

However, many times, including testing system solving software (see Section 13.5.1 on Page 255), we wish to force the existence of one solution. This can be done by choosing a $\mathbf{x}$ at random, calculating $A\mathbf{x} = \mathbf{b}$ and then discarding $\mathbf{x}$. The algorithm can then be called to try to recover $\mathbf{x}$. Since multiple solutions is an unlikely outcome, and at least one solution is forced to exist, it is *extremely likely* that exactly one solution exists.

Returning to QUAD, this is again the case. An $\mathbf{x}$ will always exist. But in QUAD, one often has $m = 2n$, and so the probability of multiple solutions is very near zero. Furthermore, the designers of QUAD state that the rank of the equations should be full-rank, and so then there is no possibility of multiple solutions. The existence of exactly one solution is important for SAT-solvers, because they only output one solution.

Meanwhile, seeing the system of equations as a map, $\mathbb{F}^n \to \mathbb{F}^m$, would result in the map being injective if it is always the case that every vector in the image $\mathbb{F}^m$ has exactly one pre-image in $\mathbb{F}^n$.

5.2.2.4 Outline of a Proof

Suppose the QUAD algorithm is run λ cycles, with a random initial key, and each cycle requires T_S time to run. Then $\lambda(m-n)$ field elements will have been outputted, and call this a sequence of type 1. Alternatively, generate the correct number of field elements ($\lambda(m-n)$ of them) by selecting each element uniformly at random from the finite field, and call this a sequence of type 2. Furthermore, the finite field in question is $\mathbb{GF}(2)$.

The four theorems of [42] then prove that if an algorithm A running in time T_A can distinguish a sequence of type 1 from a sequence of type 2, with "advantage" ε, then there is an algorithm C with the following properties.

Algorithm C is given a randomly chosen quadratic system of m polynomials in n unknowns, considered as a map $S : \mathrm{GF}(2)^n \rightarrow \mathrm{GF}(2)^m$. Furthermore, for some randomly chosen value of $\mathbf{x} \in \mathrm{GF}(2)^n$, the algorithm C is given $S(\mathbf{x})$. Algorithm C must produce a $\mathbf{y}$ such that $S(\mathbf{y}) = S(\mathbf{x})$. As stated earlier, S is almost certainly injective and so that would imply $\mathbf{y} = \mathbf{x}$. Furthermore, the success probability of Algorithm C is at least $\varepsilon/(8\lambda)$, and the running time is

$$T_C \leq \frac{2^7 n^2 \lambda^2}{\varepsilon^2} \left(2 + T_A + (\lambda + 2)T_S + \log_2 \frac{2^7 n \lambda^2}{\varepsilon^2} \right) + \frac{2^7 n \lambda^2}{\varepsilon^2} T_S$$

The proof of the Berbain-Gilbert-Patarin theorem [42] includes many vital concepts of the subject, and while it is somewhat difficult, it is recommended that the reader attempt to follow it. We will not reproduce the proof here.

Note, they write $m = kn$, where $k > 2$ is an integer. This rules out underdefined and exactly defined systems where $n \geq m$. So what is Algorithm C? It is a probabilistic pre-image finder for general quadratic systems of equations that happen to be overdefined with an "overdefinition" of $c = m/n = k$. Furthermore, the authors of QUAD recommend $k = 2$ and so $c = 2$. (see Section 13.3 on Page 248 for a description of the "overdefinition" c).

But a general pre-image finder, even if restricted to $c \geq 2$, would be a tremendous boon to all of algebraic cryptanalysis, and would likely[2] also solve all NP-Complete problems (see Section 11.5 on Page 199). Thus, before we can believe it exists, we must examine the running time, and the success probability.

5.2.2.5 Exploratory Example

First, since the success probability is $\varepsilon/(8\lambda)$ then by repeating the algorithm C, we can just keep trying until we get a right answer. The number of repetitions might be quite large, but its expected value is $8\lambda/\varepsilon$. Then we obtain an expected running time of

$$T'_C \leq \frac{2^{10} n^2 \lambda^3}{\varepsilon^3} \left(2 + T_A + \left(\lambda + 2 + \frac{1}{n} \right) T_S + \log_2 \frac{2^7 n \lambda^2}{\varepsilon^2} \right)$$

First, let us estimate T_S. The system has 320 equations and 160 unknowns, and so there are at most $\binom{160}{2} + 160 + 1 = 12881$ monomials per equation, or 4,121,920

---

[2] This point is very tricky. Such a "pre-imagine finding machine" would not solve MQ generically. It would only solve those cases where $m > n$. Only if one could prove that every SAT problem could be written as an MQ problem with this additional property ($m > n$), would it be the case that the machine could solve SAT problems even in the worse case.

Likewise, if one could prove that every instance of problem X could be written as an MQ problem with this additional property, and such that the MQ system had a size upper-bounded by some polynomial compared to the size of the instance of the problem X, where X is some NP-Complete problem, could one claim that P=NP.

possible total monomials. Since the coefficients of the system are random (generated by fair coins) then we expect 2,060,960 monomials. We can be very generous and assume 1 CPU-nanosecond per monomial, and so roughly $T_S \approx 2$ CPU-milliseconds.

Suppose the data encrypted were a gigabyte length movie. Then 2^{33} bits must be encrypted, and using the recommended $m = 2n$ (or $k = 2$) and $\mathbb{GF}(2)$, this means that $\lambda = 2^{33}/n$.

$$T_C' \leq \frac{2^{109}}{n\varepsilon^3} \left(2 + T_A + \left(\frac{2^{33}+1}{n} + 2 \right) T_S + \log_2 \frac{2^{73}}{n\varepsilon^2} \right)$$

We might also take the recommended $n = 160$. We have now assigned values to everything except T_A and ε, and T_C'.

Because T_A and ε are both variables, it appears at first glance we are stuck. Suppose we want a very good distinguisher, with advantage $\varepsilon = 1/4$, and we allow T_A to run in 10,000 CPU-years. We do not imply that we would let one CPU run for 10,000 years, but in some distributed network, such as BOINC [1] or SETI@Home [8], one obtains 100,000 volunteer CPUs and let the process run for 37 days.

Then solving the above inequality, we get $T_C' \leq 2.051 \cdots \times 10^{43}$ CPU-seconds for Algorithm C. Recall, Algorithm C is trying to find a pre-image for a 160-variable and 320-polynomial quadratic system. So then Algorithm C would be faster than a brute-force search if 2^{-160} times that running time would be the amount of time to accept or reject a brute-force guess at $\mathbf{x}$.

That, in turn, comes to 1.403×10^{-5} CPU-seconds and assuming 2 billion instructions per CPU-second, that is 28,066 instructions. This is entirely reasonable.

Therefore, the theorem tells us that if an attack on the cipher can distinguish it from random with $\varepsilon = 1/4$ in 10,000 CPU-years, and we have one 1 gigabyte of plaintext, with the recommended $n = 160$, then an algorithm which can find pre-images of a 320 polynomial 160 variable quadratic system can exist which runs in time $2.051 \cdots \times 10^{43}$ CPU-seconds or $6.499 \cdots \times 10^{35}$ CPU-years. If it had told us 10 CPU-seconds, then we would be shocked, and all agree that no such algorithm can exist, and logically we would be compelled to conclude that no such distinguisher attack on QUAD could exist either. But, the time given is not 10 CPU-seconds, but rather larger, and even a brute-force search for $\mathbf{x}$ could be done in that time. And so, in conclusion, the theorem tells us nothing for this case. Of course, finding 100,000 volunteers is extremely unlikely.

On the other hand, if $T_C' = 7.3075 \cdots \times 10^{38}$ were to come out of the equations, then we could note that this is equivalent not to 28,066 instructions per brute-force guess, but rather 1 instruction per brute-force guess. Indeed, surely that could never exist, because checking a brute-force guess would require more than one instruction. If we believe solving quadratic systems of equations to be hard, then we could believe that no such Algorithm C can exist. Then, we could solve backward for T_A and learn $T_A = 1.1152 \cdots \times 10^7$ CPU-seconds. That is 12.9 CPU-days. In other words, we have proven if no algorithm could find pre-images in time faster than 1 instruction per brute-force guess, then no distinguisher attack could run in less than

12.9 CPU-days. It is not much of a guarantee, but it is a guarantee. It would be more reassuring if the time given were in years and not days.

In a sense, the proof gives us a promise that for some parameter settings, the system is secure. But we are left without a clear idea of what those must be. The invention of the Quadratic Sieve caused similar uncertainty to users of RSA [192] at the time of the "squeamish ossifrage" paper [27].

5.2.3 The Yang-Chen-Bernstein-Chen Attack against QUAD

This attack, from [232] is against QUAD over $\mathbb{GF}(256)$ with 20 field-elements of internal state (160 bits), and 20 outputs per cycle. It is worth noting that the paper [232] has in it much more than we discuss here, and we recommend the reader, if interested in exploring QUAD further, examine it. We are also dropping the question of the semi-regularity of the equations, because the equations are generated uniformly at random.

5.2.3.1 The Combination of Wiedemann and XL-II

As described in Section 12.4 on Page 213, the XL algorithm, improved to XL-II in [80], reduces a polynomial systems of equations problem to a linear algebra problem. This brings to bear the full artillery of 50 years of linear algebra research on the difficulty of the problem. In particular, [232] applies the Block Wiedemann algorithm of Don Coppersmith to the problem [61]. For further reading on sparse matrices, see Appendix D on Page 323.

The expected number of field operations (which is not a good measurement for $\mathbb{GF}(2)$, but could be a good measurement for $\mathbb{GF}(256)$ and that is what concerns us here), for the Block-Wiedemann Algorithm is given by the following three expressions for an $n \times n$ matrix

$$\sim 3\beta n^3 \sim 3wn^2 \sim 3cn$$

where β is the sparsity (see Section 11.6 on Page 203), w is the average "weight" of a row (the number of non-zero entries in it), and c is the "content" of the matrix—the number of non-zero entries in the entire matrix. As it turns out, [232] estimates that the cost of a multiplication in this case should be $c_0 + c_1 \log n$, to represent the growing inefficiency of actually finding the entries in the sparse data structure as the matrix gets large. We do not contest that, but rather stipulate that this should be determined by experimentation.

The algorithm can be used to find vectors in the null space of the matrix, i.e. $A\mathbf{n} = \mathbf{0}$ but to solve $A\mathbf{x} = \mathbf{b}$ one can simply make a "dummy variable", and replace the constant 1 with this dummy variable. Solutions which have this dummy variable as zero are uninteresting, but those with the dummy variable equal to one are valid solutions.

Table 5.1 The number of monomials for two cases of polynomial systems over particular finite fields.

Degree	0	1	2	3	4	5	6	7	8	9	10
Monomials of this degree	1	20	210	1540	8855	42,504	177,100	657,800	2,220,075	6,906,900	20,030,010
Monomials up to this degree	1	21	231	1771	10,626	53,130	230,230	888,030	3,108,105	10,015,005	30,045,015

Above is for $GF(q)$, among 20 variables, $q > 10$.

Degree	0	1	2	3	4	5	6	7	8	9	10
Monomials of this degree	1	40	820	11,480	123,410	1,086,008	8,145,060	53,524,680	314,457,495	1,677,106,640	8,217,822,536
Monomials up to this degree	1	41	861	12341	135,751	1,221,759	9,366,819	62,891,499	377,348,994	2,054,455,634	10,272,278,170

Above is for $GF(q)$, among 40 variables, $q > 10$.

Degree	0	1	2	3	4	5	6	7	8	9	10
Monomials of this degree	1	35	630	7770	73,815	575,757	3,838,380	22,481,940	118,030,185	563,921,995	2,481,256,778
Monomials up to this degree	1	36	666	8436	82,251	658,008	4,496,388	26,978,328	145,008,513	708,930,508	3,190,187,286

Above is for $GF(q)$, among 35 variables, $q > 10$.

5.2.3.2 The Attack Itself

First, we will use Theorem 16 on Page 47 to count how many monomials there are among 20 variables in $\mathbb{GF}(256)$. This is given in Table 5.1 on Page 73. The data were found by typing the following into MAPLE [3].

```
expand((1+z+z^2+z^3+z^4+z^5+z^6+z^7+z^8+z^9+z^10)^20);
```

and one can truncate the sum at 10 terms by taking the 11-term Taylor series. Of course, the Taylor series (evaluated about the origin) of a polynomial is itself, but the following command accomplishes the trimming

```
taylor(%, z, 11)
```

Let us verify (for practice) one entry of the table. For terms of degree five, first there are terms of type $vwxyz$, and there are $\binom{20}{5} = 15,504$ of these. Then, there are terms of type v^2wxy, and there are $20\binom{19}{3} = 19,380$ of those. Next, there are terms of type v^2w^2x and there are $\binom{20}{2}18 = 3420$ of that type. Also, we must not forget terms of type v^3wx of which there are $20\binom{19}{2} = 3420$. Next, there also those of the form v^3w^2 of which there are only $20 \times 19 = 380$. Moving up to terms with fourth degree powers, there are those of the form v^4w, of which there are again only $20 \times 19 = 380$, and finally terms of the type v^5, of which there are precisely 20. This brings the total to 42,504, as desired! This was extremely tedious, and so we recommend Section 4.4.3 on Page 47 to the reader's attention.

It will be useful to take the first n of the m equations as an operator $f : \mathbb{GF}(q)^n \rightarrow \mathbb{GF}(q)^n$, and the next $m - n$ equations as an operator $g : \mathbb{GF}(q)^n \rightarrow \mathbb{GF}(q)$. This will then have f play the role of a state-update function and g play the role of a filter function as shown in Section 5.1.1.1 on Page 55. Given any state x_i, we will calculate $g(x_i) = k_i = P_i + C_i$ and $g(f(x_i)) = k_{i+1} = P_{i+1} + C_{i+1}$. Note, that this means that the attack merely requires two adjacent plaintext-ciphertext pairs anywhere in the message stream (only 320 bytes of known plaintext, but adjacent).

The equations of the form $g(x_i) = k_i$ are 20 equations, and of degree 2. Each of these has at most 231 terms in it, because that is the maximum possible for degree 2 as listed in the table. The equations of the form $g(f(x_i)) = k_{i+1}$ are again 20 equations, but they are of degree 4. Each of these has at most 10,626 terms in it—again this is the maximum possible as listed in the table.

The attack uses the XL-II algorithm, which is a variant of the XL algorithm (see Section 12.4 on Page 213), given in the paper [80]. We will elect to use "operating degree" 10 for the XL-step. This means the degree 2 equations will be multiplied by all possible monomials of degree 8 or lower (3,108,105 of those). That produces 62,162,100 equations with 231 terms in each one. Meanwhile, the other equations (which were degree 4) will be multiplied by all possible monomials of degree 6 or lower (230,230 exist). That produces 4,604,600 equations with 10,626 terms in each one. The grand total in the final system of equations has 66,766,700 equations and a total of 63,287,924,700 terms (among all equations). The total number of terms of degree 10 that are possible would be 30,045,015, the number of columns of the

XL matrix. Note, at the time of this writing, there are single PCs available with 128 gigabytes of RAM, and so storage (while expensive) would be entirely feasible even on one machine, let alone a cluster.

Therefore, we have a matrix with 66,766,700 rows and 30,045,015 columns, with content 63,287,924,700 non-zero entries (average row weight is 948 entries). The density is $\beta = 3.15492\cdots \times 10^{-5}$, which is rather sparse.

Trimming this to a square matrix of dimension 30,045,015, we would expect roughly 2.5670×10^{18} multiplications. Various machines tried in [232] had 8–14 cycles per multiplication. Using the 8 cycles per multiplication, this is 2.0536×10^{19} cycles, or at 2 billion instructors per second per core that is 325.37 core-years. A cluster of 1000 PCs would do this in under 119 days if a sufficiently parallel algorithm exists.

The internal state, of course, is 160 bits, and this attack is faster than brute force if checking one potential key takes more than 1.4051×10^{-29} core cycles, which is obviously the case. Alternatively, to compare it to a 80 bit secret key (ignoring the 80 bit initialization vector), then a verification of a potential key would have to take place in more than 1.6987×10^{-5} core cycles, which again is obviously the case.

Note, we described earlier in Section 5.1.1.4 on Page 60 why we believe the correct measurement is to compare to brute-forcing the secret key alone, and not the secret key as well as the initialization vector taken together.

5.2.4 Extending to $\mathbb{GF}(16)$

5.2.4.1 An Exercise

As an exercise, we invite the reader to shut the book, and with a blank piece of paper, replicate the above calculations for $\mathbb{GF}(16)$. There is no need to repeat the tedious calculation of Table 5.1 on Page 73. Note that to produce 160 bits of internal state, there will be 40 variables. Also, attempt the attack directly, as well as in the case of doing a "guess-and-determine" (see Section 11.7 on Page 206) attack and Fix-XL with 5 variables guessed, and 35 remaining to be solved for.

5.2.4.2 The Solution: Direct Version

We will proceed to do the example below. As noted, to achieve 160 bits of internal state, we will have 40 variables instead of 20 variables. Let us examine operating degree 8.

The equations of type $g(x_i) = k_i$ would be 40 equations, and they would be quadratic, thus have at most 861 terms. We would be multiplying by all possible monomials of degree 6. There are 9,366,819 of those. This results in 374,672,760 equations of 861 terms each, or total monomial count of 8,064,831,159.

The equations of type $g(f(x_i)) = k_{i+1}$ would be quartic, and there would be 40 of these as well. Instead of monomials of degree 6, here we will multiply by all monomials of degree 4 to bring the total to degree 8. There are 135,751 of those. This results in 5,430,040 equations of at most 135,751 terms each, or a total monomial count of 737,133,360,040.

There are now a total of 380,102,800 equations. But there are only 377,348,994 monomials of degree 8 or fewer. And therefore, we can expect an answer.

Now let us calculate how many field multiplications are expected. The total number of monomials is 745,198,191,199 (note, unlike the previous example, storage will be a serious question here). This is an average weight of $1960.52\cdots$ monomials per equation, or a $\beta = 5.1955 \times 10^{-6}$, very sparse indeed! Recall that $\sim 3\beta n^3$ is the expected number of field multiplications. Taking $n = 377348994$, we get $8.3749\cdots \times 10^{20}$ field multiplications. This is very much within the "faster than brute force" range but in order to claim feasibility, we should try to address the storage question.

5.2.4.3 Another look at Fix-XL

Let us repeat the above example but after fixing 5 of the 40 variables. This means that we will simply guess them, and try to solve the problem. If we guess correctly, with probability $16^{-5} = 2^{-20}$, then we will get the correct answer, otherwise we must repeat the attack with probability $1 - 2^{-20}$. We must expect 2^{19} trials on average. Note, this is yet another example of a guess-and-determine attack (see Section 11.7 on Page 206).

Formerly, we had 40 equations of degree two from $g(x_i) = k_i$, and since the degree was 2, we assumed they had 861 terms, as this was the worse possible. After filling-in the guessed values, we will still have 40 equations. But we will have at most 666 terms, as that is the maximum possible.

Likewise, we had 40 equations of degree four from $g(f(x_i)) = k_{i+1}$, and we again assumed the worse possible number of monomials, namely 135,751. But now, there are at most 82,251 monomials of degree 4 among 35 variables.

Let us try operating degree 7. This means that all the equations in the first group will be multiplied by all possible monomials of degree 5 or lower. There are 658,008 of those, for a total of 26,320,320 equations, with 666 monomials each, or a grand total of 438,233,328 total monomials.

Next, all the equations of the second group will be multiplied by all possible monomials of degree 3 or lower. There are 84,36 of them, for a total of 337,440 equations. They shall have 82,251 monomials each, or a grand total of 27,754,777,440 monomials.

Therefore the entire system has 26,657,760 equations. There are 26,978,328 possible monomials of degree 7, leaving us 320,568 short of the required amount. How unfortunate! This does show the importance of being exact, as otherwise small errors in approximation might have lead us to believe that we had indeed solved the problem.

Let us now try operating degree 8. The equations from the first part will be multiplied by all possible monomials of degree 6 or less. There are 4,496,388 of those. This will yield 179,855,520 total equations. Yet, there are only 145,008,513 possible monomials of degree up to 8 among 35 variables. Since the original equations had up to 666 monomials each, so will ours.

This means that the matrix will be of dimension 145,008,513 and will have only 666 entries per row, or $\beta = 4.5928 \cdots \times 10^{-6}$, which is very sparse indeed! Then the expect number of field operations will be 4.2013×10^{19}, considerably better than the previous attempt without fixing any variables.

Nonetheless, we must measure the amount of memory used. Multiplying the number of monomials per row by the dimension of the matrix, we obtain 96,575,669,658 monomials. This is well within the threshold of 128GB of some PCs at the time of this writing. That is the main advantage of guessing 5 variables in this case.

Finally, we adjust for the expected number of executions (2^{19} of them) to get a running time of 2.2027×10^{25} field operations. Previously we saw that 8–14 cycles was sufficient for a $\mathbb{GF}(256)$ multiplication. We shall make the horridly conservative estimate that the required time for a $\mathbb{GF}(16)$ multiplication is the same. Of course, it will be less, because the field is simpler, but continuing with 14 cycles per multiply, this comes to 3.0838×10^{26} cycles.

This is faster than brute force guessing all the keys if and only if checking one key is faster than 255.083 core cycles. Given our conservatism on the time for one multiplication, and the fact that 40 equations with 861 monomials would have to be evaluated, it is unclear if that this is the case.

As stated earlier, some prefer to compare against the total cost of the key and the initialization vector, in which case this attack is satisfactory if checking one potential internal state takes longer than 2.1100×10^{-22} core cycles, which is almost certainly the case.

Finally, note that the attack is not time-feasible, as it would require 4.886×10^9 core-years (in other words, with one million PCs each with 4 cores, it would take 1222 years). But the standard in cryptography is that we compare the attack to brute force, regardless if the time required is enormous.

The reader might want to recalculate the above if we guess merely 4 variables, instead of 5.

5.2.5 For Further Reading

For further reading, the following may be useful

- "QUAD: A Practical Stream Cipher with Provable Security", by Côme Berbain, Henri Gilbert, and Jacques Patarin [42], published at Eurocrypt 2006.
- "QUAD: A Multivariate Stream Cipher with Provable Security", by Côme Berbain, Henri Gilbert, and Jacques Patarin [42], published in the Journal of Symbolic Computation.

- "Analysis of QUAD", by Bo-Yin Yang, Owen Chia-Hsin Chen, Daniel J. Bernstein, and Jiun-Ming Chen [232], published at Fast Software Encryption in 2007.
- "Compact FPGA implementations of QUAD", by David Arditti, Côme Berbain, Olivier Billet and Henri Gilbert [19], published in the Proceedings of the 2007 ACM Symposium on Information, Computer and Communications Security.
- "QUAD: Overview and Recent Developments", by David Arditti, Côme Berbain, Olivier Billet, Henri Gilbert and Jacques Patarin [20], published in the Proceedings of Symmetric Cryptography 2007.

5.3 Conclusions for QUAD

No one can question that the Berbain-Gilbert-Patarin theorem is a theoretical work of great depth and also provides a security guarantee for $GF(2)$ with settings above a certain size. It does not seem possible, however, to determine what that size threshold happens to be. Without a proof of security for extension fields, one can only guess what is secure, and we have shown here (via [232]) several attacks on extension field instantiations. For $GF(2)$, the attack for $n = 160$ should leave one to believe that a minimum setting would be $n = 320$. The performance penalty one pays at that point might be too much to bear, compared to the AES. But as architectures and computing evolve, perhaps that comparison might change.

For the cryptanalyst, however, QUAD has now given us an infinite variety of ciphers against which to hone our tools, and it is for this reason that we have selected it to appear here.

Part II
Linear Systems Mod 2

Chapter 6
Some Basic Facts about Linear Algebra over $\mathbb{GF}(2)$

The purpose of this chapter is to identify some facts about $\mathbb{GF}(2)$-vector spaces and about matrices over $\mathbb{GF}(2)$. To emphasize the differences between matrices over $\mathbb{R}$, or $\mathbb{C}$, and matrices over $\mathbb{GF}(2)$, we note several interesting phenomena. The contents of this chapter are already known, but we present them here as a "warm up." They are stated here so that they can be used elsewhere, and for background.

6.1 Sources

Normally, we would cite a series of useful textbooks with background information but amazingly there is no text for finite field linear algebra. We do not know why this is the case. The algorithms book [13, Ch. 6] mentions algorithms for finite field linear algebra, but it has been out of print for many years. There are a few pages in [164, Ch. 7] that deal with this topic, there named "Linear Modular Systems." A linear modular system is a finite state machine that happens to have a state-transition function given as a matrix. The dynamics of such systems can be studied to much greater extent than those with higher degree functions for state-transitions. Also, Krishnamurthy's work [154, Ch. 2], discusses linear algebra over the integers, a related topic. The studies [133] and [117] appear highly cited and relevant but we have been unable to obtain a copy of either one. For solving linear sparse systems over finite-fields, and applications to cryptography, see [156]. For other topics in finite fields, [165] is an excellent desk-reference.

6.2 Boolean Matrices vs $\mathbb{GF}(2)$ Matrices

In graph theory, a particular ring-like object is often used. Its elements are "true" and "false"; multiplication is logical-AND and addition is logical-OR. The identity element for addition is "false." But then clearly, this algegbraic object has no addi-

G.V. Bard, *Algebraic Cryptanalysis*, DOI: 10.1007/978-0-387-88757-9_6

tive inverse for "true." Thus it is a commutative semigroup on both operations (as well as a monoid on both operations). The distributive law is easy to verify.

The name for this bizarre arrangement is a "commutative semiring." It turns out that linear algebra can be done in this world, in the sense of matrix multiplication and matrix squaring for calculating the transitive closures of digraphs. Matrices filled with elements from this semiring are called boolean matrices.

Therefore, to distinguish between those matrices and matrices from $\mathbb{GF}(2)$, we will use the term "boolean matrix" for the former and "$\mathbb{GF}(2)$-matrices" for the latter. For example, the Method of Four Russians for Multiplication was designed for boolean matrices, but as will be shown in Section 9.3 on Page 137, we have adapted it for $\mathbb{GF}(2)$-matrices.

6.2.1 Implementing with the Integers

When actually operating in the boolean semiring, the following technique is very useful. Let zero represent zero, but any positive number represent one. Then since "positive + positive" is positive, and "zero + positive" is "positive", while "zero + zero" is "zero", we see that the addition property works as desired. So does the multiplication.

One can therefore perform matrix multiplications in this way, by initially filling with zeroes and ones, and operating in the integers. Periodically, one must scan the matrix and reset all the positive values back to one to avoid overflow. But, this is a quadratic and therefore not cheap operation. If this "sweep" is also performed at the end of the matrix multiplication, one obatins the desired answer in the ring of matrices over the semiring.

6.3 Why is $\mathbb{GF}(2)$ Different?

This section contains three very basic observations that are intended to remind the reader that $\mathbb{GF}(2)$-vector spaces are different from $\mathbb{R}$-vector spaces, such as $\mathbb{R}^3$. The author assumes these examples have been known for quite some time, but they serve to remind the reader of some crucial differences, and will be touched on later in the book as facts in their own right.

6.3.1 There are Self-Orthogonal Vectors

Consider the ordinary dot product,

$$< \mathbf{x}, \mathbf{y} >= \sum_{i=1}^{i=n} x_i y_i$$

Surely in $\mathbb{GF}(2)$, one can see that

$$< (0,1,1), (0,1,1) >= 0 + 1 + 1 = 0$$

and thus there exist non-zero vectors which are orthogonal to themselves. These vectors are called "self-orthogonal." In $\mathbb{R}$, $\mathbb{C}$, and in $\mathbb{Q}$, or any field of characteristic zero, only the zero vector is self-orthogonal. Note that in $\mathbb{C}$

$$< \mathbf{x}, \mathbf{y} >= \mathbf{x}^\dagger \mathbf{y}$$

where $\mathbf{x}^\dagger$ is the complex conjugate of $\mathbf{x}^T$, the transpose of $\mathbf{x}$. This is not needed in any real or finite field.

6.3.2 Something that Fails

Consider the Gram-Schmidt algorithm, a very well-understood linear algebraic technique. Given a set S of vectors, the algorithm computes B, an orthonormal basis for the span of S. The algorithm is given in Algorithm 4 on Page 83.

INPUT: A set of vectors S.
OUTPUT: An orthonormal basis B for the span of S.
1: $B \leftarrow \{\}$
2: For each $s_i \in S$ do

   1: For each $b_j \in B$ do
      1: $s_i \leftarrow s_i - < s_i, b_j > b_j$
   2: If $s_i \neq 0$ then
      1: $s_i \leftarrow \frac{1}{||s_i||} s_i$
      2: Insert s_i into B.

3: Return B.

Algorithm 4: Gram-Schmidt, over a field of characteristic zero. [Jørgen Pedersen Gram and Erhard Schmidt]

The first problem is that the normalization step (second-to-last step) requires a norm. If the usual norm based on the inner product $||x|| = \sqrt{< x,x >}$ is used, then self-orthogonal vectors will result in division by zero. If the Hamming norm is used, then perhaps one would have to compute $1/3$ or $1/4$ times a $\mathbb{GF}(2)$-vector, which is meaningless.

However, we can drop the second-to-last step, and simply hope to create an orthogonal basis instead of an orthonormal basis (i.e. it will not necessarily be the case that the output vectors will all have norm one, but there will still be a basis and all vectors will be orthogonal to each other).

Now consider the vectors $S = \{(1,0,1,0); (1,1,1,0); (0,0,1,1)\}$. The output is $B = \{(1,0,1,0); (1,1,1,0); (0,1,1,1)\}$. Cleary the first and last vector of B are not orthogonal, as their dot-product is 1. Thus the algorithm fails. Note that the first input vector was a self-orthogonal vector.

To see why this is important, consider this basic use of an orthonormal basis. Given such a basis $B = \{\mathbf{b_1}, \ldots, \mathbf{b_n}\}$, one can write a vector $\mathbf{v} \in span(B)$ as a linear combination of the basis vectors. Let $c_i = < \mathbf{v}, \mathbf{b_i} >$, and then $\mathbf{v} = \sum c_i \mathbf{b_i}$.

In the example above, consider $(0,1,0,0)$. In this case $c_1 = 0$, $c_2 = 1$, $c_3 = 0$, by the above method. But $0\mathbf{b_1} + 1\mathbf{b_2} + 0\mathbf{b_3} = (1,1,1,0) \neq (0,1,0,0)$. Instead, a better choice would have been $c_1 = 1$, $c_2 = 1$, $c_3 = 0$, which produces the correct answer. Note the only coefficient that is wrong is the one computed for the only self-orthogonal vector. Since Gram-Schmidt is crucial in the QR-factorization algorithm, this problem rules out doing QR in $\mathbb{GF}(2)$-vector spaces, at least without serious modification to the algorithm.

6.3.3 The Probability a Random Square Matrix Singular or Invertible

The following theorem is very interesting.

Theorem 25. *A random $n \times n$ matrix over $\mathbb{GF}(2)$, filled with independent and identically distributed fair coins, is non-singular with probability:*

$$\prod_{i=1}^{i=n} 1 - 2^{i-1-n}$$

Consider the set of $n \times n$ matrices over $\mathbb{GF}(2)$. Suppose we wish to calculate the probability that a random matrix (one filled with the output of random fair coins), is singular or invertible. The ratio of invertible $n \times n$ matrices (i.e. $|GL_n(\mathbb{GF}(2))|$), to all $n \times n$ matrices (i.e. $|M_n(\mathbb{GF}(2))|$), will give us that probability. The latter calculation is trivial. Each matrix has n^2 entries and so there are 2^{n^2} such matrices.

Now for the former, consider the first column. It can be anything except the column of all zeroes, or $2^n - 1$ choices. The second column can be anything except the first column, or all zeroes, thus $2^n - 2$ choices. The third column cannot be all zeroes, the first column, the second column, or their sum, or $2^n - 4$ choices. It is clear that the ith column cannot contain a vector in the subspace generated by the previous $i - 1$ columns, which are linearly independent by construction. This subspace has 2^{i-1} elements. Thus the ith column has $2^n - 2^{i-1}$ choices.

This results in the following expression for the probability

$$\frac{\Pi_{i=1}^{i=n} 2^n - 2^{i-1}}{2^{n^2}} = \prod_{i=1}^{i=n} 1 - 2^{i-1-n}$$

The latter is obviously just a rational number. For any particular value of n, it can be calculated. But as $n \to \infty$, the product converges toward $0.28879\ldots$, a positive real number. (This is also a good approximation for $n > 10$). This value is very close to $\sqrt{1/12}$.

While this result is well known [202] or [166, Ch. 16] this is still a surprise, because for any real random variable with a continuous probability distribution function, filling a matrix with independent and identically distributed values will produce a singular matirx with probability zero.

Theorem 26. *Let A be a matrix filled with entries from $\mathbb{R}$ being given by some independantly but identically distributed continuous random function. The probability A is singular is zero.*

Proof. The determinant of an $n \times n$ matrix can be thought of as a function with n^2 variables. It is a polynomial, mapping from $\mathbb{R}^{n^2}$ to $\mathbb{R}$. Therefore, the pre-image of zero (the set of input vectors so that the determinant is zero), is the set of zeros of a multivariate polynomial, which means it is a closed set in the Zariski topology. As such, since it is not the whole domain, it therefore is of measure zero in the entire space $\mathbb{R}^{n^2}$. □

6.4 Null Space from the RREF

Given a small matrix in Reduced Row Echelon Form, surely the reader knows how to recover the null space of the matrix. (A review of null spaces can be found in Section 10.5.5 on Page 174.) However, the algorithm to do so efficiently is a bit different from the standard method by hand, because normally by hand the user has a 4×4 matrix or smaller, and the null space therefore cannot be very complicated. The algorithm has given students of mine some trouble, so we will derive it here. The author apologizes in advance if the reader's intelligence is insulted, but a review of null spaces can be found in Section 10.5.5 on Page 174.

For the moment, we will switch back to a general field $\mathbb{F}$. Since the null space of a matrix is the set of vectors **n** such that $A\mathbf{n} = \mathbf{0}$, we find the null space by solving precisely that system. There is no need to adjoin the zero-vector **0** to A prior to doing Gaussian Elimination into RREF, because that new column will remain all zeroes at all times.

Suppose in the RREF, there are r non-zero rows. This r is the rank of the matrix. Call the left-most non-zero entry in each row a pivot element. If a column has a pivot element in it, the variable associated with that column will be called dependent, and if a column has no pivot element, that variable will be called free.

Let one non-zero row in the RREF be row i, and suppose its left-most non-zero entry is A_{ij}. This should be 1 if the standard definition of RREF is used. Suppose the other non-zero entries are A_{ik}, $A_{i\ell}$, and A_{im}. Then we have

$$A_{ij}x_j + A_{ik}x_k + A_{i\ell}x_\ell + A_{im}x_m = 0$$

and because $A_{ij} = 1$, this can easily be thought of as

$$x_j = -A_{ik}x_k - A_{i\ell}x_\ell - A_{im}x_m$$

which can permit us to specialize to the case where all the free xs are zero, except perhaps x_ℓ, which we will set to negative one. Then we substitute and obtain $A_{ij}x_j = x_j = A_{i\ell}$.

Thus, for the rows $i \in \{1, 2, \ldots, r\}$ let the left-most non-zero entry be $A_{i,f(i)}$ and then we would have $x_{f(i)} = A_{i\ell}$. This can be written for all the dependent variables. The free variables have all been set to zero, except for x_ℓ which is negative one. Thus, all the variables have been accounted for. In summary, the free variables are all zero except the one we are working on, which gets -1, and the dependent variables come right out of the matrix A.

Just as we did this for x_ℓ, we should do this for each free variable. If A has n columns, then there will be $n - r$ free variables. This gives rise to the algorithm stated here as Algorithm 5 on Page 87.

Furthermore, it is easy to see that Algorithm 5 on Page 87 runs in quadratic time, because The first outer loop runs $r \leq n$ times, and the second outer loop runs $n - r$ times. If one had to scan the row i in the first loop entry by entry to find the first non-zero entry (which would be very bad programming) then it is at most $\Theta(n)$ reads, for $\Theta(rn)$ steps. The second outer loop has an inner loop which will run r times, and uses a constant number of operations. Thus the second outer loop accounts for $\Theta(r)$ steps each time it runs, or $\Theta((n-r)r) = \Theta(nr - r^2)$ steps total. The total running time is thus $\Theta(nr - r^2)$. Since $r \leq n$, this will never be negative. Thus if the rank is near 0, or near n, the algorithm is very fast.

6.5 The Number of Solutions to a Linear System

Surely over an infinite field, either $Ax = b$ has no solutions, one solution, or infinitely many solutions. To show this fact, assume that $Ax = b$ has two solutions, x and x' with $x \neq x'$, and we will produce infinitely many. First, note that $Ax - Ax' = b - b = 0$. Thus $A(x - x') = 0$ and since $x \neq x'$ then $(x - x') \neq 0$. Also, for all non-zero field elements k, then $A(k(x - x')) = kA(x - x')$. Thus A has an infinite number of vectors n such that $An = 0$, where each $n = k(x - x')$ for some k. Formally, A has an infinite null space. But observe,

$$A(x + n) = Ax + An = Ax + 0 = Ax = b$$

each null-space vector produces a new solution.

INPUT: A matrix A in reduced row-echelon form, of dimension $m \times n$.
OUTPUT: A set of vectors $\mathcal{N}$ that is a basis for the null space of A.
1: Let the number of non-zero rows in A be r.
2: Let $\mathcal{F} \leftarrow \{1, 2, 3, 4, \ldots, n\}$
3: Let $\mathbf{p}$ be an r-dimensional vector, initially the zero vector.
4: For $i = 1$ to r do

- Let A_{ij} be the left-most non-zero entry in row i.
  Note: Since A_{ij} is a pivot-element, then x_j is not a free variable.
- Remove j from $\mathcal{F}$.
- $\mathbf{p}_i \leftarrow j$

Note: Now the set $\mathcal{F}$ is the set of free variables.
5: For each $f \in \mathcal{F}$ do

  Note: We now compute the case of all free variables equal to zero except $x_f = -1$.
  1: Let $\mathbf{n}$ be an n-dimensional vector, initially the zero vector.
  2: $\mathbf{n}_f \leftarrow -1$
  3: For $i = 1$ to r do
   - $j \leftarrow \mathbf{p}_i$
     Note: This means A_{ij} is the left-most non-zero entry in row i.
     Note: The ith equation now reads $x_j = A_{if}$.
   - $\mathbf{n}_j \leftarrow A_{if}$
  4: Insert $\mathbf{n}$ into the set $\mathcal{N}$.

6: Output $\mathcal{N}$.

Algorithm 5: Finding the Null Space from an RREF [Classic]

Thus, in the case of $\mathbb{R}$, $\mathbb{C}$, or $\mathbb{Q}$, since there are infinitely many n, there are infinitely many solutions.

Likewise, the number of solutions in the finite field case is exactly equal to the size of the null space. We have already shown a map, $f(\mathbf{n}) = \mathbf{x} + \mathbf{n}$, that produces a solution to the system $A\mathbf{x} = \mathbf{b}$ for any vector n in the null space of A. This f works over finite fields as well, and is an injection from the null space of A into the solution space of $A\mathbf{x} = \mathbf{b}$.

There is an injection going the other way. Suppose $A\mathbf{x} = \mathbf{b}$ has at least one solution, and denote it $\mathbf{x}'$. For any solution $\mathbf{x}$, we have already shown that $(\mathbf{x} - \mathbf{x}')$ is in the null space. Thus $g(\mathbf{x}) = \mathbf{x} - \mathbf{x}'$ is a map from the solution space of $A\mathbf{x} = \mathbf{b}$ into the null space of A and it too is injective. Since the sets involved are finite (when working over finite fields), injection implies bijection, and both f and g are bijections. Thus we have proven

Theorem 27. *The number of solutions to $A\mathbf{x} = \mathbf{b}$, is either zero, or if a solution exists, then equal to the size of the null space of A.*

The null space of A is a subspace and so it has a basis and thus a dimension. This is called the nullity of A. For a square matrix, the nullity plus the rank of A is the dimension of A. For a rectangular matrix, the nullity plus the rank is the number of columns.

Actually in either case, the rank of A plus the nullity of A is the number of columns of A. Thus if the rank is r, and A is an $m \times n$ matrix, we have a nullity of $n - r$, or the null space has dimension $n - r$. Of course, if there are q elements in the finite field, then a subspace of dimension d has q^d elements in it. Therefore, we have the following corollary:

Corollary 28. *If A is an $m \times n$ matrix with rank r, over the field of size q, then the linear system of equations given by $A\mathbf{x} = \mathbf{b}$ has either no solutions, or q^{n-r} solutions.*

Thus over $\mathbb{GF}(2)$, we can have a system of equations with 0, 1, 2, 4, 8, 16,... solutions, not just 0, 1, or infinity solutions, as over $\mathbb{R}$, $\mathbb{C}$, or $\mathbb{Q}$.

Chapter 7
The Complexity of $\mathbb{GF}(2)$-Matrix Operations

Here, we propose a new model, counting matrix-memory operations instead of field operations, for reasons to be discussed. It turns out this model describes reality only partially—but we will explicitly discuss the circumstances in which the model is descriptive and in which it fails, see Section 7.1.4 on Page 92. The complexity expressions are summarized in Table 7.1 on Page 105. Also of interest are certain data structure choices that we made in arranging our linear algebra library, see Section 9 on Page 133. This library was used by Nicolas Courtois in his cryptographic research, as well as by the author, and now forms part of the $\mathbb{GF}(2)$ linear algebra suite of SAGE [7], an open source competitor to MAGMA [2], MATLAB [5], MAPLE [3], and MATHEMATICA[4]. These are described in Section 7.4 on Page 94.

7.1 The Cost Model

In papers on matrix operations over the real or complex numbers, the number of floating point operations is used as a measure of running time. This removes the need to account for assembly language instructions needed to manipulate index pointers, iteration counters, discussions of instruction set, and measurements of how cache coherency or branch prediction will impact running time. In this dissertation, floating point operation counts are meaningless, for matrices over $\mathbb{GF}(2)$ do not use floating point operations. Therefore, we propose that matrix entry reads and writes be tabulated, because addition (XOR) and multiplication (AND) are single instructions, and can even be aggregated (see Section 9.5.4 on Page 149) while reads and writes on rectangular arrays are much more expensive. Clearly these data structures are non-trivial in size, so memory transactions will be the bulk of the time spent.

From a computer architecture viewpoint in particular, the matrices required for cryptanalysis cannot fit in the cache of the microprocessor, so the fetches to main memory are a bottleneck. Even if exceptionally careful use of temporal and spatial locality guarantees effective caching (and it is not clear that this is even possible),

G.V. Bard, *Algebraic Cryptanalysis*, DOI: 10.1007/978-0-387-88757-9_7
© Springer Science + Business Media, LLC 2009

the data must still travel from memory to the processor and back. The bandwidth of buses has not increased proportionally to the rapid increase in the speeds of microprocessors. Given the relatively simple calculations done once the data is in the microprocessor's registers (i.e. single instructions), it is extremely likely that the memory transactions are the rate-determining step.

When attempting to convert these memory operation counts into CPU cycles, one must remember that other instructions are needed to maintain loops, execute field operations, and so forth. Also, memory transactions are not one cycle each, but can be pipelined. Thus we estimate that about 4–10 CPU cycles are needed per matrix-memory operation.

7.1.1 A Word on Architecture and Cross-Over

Often, there is an asymptotically fast algorithm for some problem, and then other algorithms which are better for small and medium-sized versions of the problem. The points at which one algorithm ceases to dominate another is called the "cross over" between those two algorithms. Calculating the cross-over point, at least approximately, is of importance, so that when presented with a specific problem instance, one knows exactly which algorithm to run on it.

Due to the variations of computer architectures, the coefficients given here may vary slightly. In particular, on some machines, 32 $\mathbb{GF}(2)$ addition operations can be a single instruction, and on others, 64. Slight variations in the coefficients might appear to be of little interest, but when comparing two algorithms (e.g. M4RM and Strassen's Matrix Multiplication Algorithm), we must consider the cross-over time. In this case, it would be given by

$$c_1 \frac{n^3}{\log n} = c_2 n^{2.807\cdots}$$

and one can see that time variations in c_1 or c_2 are very important, because $n^{0.193}/\log n = \frac{c_2}{c_1}$, or neglecting the $\log n$, roughly $n \sim (c_2/c_1)^5$. Certainly, changes in cache sizes on different machines that are otherwise identical can also change the cross-over. For this reason, the BLAS (Basic Linear Algebra System) called ATLAS (Automatically Tuned Linear Algebra System) [229] is very exciting. It automatically computes the precise cross-over sizes exactly on the machine during a tuning stage while being installed. Therefore, the algorithms always perform optimally.

On the other hand, by deriving running times mathematically rather than experimentally, one need not worry about artifacts of particular architectures or benchmarks skewing the results.

7.1.2 Is the Model Trivial?

A minor technicality is defining what regions of memory the reads and writes should count. Clearly registers do not count and the original matrix should. The standard we set is that a read or write counts unless it is to a "scratch" data structure. We define a data structure to be "scratch" if and only if it size is bounded by a constant.

For example, consider the following three step algorithm of inverting a non-singular $n \times n$ matrix, in $\sim 2n^2$ time.

1. Read in a matrix. (n^2 reads).
2. Invert the matrix. (No reads or writes).
3. Write the output matrix. (n^2 writes).

This is not allowed (or rather, we would not tabulate Step 2 as zero cost) because the temporary storage of the matrix requires n^2 field elements, and this is not upper-bounded by a constant.

7.1.3 Counting Field Operations

Traditionally, in a linear algebra text, one counts the number of field operations. Or alternatively, one can count the number of multiplications or divisions. However, as stated earlier, that does not make sense for $\mathbb{GF}(2)$(see [31, App. B]).

It is easy to see that counting field multiplications only versus counting field multiplications and additions produces two distinct tabulations in almost all cases. It is also easy to imagine that counting field multiplies and reads/writes will result in distinct tabulations.

An interesting question is if counting reads/writes is distinct from counting field multiplications and additions. In Gaussian Elimination, the answer is yes, because of "if" operations. If a row contains a zero in the pivot column, it is read but never operated upon.

The follow-up question is if counting reads/writes is distinct from counting field multiplications, additions, and conditionals (if's). After all, the latter three operations are all single logic gates.

In this case consider a one by one matrix multiplication, or one-dimensional dot-product. It requires one arithmetic operation, and three reads/writes. A two-dimensional dot product requires four reads and one write, versus two multiplications and one addition. An n-dimensional dot-product requires $2n + 1$ reads/writes but $2n - 1$ field operations, for a ratio of $\frac{2n+1}{2n-1}$. While this is ~ 1, the ratio is changing. Note it is important to have very close estimates of the coefficient when performing cross-over analysis.

An interesting distinction is between Gaussian Elimination with full-pivoting versus partial-pivoting [222, Ch. 1.8]. In the former, during iteration i, one searches $A_{ii}, \ldots, A_{mn}$ for a non-zero element (often of largest absolute value) to pivot upon. In the later, one mrely searches $A_{ii}, \ldots, A_{mi}$, the active column. The difficulty is not doing both column-swap and row-swap (full pivoting) versus row-swap alone (partial pivoting). Rather, the difficulty is that one must make roughly $(n-i+1)(m-i+1)$ comparisons versus $(n-i+1)$ of them. And in practice, partly because it obliterates any spatial locality for caching, full pivoting is extremely expensive, and is only used for the most sensitive calculations.

7.1.4 Success and Failure

The model described above has had some success. When actually implementing the algorithms in code, and performing timing experiments, the observed exponents have always been correct. When comparing different variants of the same algorithm (e.g. triangular versus complete Gaussian Elimination), the coefficients have been correct to about 2%.

However, when comparing different algorithms (e.g. MAGMA's Strassen-naïve matrix multiplication vs M4RM, or M4RM vs naïve matrix multiplication) the coefficients sometimes give ratios that are off by up to 50%. This inaccuracy above is probably due to the role of caching. Some algorithms are more friendly toward cached memory than others. It is notoriously hard to model this.

Another reason is that MAGMA has been hand-optimized for certain processors at the assembly language level, and the author's library has been written in C (though compiled with all optimization settings turned on).

In calculating the number of times a subroutine will be called (i.e. How many times do you use the black-box $n_0 \times n_0$ matrix multiply when inverting a much larger matrix?), the model is exact. Presumably because nearly all the time is spent in the black box, and it is the same single black box routine in all cases, the number of calls to the black box is all that matters. Since this is an integer, it is easy to measure if one is correct.

7.2 Notational Conventions

Precise performance estimates are useful, so rather than the usual five symbols $O(n)$, $o(n)$, $\Omega(n)$, $\omega(n)$, $\Theta(n)$, we will use $f(n) \sim g(n)$ to indicate that

$$\lim_{n\to\infty} \frac{f(n)}{g(n)} = 1$$

in the case that an exact number of operations is difficult to state. While $O(n)$ statements are perfectly adequate for many applications, coefficients must be known to determine if algorithms can be run in a reasonable amount of time on particular target ciphers.

Let $f(n) \leq\sim g(n)$ signify that there exists an $h(n)$ and n_0 such that $f(n) \leq h(n)$ for all $n > n_0$, and $h(n) \sim g(n)$. Equivalently, this means $\limsup f(n)/g(n) \leq 1$ as $n \to \infty$.

Matrices in algebraic cryptanalysis are over $\mathbb{GF}(2)$ unless otherwise stated, and are of size m rows and n columns. Denote ℓ as the lesser of n and m. If $n > m$ or $\ell = m$ the matrix is said to be underdefined, and if $m > n$ or $\ell = n$ then the matrix is said to be overdefined. Also, β is the fraction of elements of the matrix not equal to zero.

7.3 To Invert or to Solve?

Generally, four basic options exist when presented with solving systems of equations over the reals as defined by a square matrix. First, the matrix can be inverted, but this is the most computationally intensive option. Second, the system can be adjoined by the vector of constants, and the matrix reduced into a triangular form so that the unknowns can be found via back-substitution (see Section 7.5.4 on Page 98). Third, the matrix can be factored into LU-triangular form, or other forms. Fourth, the matrix can be operated upon by iterative methods, to converge to a matrix near to its inverse. Unfortunately, in finite fields concepts like convergence toward an inverse do not have meaning. This rules out option four. The second option is unattractive, because solving the same system for two sets of constants requires twice as much work, whereas in the first and third case, if the quantity of additional sets of constants is small compared to the dimensions of the matrices, trivial increase in workload is required.

Among these two remaining strategies, inversion is almost strictly dominated by LUP-factorization. The LUP-factorization is $A = LUP$, where L is lower unit triangular, U is upper unit triangular, and P is a permutation matrix. There are other factorizations, like the QR [217, Lec. 7], which are not discussed here because no one (to the author's knowledge) has proposed how to do them over $\mathbb{GF}(2)$. (For example, the QR depends on the complexity of Gram-Schmidt, but Gram-Schmidt fails over $\mathbb{GF}(2)$, see Section 6.3.2 on Page 83). While the LUP-factorization results in three matrices, and the inverse in only one, the storage requirements are about the same. This is because, other than the main diagonal, the triangular matrices have half of their entries forced at zero by definition. Also, since the main diagonal can have only units, and the only unit in this field is 1, the main diagonal of the triangular matrices need not be stored. The permutation matrix can be stored with n entries, rather than n^2, as is explained in Section 7.4 on Page 94.

Calculating the inverse is always (for all methods discussed in this book) more work than the LUP-factorization but by a factor that varies depending on which algorithm is used. Also the LUP-factorization allows the determinant to be calculated, but for all non-singular $\mathbb{GF}(2)$ matrices the determinant is 1. And for singular matrices it is zero, so this is not informative. Also, multiplying a matrix by a vector requires $\sim 3n^2$ matrix-memory operations (a read-read-write for each field operation, with $\sim n^2$ field operations). For back-substitution in the LUP-case, one must do it twice, for L and for U. The back-substitution requires $\sim n^2/2$ field operations, or $\sim (3/2)n^2$ matrix-memory operations, so this ends up being equal also.

7.4 Data Structure Choices

The most elementary way to store a matrix is as an array of scalars. Two-dimensional arrays are often stored as a series of one-dimensional arrays in sequence, or as an array of pointers to arrays (one for each row, called a "ragged array"). In either case, it is not obvious if the linear arrays should be rows or columns. For example, in a matrix multiplication AB with the naïve algorithm, spatial locality will be enhanced if A's rows and B's columns are the linear data structure. Those options are called row-major and column-major respectively. More information on the row-major versus column-major tradeoff can be found in [190].

Two data structures that we used are proposed and described below.

7.4.1 Dense Form: An Array with Swaps

For dense matrices, we present a method of storing the matrix as an array but with very fast swaps. The cells of the matrix are a two-dimensional array, with the rows being the linear data structure, since more of the work in the algorithms of this dissertation is performed upon rows than upon columns. Additionally, two one-dimensional arrays called row-swap and column-swap are used. Initially these are filled with the numbers $1, 2, \ldots m$ and $1, 2, \ldots n$. When a swap of rows or columns is called for, the numbers in the cells of the row-swap array or column-swap array corresponding to those rows are exchanged. When a cell a_{ij} is called for, the result returned is a_{r_i, c_j}, with r_i representing the ith entry of the row-swap array, and c_j the jth entry of the column-swap array. In this manner, row and column swaps can be executed in constant time, namely two read/writes each.

For example, a 5×5 matrix with rows 1 and 2 being exchanged, and then rows 4 and 2 being exchanged, would cause the matrix to have $\{2, 4, 3, 1, 5\}$ as its row-swap array.

7.4.2 Permutation Matrices

The definition of a permutation matrix at first seems arbitrary.

Definition 29. A matrix with exactly one entry in each row equal to 1, and exactly one entry in each column equal to 1, but zero everywhere else, is called a permutation matrix.

However, the cause of this definition is that multiplication by a permutation matrix on the left causes a swapping of rows, and on the right causes a swapping of columns. In fact, the permutation matrix itself is just an identity matrix, after some rows and columns have been swapped. We propose an efficient scheme for storing and performing operations on permutation matrices.

It is only necessary to store a row-swap array and column-swap array as before, not the body of the matrix. The row-swap and column-swap arrays allow a quick look-up, by calculating $a_{ij} = 1$ if and only if $r_i = c_j$ (i.e. the cell is on the main diagonal after swapping), and returning $a_{ij} = 0$ if $r_i \neq c_j$.

In linear time one can compose two permutations (multiply the matrices) or invert the permutation (invert the matrix). The algorithms for this are given in Algorithm 6 on Page 95 and Algorithm 7 on Page 95. Note that the algorithms should be called twice, once for row permutations and once for columns. Alternatively, the transpose of a permutation matrix is its inverse.

INPUT: Two permutation matrices or row-swap arrays, in the form $r_1, \ldots, r_n$, and $s_1, \ldots, s_n$.
OUTPUT: Their product $t_1, \ldots, t_n$.
1: For $i = 1$ to n do

    1: $temp \leftarrow r_i$
    2: $t_i \leftarrow s_{temp}$

NOTE: for column-swap arrays, the algorithm is identical.

Algorithm 6: To compose two permutations or row-swap arrays. [Classic]

INPUT: A permutation matrix or row-swap array, in the form $r_1, \ldots, r_n$.
OUTPUT: The inverse permutation matrix or row-swap array, in the form $s_1, \ldots, s_n$.
1: For $i = 1$ to n do

    1: $temp \leftarrow r_i$
    2: $s_{temp} \leftarrow i$

NOTE: for column-swap arrays, the algorithm is identical.

Algorithm 7: To invert a permutation matrix or row-swap array. [Classic]

It is trivial to see that a permutation can be applied to a vector in linear time, by simply moving the values around in accordance with the row-swap array. To multiply a matrix by a permutation is also a linear time operation, because one only need apply the permutation's row-swap array to the matrix's row-swap array (as in composing two permutations, in Aglorithm 6 on Page 95).

7.5 Analysis of Classical Techniques with our Model

7.5.1 Naïve Matrix Multiplication

For comparison, we calculate the complexity of the naïve matrix multiplication algorithm, for a product $AB = C$ with dimensions $a \times b$, $b \times c$ and $a \times c$, respectively.

INPUT: Two matrices, A of dimension $a \times b$, and B of dimension $b \times c$.
OUTPUT: A matrix $C = AB$, of dimension $a \times c$.
1: for $i = 1, 2, \ldots, a$

    1: for $j = 1, 2, \ldots, c$
       1: Calculate $C_{ij} \leftarrow A_{i1}B_{1j} + A_{i2}B_{2j} + \cdots A_{ib}B_{bj}$. (Costs $2b + 1$ reads/writes).

Algorithm 8: Naïve Matrix Multiplication [Classic]

From the algorithm given in Algorithm 8 on Page 96, this clearly requires $2abc + ac$ operations, or for square matrices $2n^3 + n^2$ operations. This reduces to $\sim 2abc$ or $\sim 2n^3$, when $a = b = c = n$.

7.5.2 Matrix Addition

If adding $A + B = C$, obviously $c_{ij} = a_{ij} + b_{ij}$ requires two reads and one write per matrix entry. This yields $\sim 3mn$ matrix memory operations overall, if the original matrices are $m \times n$.

7.5.3 Dense Gaussian Elimination

The algorithm known as Gaussian Elimination is very familiar. It has many variants, but three are useful to us. As a subroutine for calculating the inverse of a matrix, we refer to adjoining an $n \times n$ matrix with the $n \times n$ identity matrix to form an $n \times 2n$ matrix. This will be processed to output the $n \times n$ identity on the left, and A^{-1} on the right. The second is to solve a system directly, in which case one column is adjoined with the constant values. This is "full Gaussian Elimination" and is found in Algorithm 9 on Page 97. Another useful variant, which finishes with a triangular rather than identity submatrix in the upper-left, is listed in Algorithm 10 on Page 98, and is called "Triangular Gaussian Elimination." (That variant requires 2/3 as much time for solving a system of equations, but is not useful for finding matrix inverses). Since Gaussian Elimination is probably known to the reader, it is not described here, but it has the following cost analysis.

INPUT: A matrix A of dimension $m \times n$. In the special case for matrix inverses, this is formed by an $m \times m$ matrix B and an $m \times m$ identity matrix being adjoined to form a $m \times 2m$ matrix. In the special case for solving the system $A\mathbf{x} = \mathbf{b}$, this is formed by an $m \times n$ matrix A, and an $m \times 1$ matrix (i.e. an m-dimensional vector) $\mathbf{b}$ being adjoined.

OUTPUT: The reduced row-echelon form of the input matrix A. In the special case of a matrix inverse, if $\det B \neq 0$, this is an $m \times m$ identity matrix adjoined to a $m \times m$ matrix which is equal to B^{-1}. In the special case for solving the system $A\mathbf{x} = \mathbf{b}$, if A is full-rank, then the right-hand column is the solution vector $\mathbf{x}$, possibly with zeroes concatenated at the end.

1: Let $\ell \leftarrow \min(m,n)$†
2: For $i = 1, 2, \ldots, \ell$

   1: Search for a non-zero entry in region $a_{ii} \ldots a_{mn}$† (Expected cost is 2 reads). Call this entry a_{xy}.
   2: Exchange rows i and x via row-swap array, exchange columns i and y via column-swap array. (Costs 4 writes).
   3: For each row $j = 1, 2, \ldots, m$, but not row i
      1: If $a_{ji} = 1$ (Costs 1 read) then for each column $k \in i, i+1, \ldots, n$
         1: Calculate $a_{jk} \leftarrow a_{jk} + a_{ik}$. (Costs 2 reads, 1 write).

Note: † signifies that the value of n in these steps, in the special case when solving $A\mathbf{x} = \mathbf{b}$, refers to the number of columns of A before the adjoining, not afterward. In all other cases, n refers to the dimensions of A at that point in the algorithm.

Algorithm 9: Dense Gaussian Elimination, for Inversion [Carl Friedrich Gauss and Wilhelm Jordan]

The search for a non-zero element in a certain region in Step 2.1 above might seem a bit odd. The possibility of an entry being zero is far more common in $\mathbb{GF}(2)$ than otherwise. This process of searching for a non-zero pivot element (i.e. a one) is akin to Gaussian Elimination "with full-pivoting" (see Section 7.1.3 on Page 91). However, this algorithm here is for dense matrices. Therefore, we make the "fair coin assumption" (see Section 9.0.4 on Page 134), and therefore we would have an expected value of two entries to check before we find a one.

The total number of expected reads and writes is given by

$$= \sum_{i=1}^{i=\ell} 6 + (m-1)(1 + 0.5(3)(n-i+1))$$

$$= 1.5nm\ell - 0.75m\ell^2 + 1.75m\ell - 1.5n\ell + 0.75\ell^2 + 4.25\ell$$

$$\sim 1.5nm\ell - 0.75m\ell^2$$

Note that the 0.5 in the first line is due to the fair-coin assumption also. At worst, all columns will have a non-zero entry in column i, and at best, none will. The value 0.5 would change to 1 or 0, respectively, in those cases.

Thus for the overdefined case ($\ell = n$) one obtains $1.5n^2m - 0.75mn^2$, and for underdefined ($\ell = m$) the total is $1.5nm^2 - 0.75m^3$. For a square matrix this is $0.75n^3$.

The alternative form of the Gaussian Elimination algorithm, which outputs an upper-triangular matrix rather than the identity matrix in the upper-left $\ell \times \ell$ submatrix, is found in Algorithm 10 on Page 98. This is not useful for finding the inverse of a matrix, but is useful for LU-factorization or solving a system of m equations in n unknowns. Here it is assumed that one column is adjoined that contains the constants for a system of linear equations.

INPUT: A matrix A of dimension $m \times n$. In the special case for solving the system $A\mathbf{x} = \mathbf{b}$, this is formed by an $m \times n$ matrix A, and an $m \times 1$ matrix (i.e. an m-dimensional vector) $\mathbf{b}$ being adjoined.

OUTPUT: The row-echelon form of A. In the special case of solving $A\mathbf{x} = \mathbf{b}$ the system is now ready for back-solving, see Section 7.5.4 on Page 98.

1: Let $\ell = \min(m,n)$†

2: For each column $i = 1, 2, \ldots, \ell$

    1: Search for a non-zero entry in region $a_{ii} \ldots a_{mn}$† (Expected cost is 2 reads). Call this entry a_{xy}.

    2: Exchange rows i and x via row-swap array, exchange columns i and y via column-swap array. (Costs 4 writes).

    3: For each row $j \in i+1, i+2, \ldots, m$
        1: If $a_{ji} = 1$ then for each column $k \in i, i+1, \ldots, n$
            1: Calculate $a_{jk} \leftarrow a_{jk} + a_{ik}$ (Costs 2 reads, and 1 write).

Note: † signifies that the value of n in these steps, in the special case when solving $A\mathbf{x} = \mathbf{b}$, refers to the number of columns of A before the adjoining, not afterward. In all other cases, n refers to the dimensions of A at that point in the algorithm.

Algorithm 10: Dense Gaussian Elimination, for Triangularization [Carl Friedrich Gauss]

The total number of reads and writes is given by

$$\sum_{i=1}^{i=\ell} 6 + (m-i)(1 + 0.5(3)(n-i+1))$$

$$= 1.5nm\ell - 0.75m\ell^2 - 0.75n\ell^2 + 0.5\ell^3 + 1.75m\ell$$
$$- 0.5\ell^2 - 0.75n\ell + 5\ell$$
$$\sim 1.5nm\ell - 0.75m\ell^2 - 0.75n\ell^2 + 0.5\ell^3$$

Thus for the overdefined case $(\ell = n)$ one obtains $1.5n^2m - 0.75mn^2 - 0.25n^3$, and for underdefined $(\ell = m)$ the total is $0.75nm^2 - 0.25m^3$. For a square matrix this is $0.5n^3$.

7.5.4 Back-Solving a Triangulated Linear System

For dense matrices A, to solve $A\mathbf{x} = \mathbf{b}$ is a non-trivial operation, as we have already seen. However, in the special case that A is either upper-triangular or lower-triangular, then this becomes a quadratic-time operation. That is, of course, a huge savings compared to cubic or near-cubic, and so it is the motivation behind LUP-factorizations (see Section 10 on Page 98 and Section 9.5.2 on Page 147, as well as Theorem 42 on Page 116). This is sometimes called back-substitution.

Here we present the algorithm for upper-triangular matrices as Algorithm 11 on Page 99 but the one for lower-triangular is similar, with the indices changed in the obvious way.

INPUT: An $m \times n$ upper-triangular matrix U and an m-dimensional vector $\mathbf{b}$.
OUTPUT: An n-dimensional vector $\mathbf{x}$ such that $U\mathbf{x} = \mathbf{b}$.
1: Let $\mathbf{x}$ be initialized with the all-zero vector. (Cost: n writes)
2: For each row $i = \min(m,n)$ down to 1

  1: if $U_{ii} = 0$ then continue with next i value. (Cost: 1 read)
  2: Let $x_i \leftarrow b_i$. (Cost: 1 read and 1 write).
  3: For each column $j = n$ down to $i+1$
     1: $x_i \leftarrow x_i + x_j U_{ij}$ (Cost: 2 reads using an accumulator for x_i)
  4: (Cost: 1 write to copy accumulator for x_i into memory)

NOTE: For a lower-triangular matrix L, the algorithm would proceed identically, but with the indices changed in the obvious way.

Algorithm 11: Back-Solving a Triangulated System [Classic]

In the cost analysis, for simplicity, we assume $U_{ii} \neq 0$ each time. As before, $\ell = \min(m,n)$. The total cost is then

$$n + \sum_{i=1}^{i=\ell} 4 + \sum_{j=i+1}^{j=n} 2 = 2n\ell - \ell^2 + 3\ell + n$$

$$\sim 2n\ell - \ell^2$$

which for $m > n$ would be $\sim n^2$ matrix-memory operations, and for $m < n$ would be $\sim 2nm - m^2$ matrix-memory operations. Naturally for square matrices this comes to $\sim n^2$.

7.6 Strassen's Algorithms

Contrary to popular belief, Volker Strassen's famous paper [212] actually contains three algorithms. The first is the matrix multiplication algorithm given below; the second is a matrix inversion method given immediately after it here; the third is a method for taking determinants, which does not affect us.

7.6.1 Strassen's Algorithm for Matrix Multiplication

To find

$$\begin{bmatrix} a_{11} & a_{12} \\ a_{21} & a_{22} \end{bmatrix} \begin{bmatrix} b_{11} & b_{12} \\ b_{21} & b_{22} \end{bmatrix} = \begin{bmatrix} c_{11} & c_{12} \\ c_{21} & c_{22} \end{bmatrix}$$

one can use the algorithm found in Algorithm 12 on Page 100. One can see that this consists of 18 matrix additions and 7 matrix multiplications.

INPUT: Two matrices a and b, both of size 2×2. But note, usually the elements of a and b are themselves very large matrices.
OUTPUT: The produce matrix $c = ab$, also of size 2×2.
1: Calculate 10 sums, namely: $s_1 \leftarrow a_{12} - a_{22}$, $s_2 \leftarrow a_{11} + a_{22}$, $s_3 \leftarrow a_{11} - a_{21}$, $s_4 \leftarrow a_{11} + a_{12}$, $s_5 \leftarrow a_{21} + a_{22}$, $s_6 \leftarrow b_{11} + b_{22}$, $s_7 \leftarrow b_{11} + b_{22}$, $s_8 \leftarrow b_{11} + b_{12}$, $s_9 \leftarrow b_{12} - b_{22}$, and $s_{10} \leftarrow b_{21} - b_{11}$.
2: Calculate 7 products, namely: $m_1 \leftarrow s_1 s_6$, $m_2 \leftarrow s_2 s_7$, $m_3 \leftarrow s_3 s_8$, $m_4 \leftarrow s_4 b_{22}$, $m_5 \leftarrow a_{11} s_9$, $m_6 \leftarrow a_{22} s_{10}$, and $m_7 \leftarrow s_5 b_{11}$.
3: Calculate 8 sums, namely: $s_{11} \leftarrow m_1 + m_2$, $s_{12} \leftarrow -m_4 + m_6$, $s_{13} \leftarrow -m_3 + m_2$, $s_{14} \leftarrow -m_7 + m_5$, $c_{11} \leftarrow s_{11} + s_{12}$, $c_{12} \leftarrow m_4 + m_5$, $c_{21} \leftarrow m_6 + m_7$, and $c_{22} \leftarrow s_{13} + s_{14}$.

Algorithm 12: Strassen's Algorithm for Matrix Multiplication [Volker Strassen]

Note that the matrices c_{11} and c_{22} must be square, but need not equal each other in size. For simplicity assume that A and B are both $2n \times 2n$ matrices. The seven multiplications are to be performed by repeated calls to Strassen's algorithm. In theory one could repeatedly call the algorithm until 1×1 matrices are the inputs, and multiply them with a logical AND operand. However, its unlikely that this is optimal. Instead, the program should switch from Strassen's algorithm to some other algorithm below some size n_0.

As stated in Section 7.5.2 on Page 96, the $n \times n$ matrix additions require $\sim 3n^2$ matrix memory operations each, giving the following equation:

$$M(2n) = 7M(n) + 54n^2$$

allowing one to calculate, for a large matrix,

$$M(4n_0) = 7^2 M(n_0) + (4+7) \cdot 54n_0^2$$
$$M(8n_0) = 7^3 M(n_0) + (16 + 7 \cdot 4 + 7^2) \cdot 54n_0^2$$
$$M(16n_0) = 7^4 M(n_0) + (64 + 16 \cdot 7 + 7^2 \cdot 4 + 7^3) \cdot 54n_0^2$$
$$M(2^i n_0) = 7^i M(n_0) + (4^{i-1} + 4^{i-2}7 + 4^{i-3}7^2 + \cdots + 4 \cdot 7^{i-2} + 7^{i-1})54n_0^2$$
$$M(2^i n_0) \approx 7^i M(n_0) + 7^{i-1}(1 + 4/7 + 16/49 + 64/343 + \cdots)54n_0^2$$
$$M(2^i n_0) \approx 7^i M(n_0) + 7^i 18n_0^2$$

Now substitute $i = \log_2(n/n_0)$ and observe,

$$7^{\log_2 \frac{n}{n_0}} M(n_0) + 7^{\log_2 \frac{n}{n_0}} 72n_0^2$$

and since $b^{\log_2 a} = a^{\log_2 b}$, then we have

$$M(n) \approx \left(\frac{n}{n_0}\right)^{\log_2 7} [M(n_0) + 72n_0]$$

or finally $M(n) \sim (n/n_0)^{\log_2 7} M(n_0)$.

7.6.2 Misunderstanding Strassen's Matrix Inversion Formula

Strassen's matrix inversion formula sometimes called SMIF for short, is the following

$$A = \begin{bmatrix} B & C \\ D & E \end{bmatrix} \Rightarrow A^{-1} = \begin{bmatrix} B^{-1} + B^{-1}CS^{-1}DB^{-1} & -B^{-1}CS^{-1} \\ -S^{-1}DB^{-1} & S^{-1} \end{bmatrix}$$

where $S = D^{-1} - E^{-1}CB^{-1}$, the Schur complement of A with respect to B (See Definition 40 on Page 115), provides a fast way of calculating matrix inverses. However, this does not work for fields in which a singular B can be encountered. We will now visit this point, in detail.

7.7 The Unsuitability of Strassen's Algorithm for Inversion

It is important to note that Strassen's famous paper [212] has three algorithms. The first is a matrix multiplication algorithm, which we call "Strassen's Algorithm for Matrix Multiplication." The second is a method for using any matrix multiplication technique as an oracle for matrix inversion, in asymptotically equal time (in the big-Θ sense). We call this "Strassen's Formula for Matrix Inversion." The third is a method for the calculation of the determinant of a matrix, which is of no concern to us. Below, Strassen's Formula for Matrix Inversion is analyzed, by which a system of equations over a field can be solved.

Over all, the purpose of this section is to explain why these approaches to matrix inversion can work over fields like $\mathbb{C}$, $\mathbb{R}$, or $\mathbb{Q}$, as well as large finite fields, but be totally out-of-the-question for $\mathbb{GF}(2)$, without serious modification. Nonetheless, other authors have already figured out how to do the modification, and we list two successful examples of that. Thus, if one wants to use Strassen's Algorithms to get a better exponent in $\mathbb{GF}(2)$-linear algebra, one can, but only through modifications on the scale of those listed here. This will become important when we discuss "the Method of Four Russians for Matrix Inversion" in Section 9.8.1 on Page 152.

7.7.1 Strassen's Approach to Matrix Inversion

Given a square matrix A, by dividing it into equal quadrants one obtains the following inverse (A more detailed exposition is found in [63, Ch. 28], using the same notation):

$$A = \begin{bmatrix} B & C \\ D & E \end{bmatrix} \Rightarrow A^{-1} = \begin{bmatrix} B^{-1} + B^{-1}CS^{-1}DB^{-1} & -B^{-1}CS^{-1} \\ -S^{-1}DB^{-1} & S^{-1} \end{bmatrix}$$

where $S = E - DB^{-1}C$, which is the Schur Complement of A with respect to B (See Definition 40 on Page 115).

One can easily check that the product of A and the matrix formula for A^{-1} yields the identity matrix, either multiplying on the left or on the right. If an inverse for a matrix exists, it is unique, and so therefore this formula gives the unique inverse of A, provided that A is in fact invertible.

However, it is a clear requirement of this formula that B and S be invertible. Over the real numbers, or other subfields of the complex numbers, one can show that if A and B are non-singular, then S is non-singular also (See Lemma 41 on Page 115, or alternatively [63, Ch. 28]). The problem is to guarantee that the upper-left submatrix, B, is invertible. Strassen did not address this in the original paper, but the usual solution is as follows (more details found in [63, Ch. 28]). First, if A is positive symmetric definite (PSD), then all of its principal submatrices are positive symmetric definite, including B. All positive symmetric definite matrices are non-singular, so B is invertible. Now, if A is not positive symmetric definite, but is non-singular, then note that $A^T A$ is positive symmetric definite and that $(A^T A)^{-1} A^T = A^{-1}$. This also can be used to make a pseudoinverse for non-square matrices, called the Moore-Penrose Pseudoinverse [179], [189], [43]. In short, by inverting $A^T A$ instead of A, all the submatrices in the upper left will be invertible, throughout the recursive process.

However, the concept of positive symmetric definite does not work over a finite field, because these fields cannot be ordered (in the sense of an ordering that respects the addition and multiplication operations), and so it is not clear what positive is. Observe the following counterexample,

$$A = \begin{bmatrix} 1 & 0 & 0 & 0 \\ 1 & 0 & 1 & 0 \\ 0 & 1 & 0 & 0 \\ 0 & 1 & 0 & 1 \end{bmatrix} \quad A^T A = \begin{bmatrix} 0 & 0 & 1 & 0 \\ 0 & 0 & 0 & 1 \\ 1 & 0 & 1 & 0 \\ 0 & 1 & 0 & 1 \end{bmatrix}$$

Both A and $A^T A$ have det $= 1$, thus are invertible. Yet in both cases the upper-left hand 2×2 submatrices have det $= 0$, and therefore are not invertible. Thus Strassen's formula for inversion is unusable without modification. The modification below is from Aho, Hopcroft and Ullman's book [13, Ch. 6] though it first appeared in [55].

7.7.2 Bunch and Hopcroft's Solution

Consider a matrix L that is unit lower triangular, and a matrix U that is unit upper triangular. Then Strassen's Matrix Inversion Formula indicates

$$L = \begin{bmatrix} B & 0 \\ D & E \end{bmatrix} \Rightarrow L^{-1} = \begin{bmatrix} B^{-1} & 0 \\ -E^{-1}DB^{-1} & E^{-1} \end{bmatrix}$$

$$U = \begin{bmatrix} B & C \\ 0 & E \end{bmatrix} \Rightarrow U^{-1} = \begin{bmatrix} B^{-1} & -B^{-1}CE^{-1} \\ 0 & E^{-1} \end{bmatrix}$$

Note S becomes E in both cases, since either C or D is the zero matrix. Since L (or U) is unit lower (or upper) triangular, then its submatrices B and E are also unit lower (or upper) triangular, and therefore invertible. Thus Strassen's Matrix Inversion Formula over $\mathbb{GF}(2)$ will always work for unit lower or upper triangular matrices.

It is well known that any matrix over any field has a factorization $A = LUP$ where P is a permutation matrix, L is unit lower triangular and U is unit upper triangular [136, Lec. 21]. Once A is thus factored, the matrix inversion formula is sufficient to calculate A^{-1}. Aho, Hopcroft and Ullman [13, Ch. 6] give an algorithm for computing the LUP-factorization over an arbitrary field, in time equal to big-Θ of matrix multiplication, by use of a black-box matrix multiplication algorithm. We call this algorithm AHU-LUP. The algorithm is described, in mathematical form, in Section 8.2.2 on Page 111. Once the factorization of A is complete, Strassen's Matrix Inversion Formula can be applied to U and L. Note $A^{-1} = P^{-1}U^{-1}L^{-1}$, and inverting a permutation matrix is easy, because $P^{-1} = P^T$. Another method is shown in Algorithm 7 on Page 95.

7.7.3 Ibara, Moran, and Hui's Solution

In [141], Ibara, Moran, and Hui show how to perform an LQUP-factorization with black-box matrix multiplication. The LQUP-factorization is similar to the LUP-factorization, but can operate on rank-deficient matrices. Therefore, if intermediate submatrices are singular, there is no difficulty.

A factorization $A = LQUP$ has L as lower-triangular, $m \times m$, and U as upper-triangular, $m \times n$. The permutation matrix P is $n \times n$ as before. The added flexibility comes from the matrix $m \times m$ matrix Q which is zero everywhere off of the main diagonal, and whose main diagonal contains r ones followed by $m - r$ zeroes. Here r is the rank of A.

This is not to be confused with the $QLUP$ factorization, where Q is a permutation matrix, just as P is. In the $QLUP$, the P is for row swaps and the Q is for column swaps. This occurs when one does Gaussian Elimination with full pivoting, instead of Gaussian Elimination with partial pivoting, which would produce LUP. See also, Section 7.1.3 on Page 91 and Section D.2 on Page 325.

This is how a singular (or a rank-deficient) A can be represented, while L, U, and P can be kept invertible (or full-rank). The determinant of Q is zero if and only if $r < m$. The algorithm is simpler than Bunch and Hopcroft, but is less amenable to parallelization, as it requires copying rows between submatrices after cutting.

Table 7.1 Algorithms and Performance, for $m \times n$ matrices

Algorithm	Overdefined	Square	Underdefined	Derivation
Matrix Inversion				
M4RI†	$\sim (1.5n^3 + 1.5n^2m)/(\log_2 n)$	$\sim (3n^3)/(\log_2 n)$	$\sim (6nm^2 - 3m^3)/(\log_2 m)$	see Section 9.5 on Page 146
Dense Gaussian Elim.	$\sim 1.5n^2m - 0.75mn^2$	$\sim 0.75n^3$	$\sim 1.5nm^2 - 0.75m^3$	see Section 7.5.3 on Page 97
Upper-Triangularization				
Dense Gaussian Elim.	$\sim 1.5n^2m - 0.75mn^2 - 0.25n^3$	$\sim 0.5n^3$	$\sim 0.75nm^2 - 0.25m^3$	see Section 7.5.3 on Page 98
M4RM†	$\sim (4.5n^3 + 1.5n^2m)/(\log_2 r)$	$\sim (6n^3)/(\log_2 n)$	$\sim (4.5nm^2 + 1.5m^3)/(\log_2 m)$	see Section 9.5.2 on Page 147
Back-Solve	$\sim n^2$	$\sim n^2$	$\sim 2nm - m^2$	see Section 7.5.4 on Page 98

Algorithm	Rectangular $a \times b$ by $b \times c$	Square $n \times n$	Derivation
Multiplication			
M4RM†	$\sim (3b^2c + 3abc)/(\log_2 b)$	$\sim (6n^3)/(\log_2 n)$	see Section 9.3.1 on Page 138
Naïve Multiplication	$\sim 2abc$	$\sim 2n^3$	see Section 7.5.1 on Page 96
* Strassen's Algorithm	$\sim M(n_0) \left(\frac{\sqrt[3]{abc}}{n_0} \right)^{\log_2 7}$	$\sim M(n_0)(n/n_0)^{\log_2 7}$	see Section 7.6.1 on Page 101

* Here $M(n_0)$ signifies the time required to multiply an $n_0 \times n_0$ matrix in some "base-line" algorithm.
† These refer to the Method of Four Russians for Multiplication, or for Inversion.

Chapter 8
On the Exponent of Certain Matrix Operations

A great deal of research was done in the period 1969–1987 on fast matrix operations, including [185, 212, 206, 213, 62]. Various proofs showed that many important matrix operations, such as QR-decomposition, LU-factorization, inversion, and finding determinants, are no more complex than matrix multiplication, in the big-Oh sense, see [13, Ch. 6] or [63, Ch. 28].

For this reason, many fast matrix multiplication algorithms were developed. Almost all were intended to work over a general ring. However, one in particular was intended for boolean matrices, and by extension $\mathbb{GF}(2)$-matrices, which was named the Method of Four Russians, "after the cardinality and the nationality of its inventors."[1] While the Method of Four Russians was conceived as a boolean matrix multiplication tool, we show how to use it for $\mathbb{GF}(2)$ matrices and for inversion, in Section 9.3 on Page 137 and Section 9.4 on Page 141.

Of the general purpose algorithms, the most famous and frequently implemented of these is Volker Strassen's 1969 algorithm for matrix multiplication in time $n^{2.807}$. However, many algorithms have a lower exponent in their complexity expression.

8.1 Very Low Exponents

The algorithms with exponents below $O(n^{2.807})$ all derive from the following argument in so far as the author is aware. Matrix multiplication of any particular fixed dimensions is a bilinear map from one vector space to another. The input space is of matrices $\oplus$ matrices as a direct sum, and the output space is another matrix space. Therefore, the map can be written as a tensor. By finding a shortcut for a particular matrix multiplication operation of fixed dimensions, one writes an upper bound for the complexity[2] of this tensor for those fixed dimensions. Specifically, Strassen

---

[1] Quoted from Aho, Hopcroft & Ullman textbook [13, Ch. 6]. Later information demonstrated that not all of the authors were Russians.

[2] An element of a tensor space is a sum of simple tensors. Here, the complexity of a tensor is the smallest number of simple tensors required. This is often called the rank of the tensor, but

G.V. Bard, *Algebraic Cryptanalysis*, DOI: 10.1007/978-0-387-88757-9_8   107
© Springer Science + Business Media, LLC 2009

performs 2×2 by 2×2 in seven steps instead of eight [212]. Likewise, Victor Pan's algorithm performs 70×70 by 70×70 in 143,640 steps rather than 343,000, for an exponent of 2.795 [185, Ch. 1]. A summary of some of these algorithms and how they are mapped into tensors, including in particular those of Schönhage [206], is available in the form of a technical report by the author of this book [29].

One can now write upper bounds for the complexity of matrix multiplication in general by extending the shortcut. Different papers execute this extension differently, but usually the cross-over[3] can be calculated explicitly. While the actual crossover in practice might vary, these matrices have millions of rows and are infeasible. For example, for Schönhage's algorithm at $O(n^{2.70})$, the crossover is given by [206] at $n = 3^{14} \approx 4.78 \times 10^6$ rows, or $3^{28} \approx 22.88 \times 10^{12}$ entries compared to naïve dense Gaussian Elimination. The crossover would be much higher versus Strassen's Algorithm or the Method of Four Russians.

Therefore, we have essentially three choices: algorithms of complexity equal to Strassen's exponent, of complexity equal to the Method of Four Russians, and algorithms of cubic complexity. The purpose of the linear algebra part of this book is to combine these effectively.

8.2 The Equicomplexity Theorems

The following is a series of theorems which prove that matrix multiplication, inversion, LUP factorization, and squaring, are equally complex in the sense of big-Θ. This implies that there is a real number, denoted ω, ironically called the exponent of matrix multiplication considering how many operations it describes, such that all these operations are $\Theta(n^{\omega})$. Several papers have been written trying to find tighter upper bounds for this value [185, 212, 206, 213, 62]. Other work has tried to lower-bound this value but lower bounds are not discussed here (but see [198]). In theory, Coppersmith and Winograd still hold the record at $\omega \le 2.36$, while in practice $\omega = 2.807$ (Strassen's algorithm) is the fastest algorithm in use [62, 212].

The theorems in this section have been known for a long time. In fact, all of them are found in or can be derived from the union of the papers [212, 55, 180] and the book [13, Ch. 6], except Theorem 31. The author cannot remember where he read that theorem, but he is certain that he did not invent the proof. On the other hand, the proofs of Theorem 42, Corollary 43, and Theorem 45 are new.

For now, we will exclude rings that are not fields. Suppose R is a ring that is not a division ring. Then there exists an element z which has no inverse. What would the inverse of the matrix zI be? Normally, diagonal matrices with non-zero entries on

---

other authors use the word "rank" differently. The rank of the tensor is directly proportional to the complexity of the operation [213].

[3] The cross-over point is the size where the new tensor has rank (complexity) equal to the naïve algorithm's tensor. More simply, the naïve algorithm and the algorithm being discussed, for this size matrix, take very nearly the same running time. For sizes smaller than this, the naïve algorithm is better, and for larger matrices, the algorithm being measured is better.

the main diagonal have inverses. Therefore, while these questions can be answered (by excluding matrices with non-invertible determinant and other methods) we will exclude them in this book. "Skew fields" are rings that are division rings but not commutative, and thus not fields. An example is the quaternion field. These cases are also beyond the scope of this book.

A brief notational comment is needed. One can sometimes show that a particular algorithm is $\Theta(f(n))$ or if not, then $O(f(n))$. But, the complexity of a problem is defined as the complexity of the *best* algorithm for it, in terms of asymptotic running time. Therefore showing an algorithm for solving a problem is either $\Theta(f(n))$ or $O(f(n))$, in both cases only proves that the problem is $O(f(n))$.

The definitions of $\Theta(f(n))$, $O(f(n))$, and $\Omega(f(n))$ can be found on page xxxiii, but remember that any algorithm which is $\Theta(n^3)$ is also $\Omega(n^2)$ and $O(n^4)$.

8.2.1 Starting Point

Recall that the inverse or square of an $n \times n$ matrix, as well as the product of two $n \times n$ matrices, will be an $n \times n$ matrix with n^2 entries. Therefore, just outputting the answer requires $\Omega(n^2)$ time and these operations are $\Omega(n^2)$. Likewise the LUP factorization of a non-singular $n \times n$ matrix requires 3 matrices, each $n \times n$, to write down, so that problem also is $\Omega(n^2)$.

This is in stark contrast to the determinant which is a single field element. However, we show in Theorem 57 on Page 132 that a $\Theta(f(n))$ algorithm for finding the determinant, provided $f(n)$ is $\Omega(n^2)$, will produce a $\Theta(f(n))$ algorithm for matrix multiplication in time $O(f(n))$. Furthermore, it will never be the case that $f(n)$ fails to be $\Omega(n^2)$, because all the n^2 inputs of the $n \times n$ matrix must be read in order to calculate the determinant.

Because naïve matrix multiplication is $\Theta(n^3)$ (see Section 7.5.1 on Page 96) we know that matrix multiplication and squaring are both $O(n^3)$. Likewise, because Gaussian Elimination is $\Theta(n^3)$ (see Section 7.5.3 on Page 96), we know that matrix inversion or LUP factorization is $O(n^3)$ (since that algorithm can be used for both).

8.2.2 Proofs

Theorem 30. *If there exists an algorithm for matrix inversion of unit upper (or lower) triangular $n \times n$ matrices over the field F, in time $\Theta(n^\omega)$, with $\omega \leq 3$, then there is an algorithm for $n \times n$ matrix multiplication over the field F in time $\Theta(n^\omega)$.*

Proof. Let A, B be $n \times n$ matrices over the field F. Consider the matrix on the left in the formula below:

$$\begin{bmatrix} I & A & 0 \\ 0 & I & B \\ 0 & 0 & I \end{bmatrix}^{-1} = \begin{bmatrix} I & -A & AB \\ 0 & I & -B \\ 0 & 0 & I \end{bmatrix}$$

This matrix is $3n \times 3n$ upper-triangular and has only ones on the main diagonal, and is also composed of entries only from F. Therefore its determinant is one and it is non-singular. Its inverse can be calculated in time $\Theta(n^\omega)$, and then the product AB can be read in the "northeast" corner.   □

The requirement of $\omega \leq 3$ was not quite superfluous. Any real r with $\omega \leq r$ would have done. If the matrix inversion requires $f(n)$ time for an $n \times n$ matrix, we need to know that $f(n)$ is upper-bounded by a polynomial. Call the degree of that polynomial d. This means that $f(3n) \leq 3^d f(n)$ for sufficiently large n. Thus $f(3n) = \Theta(f(n))$.

For example, if $\omega = \log n$, or more precisely if $f(n) = n^{\log n}$ then this would be problematic. In that case, $f(3n) = 3^{\log 3} n^{2 \log 3} f(n)$ and therefore $f(3n) \neq \Theta(f(n))$.

Theorem 31. *If there exists an algorithm for squaring an $n \times n$ matrix over the field F in time $\Theta(n^\omega)$ with $\omega \leq 3$, then there is an algorithm for $n \times n$ matrix multiplication over the field F in time $\Theta(n^\omega)$.*

Proof. Let A, B be $n \times n$ matrices over the field F. Consider the matrix on the left in the formula below:

$$\begin{bmatrix} A & B \\ 0 & 0 \end{bmatrix}^2 = \begin{bmatrix} A^2 & AB \\ 0 & 0 \end{bmatrix}$$

This matrix is $2n \times 2n$ and is also composed of entries only from F. Its square can be calculated in time $\Theta(n^\omega)$, and then the product AB can be read in the "north east" corner.   □

Again, the $\omega \leq 3$ was useful so that, since $f(n)$ is upper-bounded by a polynomial of degree $d \leq 3$, we can say that $f(2n) \leq 2^d f(n)$ for sufficiently large n and therefore $f(2n) = \Theta(f(n))$.

Theorem 32. *If there exists an algorithm for multiplying two $n \times n$ matrices over the field F in time $\Theta(n^\omega)$ then there is an algorithm for $n \times n$ matrix squaring over the field F in time $\Theta(n^\omega)$.*

Proof. $A \times A = A^2$ □

Theorem 33. *If there exists an algorithm for multiplying two $n \times n$ matrices over the field F, in time $\Theta(n^\omega)$ then there is an algorithm for inverting an $n \times n$ unit upper (or lower) triangular matrix over the field F, in time $\Theta(n^\omega)$.*

Proof. We will do the proof for the lower triangular case. It is almost unchanged for the upper triangular case—just take the transpose of every matrix.
    Observe,

$$\begin{bmatrix} A & 0 \\ B & C \end{bmatrix}^{-1} = \begin{bmatrix} A^{-1} & 0 \\ -C^{-1}BA^{-1} & C^{-1} \end{bmatrix}$$

If the original matrix is unit lower triangular, so are A and C. Thus an $n \times n$ unit lower triangular inverse requires two $n/2 \times n/2$ matrix multiplications and two $n/2 \times n/2$ unit lower triangular matrix inverses. Let the time required for an $n \times n$ lower triangular inverse be $I(n)$ and for an $n \times n$ matrix product $M(n)$.

We have

$$
\begin{aligned}
I(n) &= 2I(n/2) + 2M(n/2) \\
&= 4I(n/4) + 4M(n/4) + 2M(n/2) \\
&= 8I(n/8) + 8M(n/8) + 4M(n/4) + 2M(n/2) \\
&= 2^i I\left(\frac{n}{2^i}\right) + 2^i M\left(\frac{n}{2^i}\right) + \cdots + 2M(n/2) \\
&= 2^i I\left(\frac{n}{2^i}\right) + 2^i k\left(\frac{n}{2^i}\right)^\omega + \cdots + 2k(n/2)^\omega \\
&= 2^i I(2^{-i}n) + kn^\omega \left(\frac{1 - (2^i)^{1-\omega}}{2^{\omega-1} - 1}\right)
\end{aligned}
$$

Now we substitute $i = \log n$, and observe that a 1×1 unit lower triangular matrix is just the reciprocal of its only entry, and calculating that requires $\Theta(1)$ time. Then, we have

$$
\begin{aligned}
&= nI(1) + kn^\omega \left(\frac{1 - n^{1-\omega}}{2^\omega - 1}\right) \\
&= nI(1) + \frac{kn^\omega - kn}{2^\omega - 1} \\
&= \Theta(n) + \Theta(n^\omega) = \Theta(n^\omega)
\end{aligned}
$$

$\square$

The following can be found in [55] but in different notation.

Lemma 34 (Bunch and Hopcroft). *Let $m = 2^t$ where t is a positive integer, and $m < n$. Calculating the LUP factorization of a full-row-rank $m \times n$ matrix can be done with two LUP factorizations of size $m/2 \times n$ and $m/2 \times n - m/2$, two matrix products of size $m/2 \times m/2$ by $m/2 \times m/2$ and $m/2 \times m/2$ by $m/2 \times n - m/2$, the inversion of an $m/2 \times m/2$ triangular matrix, and some quadratic operations. Furthermore, L, U, P will be each full-row-rank.*

Proof. Step One: Divide A horizontally into two $m/2 \times n$ pieces.

This yields $A = \begin{bmatrix} B \\ C \end{bmatrix}$.

Step Two: Factor B into $L_1 U_1 P_1$. (Note that L_1 will be $m/2 \times m/2$, U_1 will be $m/2 \times n$, and P_1 will be $n \times n$.

Step Three: Let $D = CP_1^{-1}$. Thus D is $m/2 \times n$. Recall $P^{-1} = P^T$ for any permutation matrix. This yields

$$
A = \begin{bmatrix} L_1 U_1 \\ D \end{bmatrix} P_1
$$

Step Four: Let E be the left-most $m/2$ columns of U_1, and E' the remainder. Let F be the left-most $m/2$ columns of D, and F' the remainder. Now compute E^{-1}. Since U_1 is unit upper triangular then E is therefore also unit upper triangular, and thus invertible.

This yields

$$A = \begin{bmatrix} L_1 E & L_1 E' \\ F & F' \end{bmatrix} P_1$$

which implies

$$A = \begin{bmatrix} L_1 & 0 \\ 0 & I_{m/2} \end{bmatrix} \begin{bmatrix} E & E' \\ F & F' \end{bmatrix} P_1 \qquad (8.1)$$

Step Five: Consider $T = D - FE^{-1}U_1$. This can be thought of as $G = F - FE^{-1}E = 0$ and $G' = F' - FE^{-1}E'$, with $T = G|G'$ since $D = F|F'$ and $U_1 = E|E'$, where the $|$ denotes concatenation. The matrices E', F', G' are all $n - m/2$ columns wide. In the algorithm, we need only compute $G' = F' - FE^{-1}E'$. Along the way we should store FE^{-1} which we will have need of later. We have now

$$\begin{bmatrix} I_{m/2} & 0 \\ -FE^{-1} & I_{m/2} \end{bmatrix} \begin{bmatrix} E & E' \\ F & F' \end{bmatrix} = \begin{bmatrix} E & E' \\ 0 & G' \end{bmatrix}$$

Step Six: Factor $G' = L_2 U_2 P_2$, and observe

$$\begin{bmatrix} I_{m/2} & 0 \\ -FE^{-1} & I_{m/2} \end{bmatrix} \begin{bmatrix} E & E' \\ F & F' \end{bmatrix} = \begin{bmatrix} E & E' \\ 0 & L_2 U_2 P_2 \end{bmatrix} \qquad (8.2)$$

Note that since G' was $m/2 \times n - m/2$ wide, then L_2 will be $m/2 \times m/2$ and U_2 will be $m/2 \times n - m/2$ and P_2 will be $n - m/2 \times n - m/2$.

Step Seven: Let

$$P_3 = \begin{bmatrix} I_{m/2} & 0 \\ 0 & P_2 \end{bmatrix}$$

so that P_3 is a $n \times n$ matrix.

Step Eight: Calculate $E'P_2^{-1}$. Recall $P^{-1} = P^T$ for any permutation matrix P. This enables us to write

$$\begin{bmatrix} E & E'P_2^{-1} \\ 0 & L_2 U_2 \end{bmatrix} \underbrace{\begin{bmatrix} I_{m/2} & 0 \\ 0 & P_2 \end{bmatrix}}_{=P_3} = \begin{bmatrix} E & E' \\ 0 & L_2 U_2 P_2 \end{bmatrix}$$

and breaking up the left-hand matrix yields

$$\begin{bmatrix} E & E'P_2^{-1} \\ 0 & L_2 U_2 \end{bmatrix} = \begin{bmatrix} I_{m/2} & 0 \\ 0 & L_2 \end{bmatrix} \begin{bmatrix} E & E'P_2^{-1} \\ 0 & U_2 \end{bmatrix}$$

and that yields

$$\begin{bmatrix} I_{m/2} & 0 \\ 0 & L_2 \end{bmatrix} \begin{bmatrix} E & E'P_2^{-1} \\ 0 & U_2 \end{bmatrix} P_3 = \begin{bmatrix} E & E' \\ 0 & L_2U_2P_2 \end{bmatrix}$$

Substituting Equation 8.2 into this last equation we obtain

$$\begin{bmatrix} I_{m/2} & 0 \\ 0 & L_2 \end{bmatrix} \begin{bmatrix} E & E'P_2^{-1} \\ 0 & U_2 \end{bmatrix} P_3 = \begin{bmatrix} I_{m/2} & 0 \\ -FE^{-1} & I_{m/2} \end{bmatrix} \begin{bmatrix} E & E' \\ F & F' \end{bmatrix}$$

Because

$$\begin{bmatrix} I_{m/2} & 0 \\ -FE^{-1} & I_{m/2} \end{bmatrix}^{-1} = \begin{bmatrix} I_{m/2} & 0 \\ FE^{-1} & I_{m/2} \end{bmatrix}$$

we can write

$$\begin{bmatrix} I_{m/2} & 0 \\ FE^{-1} & I_{m/2} \end{bmatrix} \begin{bmatrix} I_{m/2} & 0 \\ 0 & L_2 \end{bmatrix} \begin{bmatrix} E & E'P_2^{-1} \\ 0 & U_2 \end{bmatrix} P_3 = \begin{bmatrix} E & E' \\ F & F' \end{bmatrix}$$

and substitute this into Equation 8.1 to obtain

$$A = \begin{bmatrix} L_1 & 0 \\ 0 & I_{m/2} \end{bmatrix} \begin{bmatrix} I_{m/2} & 0 \\ FE^{-1} & I_{m/2} \end{bmatrix} \begin{bmatrix} I_{m/2} & 0 \\ 0 & L_2 \end{bmatrix} \begin{bmatrix} E & E'P_2^{-1} \\ 0 & U_2 \end{bmatrix} P_3P_1$$

This now is sufficient for the factorization:

$$A = \underbrace{\begin{bmatrix} L_1 & 0 \\ FE^{-1} & L_2 \end{bmatrix}}_{=L} \underbrace{\begin{bmatrix} E & E'P_2^{-1} \\ 0 & U_2 \end{bmatrix}}_{=U} \underbrace{P_3P_1}_{=P}$$

Since L_1 and L_2 are outputs of the factor algorithm they are unit lower triangular, as is L. Likewise E and U_2 are unit upper triangular, and so is U. The product of two permutation matrices is a permutation matrix, as is P. Thus all three have full row-rank.

Note that the matrix products and inverses involving permutation matrices are quadratic or faster, as discussed in Section 7.4.2 on Page 94, and thus negligible.

□

Lemma 35. *Let A be a non-zero $1 \times n$ matrix, with a non-zero entry at i. Then $L = [1]$, $U = [x_i, x_2, x_3, \ldots, x_{i-1}, x_1, x_{i+1}, x_{i+2}, \ldots, x_n]$ and P being the permutation matrix which swaps columns i and 1, is a factorization $A = LUP$.*

Proof. Obvious. □

Theorem 36. *If matrix multiplication of two $n \times n$ matrices is $O(n^{c_1})$ over the field F and matrix inversion of an $n \times n$ triangular matrix is $O(n^{c_2})$ over the field F then the LUP factorization of an $m \times n$ matrix, with m being a power of two and $m \leq n$, is $O(n^{\max(c_1,c_2)})$, over the field F. Let $2 \leq c_1 \leq 3$ and $2 \leq c_2 \leq 3$.*

Proof. Suppose matrix multiplication can be done in time $O(n^{c_1})$ and triangular matrix inversion in time $O(n^{c_2})$. Let $c = \max(c_1, c_2)$. For sufficiently large n, the time of either of these operations is $\leq kn^c$ for some real number k.

Also, since the time required to do an $m/2 \times n$ LUP factorization is greater than or equal to the time required to do an $m/2 \times n - m/2$ LUP factorization (because the latter matrix is smaller), we will represent both as $L(m/2, n)$, being slightly pessimistic.

Since $m < n$ in Lemma 34 the two matrix products and one triangular inversion require at most $3kn^c$ time.

From Lemma 34, we have that

$$
\begin{aligned}
L(m,n) &= 2L(m/2,n) + 3kn^c \\
&= 4L(m/4,n) + 6k(n/2)^c + 3kn^c \\
&= 8L(m/8,n) + 12k(n/4)^c + 6k(n/2)^c + 3kn^c \\
&= 16L(m/16,n) + 24k(n/8)^c + 12k(n/4)^c + 6k(n/2)^c + 3kn^c \\
&= 2^i L(m/2^i,n) + 3kn^c \left[\frac{2^i}{(2^i)^c} + \cdots + 4/4^c + 2/2^c + 1/1^c \right] \\
&= 2^i L(m/2^i,n) + 3kn^c \frac{2^c}{2^c - 2}
\end{aligned}
$$

Now let $i = \log_2 m$.

$$
L(m,n) = mL(1,n) + \frac{3kn^c 2^c}{2^c - 2}
$$

Since $L(1,n)$ is $\Theta(n)$ by Lemma 35, and that last term is $O(n^c)$ for any constant c and constant k, we obtain that $L(m,n) = O(n^c)$. $\qquad\square$

Theorem 37. *If matrix multiplication of two $n \times n$ matrices is $O(n^{c_1})$, over the field F, and triangular matrix inversion is $O(n^{c_2})$, over the field F, then the LUP factorization of an $m \times n$ matrix, with $m \leq n$, is $O(n^{\max(c_1,c_2)})$, over the field F. Let $2 \leq c_1 \leq 3$ and $2 \leq c_2 \leq 3$.*

Proof. This is an identical claim to Lemma 36 except that the requirement that m be a power of two has been dropped. If m is a power of two and $m = n$, factor as in Lemma 34. If not, let m' be the next power of two greater than or equal to both m and n.

$$
A = L_1 U_1 P_1 \Leftrightarrow
\begin{bmatrix} A & 0 \\ 0 & I_{m'-m} \end{bmatrix} =
\begin{bmatrix} L_1 & 0 \\ 0 & I_{m'-m} \end{bmatrix}
\begin{bmatrix} U_1 & 0 \\ 0 & I_{m'-m} \end{bmatrix}
\begin{bmatrix} P_1 & 0 \\ 0 & I_{m'-m} \end{bmatrix}
$$

by extending A diagonally as shown, we at most double the size of m. We therefore, at worse, increase the running time eightfold, since using Gaussian Elimination for LUP factorization is $\Theta(n^3)$. $\qquad\square$

Theorem 38. *If multiplying two $n \times n$ matrices is $O(n^c)$ over the field F, then inverting an $n \times n$ matrix is $O(n^c)$ over the field F.*

Proof. Because multiplying two $n \times n$ matrices is $O(n^c)$, we know by Theorem 33, that inverting a unit lower triangular matrix is $O(n^c)$. Then via Theorem 37, an LUP factorization can be computed in $O(n^c)$ time. If the original $n \times n$ matrix is A, then $A = LUP$ with L and U being unit lower/upper triangular. Thus we can invert them, and inverting P is a quadratic operation (See Section 7.4.2 on Page 94). Surely then $A^{-1} = P^{-1}U^{-1}L^{-1}$, and we required a constant number of $O(n^c)$ operations. Thus we have inverted A in $O(n^c)$ time.   □

Theorem 39. *If calculating the LUP factorization of a non-singular $n \times n$ matrix is $O(n^c)$ over a field F, then finding the determinant of a non-singular $n \times n$ matrix over a field F is also $O(n^c)$.*

Proof. If $A = LUP$ then $det(A) = det(L) \times det(U) \times det(P)$. Note that $det(L)$ is the product of the entries of the main diagonal, just as is $det(U)$, because both matrices are triangular. The determinant of a permutation matrix is the sign of that permutation, thus $+1$ or -1. This can be calculated in linear time by "undoing" the permutation as a series of swaps, and counting the number required x, and returning the determinant as $(-1)^x$.   □

Definition 40. If a matrix

$$A = \begin{bmatrix} W & X \\ Y & Z \end{bmatrix}$$

then the Schur Complement of A with respect to W is $Z - Y(W^{-1})X$.

Two interesting things about Schur Complements are worthy of note. First, the matrix W can be any of the $n - 1$ submatrices rooted in the "northwest" corner, ranging from the 1×1 submatrix of the upper-left most element to the $n - 1 \times n - 1$ submatrix consisting of all but the bottom row and rightmost column. The other interesting fact is the following lemma. This short proof is due to Larry Washington, though the result is certainly quite old.

Lemma 41. *If a matrix*

$$A = \begin{bmatrix} W & X \\ Y & Z \end{bmatrix}$$

is non-singular, and W is non-singular also, then the Schur Complement of A with respect to W is non-singular.

Proof. [L. Washington] Observe,

$$\begin{bmatrix} I & 0 \\ -Y & I \end{bmatrix} \begin{bmatrix} W^{-1} & 0 \\ 0 & I \end{bmatrix} \begin{bmatrix} W & X \\ Y & Z \end{bmatrix} = \begin{bmatrix} I & W^{-1}X \\ 0 & Z - YW^{-1}X \end{bmatrix}$$

and recall that the Schur complement of A with respect to W is $S = Z - YW^{-1}X$. Thus by taking the determinant of the above one obtains

$$(1)(\det W^{-1})(\det A) = (\det S)$$

Therefore $\det S$ is non-zero. $\square$

Theorem 42. *If a matrix*

$$A = \begin{bmatrix} W & X \\ Y & Z \end{bmatrix}$$

is non-singular, with W being non-singular also, then let $W = L_1 U_1 P_1$, be an LUP factorization, denote the Schur Complement of A with respect to W as S, and let $S = L_2 U_2 P_2$ be an LUP factorization. We have

$$A = \begin{bmatrix} L_1 & 0 \\ Y P_1^{-1} U_1^{-1} & L_2 \end{bmatrix} \begin{bmatrix} U_1 & L_1^{-1} X P_2^{-1} \\ 0 & U_2 \end{bmatrix} \begin{bmatrix} P_1 & 0 \\ 0 & P_2 \end{bmatrix}$$

Proof. Simply multiply those three matrices, and observe that one obtains A. $\square$

Corollary 43. *If a matrix*

$$A = \begin{bmatrix} U_1 & 0 \\ I & Z \end{bmatrix}$$

is non-singular, with U_1 being unit upper-triangular, then the Schur Complement of A with respect to U is I (See Definition 40 on Page 115); let $Z = L_2 U_2 P_2$ be an LUP factorization. Then an LUP factorization of A is

$$A = \begin{bmatrix} I & 0 \\ U_1^{-1} & L_2 \end{bmatrix} \begin{bmatrix} U_1 & 0 \\ 0 & U_2 \end{bmatrix} \begin{bmatrix} I & 0 \\ 0 & P_2 \end{bmatrix}$$

Proof. This is Theorem 42, but with $X = 0$ and $Y = I$. Furthermore, since U_1 is unit upper triangular, it is non-singular. Also, the LUP factorization of U_1 is $(I)(U_1)(I)$, as the identity matrix I is both unit lower triangular and a permutation matrix. $\square$

The inverse of a matrix (triangular or general), the product of two matrices, or the square of a matrix are all unique specific matrices. But a matrix can have many LUP factorizations. For any $k \neq 0$, one ambiguity is that if $(L)(U)(P) = A$ then surely $(k^{-1}L)(kU)(P) = A$ is also a factorization and both L and U remain triangular in the correct directions. In $\mathbb{GF}(2)$, this is not a concern, as the only scalar not equal to zero is one itself, and this presents no change to the factorization.

Outside of $\mathbb{GF}(2)$, one can divide each row i of L by L_{ii}, and make a diagonal matrix of the old L_{ii} values. Likewise this can be done with U. If the product of the diagonal matrices is D then $LDUP$ has both L and U as non-singular lower/upper-triangular matrices (as appropriate) and with 1's on the main diagonals of L and of U. It is curious to note that $\det L = \det U = |\det P|$, and so $\det D = |\det A|$, and $\det D$ is just the product of the diagonal entries.

Also, note $L_{ii} \neq 0$ because if $L_{ii} = 0$ then $\det L = 0$ and thus $\det(LUP) = 0$, or $\det A = 0$. But our original matrix is non-singular. Likewise, $U_{ii} \neq 0$.

In the case of our particular problem, we wish not to have such a D, and to have all ones on the main diagonal of L. Therefore, we should calculate D as above, and replace U with DU. This produces an LUP factorization, with precisely the required properties.

These scalar multiplies are (all together) a quadratic time operation. Recall, no matrix factorization is possible faster than in quadratic time since all n^2 entries in the original matrix must be read.

The following term, "non-permutative" is mine, but it gets us around an interesting sticking-point.

Definition 44. Call an algorithm for LUP factorization of a matrix A "non-permutative" if and only if the output $A = LUP$ will have $P = I$ in the cases of those particular A where this is possible.

This topic is covered in [126, Ch. 3.2] in great detail. Over the real numbers, all positive definite matrices have LUP factorizations with $P = I$. A simple matrix without such a factorization is

$$\begin{bmatrix} 0 & 1 \\ 1 & 0 \end{bmatrix}$$

All standard methods of calculating the LUP factorization *over a finite field or the rational numbers* are "non-permutative". Examples include Gaussian Elimination or the Bunch-Hopcroft algorithm given above, or various block methods covered in [126, Ch. 3.2], which fall back on Gaussian Elimination for small block sizes at the end of the recursion.

A notable exception is Gaussian Elimination with Partial Pivoting, which will usually produce a P. This is because, in iteration i, the pivot element is taken to be the element of the active column i with the largest absolute value, excepting the elements above A_{ii}, which belong to rows that should not be touched at that point. So unless, by coincidence, A_{ii} is indeed the largest element each time, then $P \neq I$. This is done to reduce rounding-error. For completeness, we should note that full-pivoting produces a $A = P_1 LU P_2$ factorization, and so is not relevant here.

And so we have noted that an LUP factorization is not unique and the inverse of U_1 is unique. At first this presents a paradox, because only one LUP factorization will produce the inverse. However, as we began to discuss earlier, there are only two points of ambiguity in an LUP factorization. The issue of moving constants between L and U is of no consequence, because it is resolved by using a D matrix as explained before. In the case of "non-permutative" algorithms, we are okay, because we do not need a permutation matrix in either case, and so none will be used, and the ambiguity is removed.

Theorem 45. *If there exists a non-permutative algorithm for the LUP factorization of a non-singular $n \times n$ matrix over a field F in time $\Theta(n^c)$, then there is an algorithm for inverting upper-triangular $n \times n$ matrices over the field F in time $\Theta(n^c)$.*

Proof. Given any unit upper triangular matrix U_1, one can construct the matrix A as used in the proof of Corollary 43. Then the LUP factorization of A will contain U^{-1} in the "southwest" corner of L. If U_1 is an $n \times n$ matrix, A will be $2n \times 2n$. This will take $\Theta(2^c n^c)$ time. Since $c \leq 3$ (because LUP factoring via Gaussian Elimination is $\Theta(n^3)$) and $c \geq 2$ (because all $(n^2 + n)/2$ non-zero elements of U_1 must be read into the algorithm), we can say $\Theta(2^c n^c) = \Theta(n^c)$.

Note that if U_1 were upper triangular but not unit upper triangular, then it would be singular and so it would have no inverse, and thus no algorithm could invert it.

$\square$

We are almost finished. We now have

Theorem 46. *If any of matrix inversion, LUP factorization, matrix multiplication, triangular matrix inversion, or matrix squaring, of $n \times n$ matrices over the field F, is $\Theta(n^c)$, then all of these operations are $\Theta(n^c)$. In addition, calculating the determinant is $O(n^c)$.*

Note that we will fix the determinant in a few more pages, namely Theorem 57 on Page 132.

8.3 Determinants and Matrix Inverses

In this section we will examine by far the deepest theorems in this chapter. While we will have to cover a large amount of background information, on the other hand, we will prove along the way the theorem of Baur and Strassen [39], as exposited by Morgenstern [180], that finding all the partial derivatives of a function is only 5 times as complex as finding the value of it, in terms of floating-point operations, regardless of the number of variables.

8.3.1 Background

One of the classical definitions of the determinant, as taught in many American high schools, is to "expand by cofactors." The method is due to Pierre-Simon Laplace, and so we shall call it Laplace's Formula. This is computationally very inefficient (requiring $n!$ field multiplications if implemented naïvely[4]) but it provides a useful mechanism for proofs by induction.

To explain this, we require a definition of a matrix minor. The minor M_{ij} for an $n \times n$ matrix A is an $(n-1) \times (n-1)$ matrix, whose entries are formed by deleting the ith row and the jth column of A. Let $m_{ij} = \det M_{ij}$.

Usually one is taught to expand a determinant as follows:

$$\det(A) = a_{11}m_{11} - a_{12}m_{12} + a_{13}m_{13} - a_{14}m_{14} + \cdots + (-1)^{n-1}m_{1n}$$

---

[4] If one stores the determinants of matrix minors as one computes them, then they can be reused. This is a huge savings in practice. But there are other, still faster ways of calculating the determinant.

but actually one can expand on any of the rows or any of the columns. This is important when taking the determinant by hand on an exam, especially if many of the entries in some row or column are zero.

The reason this flexibility is possible is that swapping two rows or swapping two columns only flips the sign of the determinant. We will state the following theorem, but not prove it, as it is contained in many linear algebra texts (e.g. [131, Ch. 4.5]).

Theorem 47. *[Laplace's Formula]. Let M_{ij} signify the minor of A formed by deleting row i and column j. Let $m_{ij} = \det M_{ij}$. For an $n \times n$ matrix A, if $n > 1$*

$$\det(A) = \sum_{i=1}^{i=n} (-1)^{m+i} a_{mi} m_{mi}$$

for any $1 \leq m \leq n$. If $n = 1$, i.e. A is a 1×1 matrix, then $\det(A) = A_{11}$.

Note: This enables you to expand on any particular row m. If you want to expand on a column, then note $\det(A) = \det(A^T)$, and so the process is identical.

Corollary 48. *For any n, the determinant of an $n \times n$ matrix is a polynomial in terms of the entries of the matrix.*

Proof. If this were true for any n, then Laplace's formula makes it true for $n + 1$. And for a 1×1 matrix A, recall $\det(A) = A_{11}$, which is a polynomial in terms of the entry of the matrix A. □

The determinant is a map from the set of $n \times n$ matrices over a ring R to the ring R, and in fact, we just showed that it is a polynomial. As a polynomial, it has n^2 variables, one for each entry in the $n \times n$ matrix. Therefore, we would be justified in thinking of it as a map not from $M_n(R) \rightarrow R$ but instead as a map $R^{n^2} \rightarrow R$. There would obviously be n^2 partial derivatives. Using these, we could find $\nabla \det(A)$, which might be very useful. Therefore, let us take the partial derivative of Laplace's Formula. This will allow us to calculate $(\partial \det(A)/\partial A_{ij})$, the partial derivative of the determinant of A in terms of the matrix element A_{ij}.

First assume $n > 1$. Let us consider the minors formed by expanding on row i, namely M_{i1}, M_{i2}, ..., M_{in}. Surely A_{ij} does not appear in any of them, because they are formed by deleting row i. This means that they are constants if everything in A is held fixed except for A_{ij}. With this in mind, it becomes obvious that the partial derivative in question is $\pm \det M_{ij} = \pm m_{ij}$, as can be observed from Laplace's formula. Finally, if $n = 1$ then the determinant function is the identity function, so the partial derivative in question is 1. Therefore, we have proven

Lemma 49. *For any $n \times n$ matrix A, with $n > 1$, let M_{ij} signify the minor formed by deleting row i and column j from the matrix A. Then*

$$\frac{\partial}{\partial A_{ij}} \det(A) = (-1)^{i+j} \det M_{ij}$$

While $\nabla \det(A)$ is a vector of length n^2, with entries in R, it would be more convenient to think of it as a matrix. Let us construct

Definition 50. Let A be an $n \times n$ matrix. The cofactor matrix of A is the $n \times n$ matrix given by

$$
C = \begin{bmatrix}
\frac{\partial}{\partial A_{11}} \det(A) & \frac{\partial}{\partial A_{12}} \det(A) & \cdots & \frac{\partial}{\partial A_{1n}} \det(A) \\[2ex]
\frac{\partial}{\partial A_{21}} \det(A) & \frac{\partial}{\partial A_{22}} \det(A) & \cdots & \frac{\partial}{\partial A_{2n}} \det(A) \\[2ex]
\vdots & \vdots & \ddots & \vdots \\[2ex]
\frac{\partial}{\partial A_{n1}} \det(A) & \frac{\partial}{\partial A_{n2}} \det(A) & \cdots & \frac{\partial}{\partial A_{nn}} \det(A)
\end{bmatrix}
$$

Another interesting formula in linear algebra is

Lemma 51. *For any matrix $n \times n$ matrix A,*

$$
AC^T = \det(A)I
$$

where C is the cofactor matrix of A, and I is the $n \times n$ identity matrix.

Proof. For a proof, see any standard linear algebra text, including [222, Ch. 1.8]. $\qquad\square$

Of course, this means

Corollary 52. *Given an invertible $n \times n$ matrix A, whose cofactor matrix is C,*

$$
A^{-1} = \frac{1}{\det(A)} C^T
$$

which is useful. Also note that if A were not invertible, then $\det(A) = 0$ and so its reciprocal does not exist— thus the formula does not produce a wrong answer.

The transpose of the cofactor matrix A^T in some older texts is called the "adjoint matrix." However, in physics, the adjoint of a matrix in $M_n(\mathbb{C})$ is the complex conjugate of its transpose. This is a totally unrelated concept, and so to distinguish those two ideas some authors call C^T the "classical adjoint" or the "adjugate". However, this author prefers "the transpose of the cofactor matrix." All this would seem rather empty as an exercise if it were not for the following amazing result.

8.3.2 The Baur-Strassen-Morgenstern Theorem

In this section, we will prove that if one can evaluate a rational function in f of n variables with t field operations, then one can evaluate it and all of its partial derivatives in $\leq 5t$ field operations. Notice that this is shocking, because n does not appear in the complexity bound. Intuitively, if there are very few steps compared to the number of variables, then the partial derivatives are very simple. We shall seek a rigorous proof.

This was first proven by Baur and Strassen in [39], but actually we will recount here a proof by Morgenstern from [180].

8.3.2.1 The Computational Model

We will define the computation of $f(x_1,\ldots,x_n)$, a particular rational function in n variables over the field $\mathbb{F}$ as follows. There will be s operations, whose outputs are $g_1, g_2, \ldots, g_s$, and some of which are of the type $g_k = g_i \star g_j$, with $i < k$ and $j < k$, and where $\star$ can be any of the four field operations. Alternatively, we also permit $g_k = \mathrm{LOAD}(x_i)$, where x_i is one of the n inputs to the function, and $g_k = \mathrm{CONS}(a)$ where $a \in \mathbb{F}$ is some fixed constant. The final g_s can be assumed to be the true output of the function f. (In other words, we assume we are presented with a correct algorithm for f). The issue of dividing by zero will be a concern later.

The algorithm will be defined as the sequence of $g_1, g_2, \ldots, g_z$, each of which is either a field operation, $\mathrm{LOAD}(x_i)$, or $\mathrm{CONS}(c)$ for some $c \in \mathbb{F}$. In the proof, we will actually construct a new algorithm by appending additional steps $g_{z+1}, g_{z+2}, \ldots$ at each inductive step.

We shall further specify $p_0, p_1, \ldots, p_n$ which are positive integers. By $p_i = k$ we intend to indicate that $\partial f(x_1, x_2, \ldots, x_n)/\partial x_i = g_k$, and by $p_0 = k$ we intend to indicate $f(x_1, x_2, \ldots, x_n) = g_k$. The purpose of these p_k is that our language has no output instruction.

Oddly enough, a model extremely similar to this one pops up also in cryptographic proofs. An example is [12], where it is proven that breaking RSA in a certain sense is equivalent to factoring. Several papers cited by that one also use this "generic ring model."

8.3.2.2 Theorem and Proof

This result is non-trivial. The author recommends that the reader skim the proof and the example given after it, and then read the proof carefully, and the example carefully again.

Suppose now that of the s instructions in f we see that $t \leq s$ of them are neither LOAD nor CONS. These are thus field operations. Furthermore, $h \leq t$ are either multiplications or divisions, which in many fields are very time consuming operations. (Thus, we denote them "h" for hard.)

Theorem 53. *[Baur-Strassen] If a rational function f can be calculated in s instructions, of which t are field operations and of those, h are multiplications and divisions, then f as well as all its partial derivatives can be calculated using at most $5s$ instructions, $5t$ field operations, and $3h$ multiplications and divisions.*

Proof. We shall now prove the inductive hypothesis:

Given that for all rational functions f over the field $\mathbb{F}$ having $t \leq t_{max}$ field operations, the partial derivatives of f can be calculated by an algorithm having $\leq 5t_{max}$ field operations, it is the case that a rational function f over the field $\mathbb{F}$ having $t_{max} + 1$ field operations can be calculated, along with all of its partial derivatives, by an algorithm having $\leq 5t_{max} + 5$ field operations.

Assume we have an f, a rational function over the field $\mathbb{F}$, such that it has s instructions $g_1, g_2, \ldots, g_s$ and of these $t_{max} + 1$ are not LOAD nor CONS. Let g_a be the first instruction that is neither LOAD nor CONS, if there is at least one (if there are none, we handle this as a special "base case" at the end).

Note that instruction a is either a sum, difference, product, or quotient. Therefore we write $g_a = g_i \star g_j$ so that $i < a$ and $j < a$ and also $\star$ is one of the four field operations. But we said that a was the first instruction not a LOAD nor a CONS, and so we know that g_i and g_j are each either x_k for some input variable $1 \leq k \leq n$, or some constant from the field $\mathbb{F}$.

Superfluous Operations:

First, we will remove a few annoying special cases that should not interest anyone. If both of the operands (inputs) of the field operation are constants from the field, we can remove the operation and replace it with a CONS, whose value is the answer to that operation which is being deleted. For the case of division by zero, see Section 8.3.2.4 on Page 131. The resulting computation is totally unchanged, but has only t_{max} operations which are neither LOAD nor CONS, and so by the inductive hypothesis, there exists an algorithm using $5t_{max}$ field operations that computes it and its partial derivatives. Therefore we are done.

Second, perhaps both operands are input variables x_i and x_j, meaning that $g_i = \text{LOAD}(x_i)$ and $g_j = \text{LOAD}(x_j)$. Now consider the possibility that $i = j$. If we have $g_a = x_i - x_i$ then replace it with $g_a = \text{CONS}(0)$ and if we have $g_a = x_i/x_i$ then replace it with $g_a = \text{CONS}(1)$. The resulting computation is totally unchanged, but has only t_{max} operations which are neither LOAD nor CONS, and so by the inductive hypothesis, there exists an algorithm using $5t_{max}$ field operations that computes it and its partial derivatives. Therefore we are done. We are forced to consider $g_a = x_i + x_i$ (in fields of characteristic not 2) and $g_a = x_i x_i$ (in all fields) as legitimate cases.

Now consider fields of characteristic two. In these fields, $g_a = x_i + x_i$ is replaceable by $g_a = \text{CONS}(0)$, and so can be removed just like $g_a = x_i - x_i$.

Useful Operations:

Now define $\overline{f}$ to be a rational function of $n + 1$ variables. Let its instructions be identical to those of f, with the exception that g_a is replaced by $g_a = \text{LOAD}(x_{n+1})$.

Note that

$$f(x_1, \ldots, x_n) = \overline{f}(x_1, \ldots, x_n, G(x_1, \ldots, x_n))$$

where G is some rational function to be defined momentarily, will render $\overline{f}$ equivalent to f provided G is correctly defined. In fact, G will be the operation we are removing. And here, equivalent means that the two functions have the same output for all possible $x_1, x_2, \ldots, x_n$.

Note further that

$$\frac{\partial f}{\partial x_i} = \frac{\partial \overline{f}}{\partial x_i} + \frac{\partial \overline{f}}{\partial x_{n+1}} \frac{\partial G}{\partial x_i}$$

via the multivariate chain rule.

Of course, $\overline{f}$ has only t_{max} field operations and so by the inductive hypothesis there exists an algorithm $\nabla \overline{f}$ such that the value of $\overline{f}$ is calculated, as well as all of the partial derivatives of $\overline{f}$, using $\leq 5t_{max}$ field operations. Using $\nabla \overline{f}$ we will construct ∇f by only appending instructions to $\nabla \overline{f}$.

Notation

For notational simplicity, let g_z be the final instruction of $\nabla \overline{f}$. Let $p_0, \ldots, p_n$ refer to the pointers to the outputs of $\nabla \overline{f}$ as discussed earlier, and let $p'_0, \ldots, p'_n$ refer to the pointers to the outputs of ∇f as we construct it.

Case 1: Sum of Two Distinct Variables

We have $g_a = x_i + x_j$, and so $G(x_1, \ldots, x_n) = x_i + x_j$. In this case,

$$\frac{\partial f}{\partial x_i} = \frac{\partial \overline{f}}{\partial x_i} + \frac{\partial \overline{f}}{\partial x_{n+1}}; \qquad \frac{\partial f}{\partial x_j} = \frac{\partial \overline{f}}{\partial x_j} + \frac{\partial \overline{f}}{\partial x_{n+1}}$$

and so we add the following instructions:

$$g_{z+1} = g_{p_{n+1}} + g_{p_i}$$
$$g_{z+2} = g_{p_{n+1}} + g_{p_j}$$

and write $p'_i = z+1$, $p'_j = z+2$, as well as $p'_k = p_k$ for all other p'.

Since $\nabla \overline{f}$ has $\leq 5t_{max}$ operations, then ∇f has $\leq 5t_{max} + 2 \leq 5(t_{max} + 1)$ operations, and we are done.

Case 2: Sum of a Variable with Itself

We have $g_a = x_i + x_i$, and so $G(x_1, \ldots, x_n) = 2x_i$. Either the field has characteristic two, or it does not. If it does, this was handled in "superfluous operations" above. This tedious detail now resolved, for all other fields

$$\frac{\partial f}{\partial x_i} = \frac{\partial \overline{f}}{\partial x_i} + 2\frac{\partial \overline{f}}{\partial x_{n+1}}$$

and so we add the following instructions:

$$g_{z+1} = g_{p_{n+1}} + g_{p_{n+1}}$$
$$g_{z+2} = g_{z+1} + g_{p_i}$$

and write $p'_i = z + 2$, as well as $p'_k = p_k$ for all other p'.

Since $\nabla \overline{f}$ has $\leq 5t_{max}$ operations, then ∇f has $\leq 5t_{max} + 2 \leq 5(t_{max} + 1)$ operations, and we are done.

Case 3: Sum of a Variable and a Constant

We have $g_a = x_i + c$, and so $G(x_1, \ldots, x_n) = x_i + c$. In this case,

$$\frac{\partial f}{\partial x_i} = \frac{\partial \overline{f}}{\partial x_i} + \frac{\partial \overline{f}}{\partial x_{n+1}}$$

and so we add the following instruction:

$$g_{z+1} = g_{p_{n+1}} + g_{p_i}$$

and write $p'_i = z + 1$, as well as $p'_k = p_k$ for all other p'.

Since $\nabla \overline{f}$ has $\leq 5t_{max}$ operations, then ∇f has $\leq 5t_{max} + 1 \leq 5(t_{max} + 1)$ operations, and we are done.

Case 4: Sum of a Constant and a Variable

Since addition is commutative in a field, then $g_a = c + x_i$ can be handled identically to the previous case, $g_a = x_i + c$.

Case 5: Difference of Two Distinct Variables

We have $g_a = x_i - x_j$, and so $G(x_1, \ldots, x_n) = x_i - x_j$. In this case,

$$\frac{\partial f}{\partial x_i} = \frac{\partial \overline{f}}{\partial x_i} + \frac{\partial \overline{f}}{\partial x_{n+1}}; \qquad \frac{\partial f}{\partial x_j} = \frac{\partial \overline{f}}{\partial x_j} - \frac{\partial \overline{f}}{\partial x_{n+1}}$$

and so we add the following instructions:

$$g_{z+1} = g_{p_i} + g_{p_{n+1}}$$
$$g_{z+2} = g_{p_j} - g_{p_{n+1}}$$

and write $p_i' = z + 1$, $p_j' = z + 2$, as well as $p_k' = p_k$ for all other p'.

Since $\nabla \overline{f}$ has $\leq 5t_{max}$ operations, then ∇f has $\leq 5t_{max} + 2 \leq 5(t_{max} + 1)$ operations, and we are done.

Case 6: Deducting a Constant from a Variable

We have $g_a = x_i - c$, and so $G(x_1, \ldots, x_n) = x_i - c$. In this case,

$$\frac{\partial f}{\partial x_i} = \frac{\partial \overline{f}}{\partial x_i} + \frac{\partial \overline{f}}{\partial x_{n+1}}$$

and so we add the following instruction:

$$g_{z+1} = g_{p_{n+1}} + g_{p_i}$$

and write $p_i' = z + 1$, as well as $p_k' = p_k$ for all other p'.

Since $\nabla \overline{f}$ has $\leq 5t_{max}$ operations, then ∇f has $\leq 5t_{max} + 1 \leq 5(t_{max} + 1)$ operations, and we are done.

Case 7: Deducting a Variable from a Constant

We have $g_a = c - x_i$, and so $G(x_1, \ldots, x_n) = c - x_i$. In this case,

$$\frac{\partial f}{\partial x_i} = \frac{\partial \overline{f}}{\partial x_i} - \frac{\partial \overline{f}}{\partial x_{n+1}}$$

and so we add the following instruction:

$$g_{z+1} = g_{p_i} - g_{p_{n+1}}$$

and write $p_i' = z + 1$, as well as $p_k' = p_k$ for all other p'.

Since $\nabla \overline{f}$ has $\leq 5t_{max}$ operations, then ∇f has $\leq 5t_{max} + 1 \leq 5(t_{max} + 1)$ operations, and we are done.

Case 8: Product of Two Distinct Variables

We have $g_a = x_i x_j$, and so $G(x_1, \ldots, x_n) = x_i x_j$. In this case,

$$\frac{\partial f}{\partial x_i} = \frac{\partial \overline{f}}{\partial x_i} + x_j \frac{\partial \overline{f}}{\partial x_{n+1}}; \qquad \frac{\partial f}{\partial x_j} = \frac{\partial \overline{f}}{\partial x_j} + x_i \frac{\partial \overline{f}}{\partial x_{n+1}}$$

and so we add the following instructions:

$$g_{z+1} = g_{p_{n+1}} g_j$$

$$g_{z+2} = g_{p_i} + g_{z+1}$$
$$g_{z+3} = g_{p_{n+1}} g_i$$
$$g_{z+4} = g_{p_j} + g_{z+3}$$

and write $p'_i = z + 2$, $p'_j = z + 4$, as well as $p'_k = p_k$ for all other p'.

Since $\nabla \overline{f}$ has $\leq 5 t_{max}$ operations, then ∇f has $\leq 5 t_{max} + 4 \leq 5(t_{max} + 1)$ operations, and we are done.

Case 9: Product of a Variable with Itself

We have $g_a = x_i x_i$, and so $G(x_1, \ldots, x_n) = x_i^2$. Thus we know

$$\frac{\partial f}{\partial x_i} = \frac{\partial \overline{f}}{\partial x_i} + 2x_i \frac{\partial \overline{f}}{\partial x_{n+1}}$$

and so we add the following instructions:

$$g_{z+1} = g_i + g_i$$
$$g_{z+2} = g_{z+1} g_{p_{n+1}}$$
$$g_{z+3} = g_{z+2} + g_{p_i}$$

and write $p'_i = z + 3$, as well as $p'_k = p_k$ for all other p'.

Since $\nabla \overline{f}$ has $\leq 5 t_{max}$ operations, then ∇f has $\leq 5 t_{max} + 3 \leq 5(t_{max} + 1)$ operations, and we are done.

Case 10: Product of a Constant and a Variable

We have $g_a = c x_j$, and so $G(x_1, \ldots, x_n) = c x_j$, where c is g_i, the result of a CONS instruction at i. In this case,

$$\frac{\partial f}{\partial x_j} = \frac{\partial \overline{f}}{\partial x_j} + c \frac{\partial \overline{f}}{\partial x_{n+1}}$$

and so we add the following instructions:

$$g_{z+1} = g_{p_{n+1}} g_i$$
$$g_{z+2} = g_{z+1} + g_{p_j}$$

and write $p'_j = z + 2$, as well as $p'_k = p_k$ for all other p'.

Since $\nabla \overline{f}$ has $\leq 5t_{max}$ operations, then ∇f has $\leq 5t_{max} + 2 \leq 5(t_{max} + 1)$ operations, and we are done.

Case 11: Product of a Variable and a Constant

Since multiplication is commutative in a field, then $g_a = x_i c$ can be handled identically to the previous case, $g_a = cx_i$.

Case 12: Quotient of Two Distinct Variables

We have $g_a = x_i/x_j$, and so $G(x_1, \ldots, x_n) = x_i/x_j$. In this case,

$$\frac{\partial f}{\partial x_i} = \frac{\partial \overline{f}}{\partial x_i} + \frac{\partial \overline{f}}{\partial x_{n+1}} \frac{1}{x_j}; \qquad \frac{\partial f}{\partial x_j} = \frac{\partial \overline{f}}{\partial x_j} - \frac{x_i}{x_j^2} \frac{\partial \overline{f}}{\partial x_{n+1}}$$

and so we add the following instructions:

$$g_{z+1} = g_{p_{n+1}}/g_j$$
$$g_{z+2} = g_{p_i} + g_{p_{z+1}}$$
$$g_{z+3} = g_{z+1}/g_j$$
$$g_{z+4} = g_{z+3}g_i$$
$$g_{z+5} = g_{p_j} - g_{z+4}$$

and write $p_i' = z + 2$, $p_j' = z + 5$, as well as $p_k' = p_k$ for all other p'.

Since $\nabla \overline{f}$ has $\leq 5t_{max}$ operations, then ∇f has $\leq 5t_{max} + 5 \leq 5(t_{max} + 1)$ operations, and we are done.

Case 13: A Constant Divided by a Variable

We have $g_a = c/x_j$, and so $G(x_1, \ldots, x_n) = c/x_j$, again note c is the output of a CONS instruction at step i. In this case,

$$\frac{\partial f}{\partial x_j} = \frac{\partial \overline{f}}{\partial x_j} - \frac{c}{x_j^2} \frac{\partial \overline{f}}{\partial x_{n+1}}$$

and so we add the following instructions:

$$g_{z+1} = g_{p_{n+1}}/g_j$$
$$g_{z+2} = g_{z+1}/g_j$$
$$g_{z+3} = g_{z+2}g_i$$
$$g_{z+4} = g_{p_j} - g_{z+3}$$

and write $p'_j = z + 4$, as well as $p'_k = p_k$ for all other p'.

Since $\nabla \overline{f}$ has $\leq 5t_{max}$ operations, then ∇f has $\leq 5t_{max} + 4 \leq 5(t_{max} + 1)$ operations, and we are done.

Case 14: A Variable Divided by a Constant

We have $g_a = x_i/c$, and so $G(x_1, \ldots, x_n) = x_i/c$. As before, c is the output of a CONS instruction at step j. In this case,

$$\frac{\partial f}{\partial x_i} = \frac{\partial \overline{f}}{\partial x_i} + \frac{\partial \overline{f}}{\partial x_{n+1}} \frac{1}{c}$$

and so we add the following instructions:

$$g_{z+1} = g_{p_{n+1}}/g_j$$
$$g_{z+2} = g_{p_i} + g_{p_{z+1}}$$

and write $p'_i = z + 2$, as well as $p'_k = p_k$ for all other p'.

Since $\nabla \overline{f}$ has $\leq 5t_{max}$ operations, then ∇f has $\leq 5t_{max} + 2 \leq 5(t_{max} + 1)$ operations, and we are done.

The Base Case

The base case is zero field operations. Then the last operation of f is either LOAD or CONS. By convention, we stated that the last operation of f is its output. Suppose that the last instruction is step z.

If it is LOAD, then $f(x_1, \ldots, x_n) = x_i$ for some $1 \leq i \leq n$. Then ∇f can be defined as appending steps $z + 1$ and $z + 2$ as follows

$$g_z = \text{LOAD}(x_i)$$
$$g_{z+1} = \text{CONS}(1)$$
$$g_{z+2} = \text{CONS}(0)$$

with $p_0 = z$, $p_i = z + 1$ and $p_k = z + 2$ for all other p.

Here, f has 0 field operations, and ∇f has 0 field operations, and so the inequality on t holds trivially.

If it is CONS, then $f(x_1, \ldots, x_n) = c$ for some $c \in \mathbb{F}$. Then ∇f can be defined as appending step $z + 1$ as follows

$$g_z = \text{CONS}(c)$$
$$g_{z+1} = \text{CONS}(0)$$

with $p_0 = z$, and $p_k = z + 1$ for all other p.

Again, f has 0 field operations, and ∇f has 0 field operations, and so the inequality on t holds trivially.

Considering s and h also

At all stages of the inductive proof, we maintained that if $\overline{f}$ has at most t field operations and $\nabla \overline{f}$ has at most $5t$ field operations. Because we add at most only 5 field operations (e.g. in case 12), then ∇f has at most $5t + 5 = 5(t + 1)$ field operations. This is satisfactory because f has $t + 1$ field operations.

Likewise, assume that $\overline{f}$ has at most s steps, and $\nabla \overline{f}$ has at most $5s$ steps. Because we add at most only 5 steps, then ∇f has at most $5s + 5 = 5(s + 1)$ steps.

Finally, if the instruction a were not a multiply/divide, (cases 1 to 7), then we add no multiplies or divides. If a were a multiply/divide (cases 8 to 14), then we add at most 3 multiplies/divides (e.g. in case 12). Thus by the above argument, if f has $\leq h$ multiplies/divides, then ∇f will have $\leq 3h$.

$\square$

8.3.2.3 A Running Example

Suppose we have the function f_0 computed by $g_1 = \text{LOAD}(x_1)$; $g_2 = \text{LOAD}(x_2)$; $g_3 = \text{LOAD}(x_3)$; $g_4 = \text{LOAD}(x_4)$; $g_5 = g_1 g_2$; $g_6 = g_5 + g_3$; $g_7 = g_6/g_4$ which calculates the rational function

$$f_0(x_1, x_2, x_3, x_4) = \frac{x_1 x_2 + x_3}{x_4}$$

Then the induction will proceed to consider the function f_1 computed by $g_1 = \text{LOAD}(x_1)$; $g_2 = \text{LOAD}(x_2)$; $g_3 = \text{LOAD}(x_3)$; $g_4 = \text{LOAD}(x_4)$; $g_5 = \text{LOAD}(x_5)$; $g_6 = g_5 + g_3$; $g_7 = g_6/g_4$ which calculates the rational function

$$f_1(x_1, x_2, x_3, x_4, x_5) = \frac{x_5 + x_3}{x_4}$$

Now the induction will proceed to consider f_2 computed by $g_1 = \text{LOAD}(x_1)$; $g_2 = \text{LOAD}(x_2)$; $g_3 = \text{LOAD}(x_3)$; $g_4 = \text{LOAD}(x_4)$; $g_5 = \text{LOAD}(x_5)$; $g_6 = \text{LOAD}(x_6)$; $g_7 = g_6/g_4$ which calculates the rational function

$$f_2(x_1, x_2, x_3, x_4, x_5, x_6) = \frac{x_6}{x_4}$$

Finally the induction will consider f_3 computed by $g_1 = \text{LOAD}(x_1)$; $g_2 = \text{LOAD}(x_2)$; $g_3 = \text{LOAD}(x_3)$; $g_4 = \text{LOAD}(x_4)$; $g_5 = \text{LOAD}(x_5)$; $g_6 = \text{LOAD}(x_6)$; $g_7 = \text{LOAD}(x_7)$ which calculates the rational function

$$f_3(x_1, x_2, x_3, x_4, x_5, x_6, x_7) = x_7$$

Using the base case of the induction, we have ∇f_3 computed by $g_1 = \text{LOAD}(x_1)$; $g_2 = \text{LOAD}(x_2)$; $g_3 = \text{LOAD}(x_3)$; $g_4 = \text{LOAD}(x_4)$; $g_5 = \text{LOAD}(x_5)$; $g_6 = \text{LOAD}(x_6)$; $g_7 = \text{LOAD}(x_7)$; $g_8 = \text{CONS}(1)$; $g_9 = \text{CONS}(0)$ along with $p_0 = 7$, $p_7 = 8$ and $p_k = 9$ for all other k. This computes the vector

$$\nabla f_3 = (0,0,0,0,0,0,1)$$

along with the value of f_3.

Now we must refer to case 12, the worst of all 14 cases. We now have ∇f_2 computed by $g_1 = \text{LOAD}(x_1)$; $g_2 = \text{LOAD}(x_2)$; $g_3 = \text{LOAD}(x_3)$; $g_4 = \text{LOAD}(x_4)$; $g_5 = \text{LOAD}(x_5)$; $g_6 = \text{LOAD}(x_6)$; $g_7 = g_6/g_4$; $g_8 = \text{CONS}(1)$; $g_9 = \text{CONS}(0)$; $g_{10} = g_8/g_4$; $g_{11} = g_9 + g_{10}$; $g_{12} = g_{10}/g_4$; $g_{13} = g_{12}g_6$; $g_{14} = g_9 - g_{13}$ along with $p_0 = 7$, $p_6 = 11$, $p_4 = 14$ and $p_k = 9$ for all other k. Note p_7 is no longer useful because x_7 no longer exists. This computes the vector

$$\nabla f_2 = (0,0,0,\frac{-x_6}{x_4^2},0,\frac{1}{x_4})$$

along with the value of f_2.

The next case is much more merciful, using case 1. This yields ∇f_1 computed by $g_1 = \text{LOAD}(x_1)$; $g_2 = \text{LOAD}(x_2)$; $g_3 = \text{LOAD}(x_3)$; $g_4 = \text{LOAD}(x_4)$; $g_5 = \text{LOAD}(x_5)$; $g_6 = g_5 + g_3$; $g_7 = g_6/g_4$; $g_8 = \text{CONS}(1)$; $g_9 = \text{CONS}(0)$; $g_{10} = g_8/g_4$; $g_{11} = g_9 + g_{10}$; $g_{12} = g_{10}/g_4$; $g_{13} = g_{12}g_6$; $g_{14} = g_9 - g_{13}$; $g_{15} = g_{11} + g_9$; $g_{16} = g_{11} + g_9$ along with $p_0 = 7$, $p_4 = 14$, $p_5 = 15$, $p_3 = 16$ and $p_k = 9$ for all other k. Note p_6 is no longer useful because x_6 no longer exists. This computes the vector

$$\nabla f_1 = (0,0,\frac{1}{x_4},\frac{-x_5 - x_3}{x_4^2},\frac{1}{x_4})$$

along with the value of f_1.

At long last, we nearing the end of our computation. We must use case 8, and thus have ∇f_0 computed by $g_1 = \text{LOAD}(x_1)$; $g_2 = \text{LOAD}(x_2)$; $g_3 = \text{LOAD}(x_3)$; $g_4 = \text{LOAD}(x_4)$; $g_5 = g_1g_2$; $g_6 = g_5 + g_3$; $g_7 = g_6/g_4$; $g_8 = \text{CONS}(1)$; $g_9 = \text{CONS}(0)$; $g_{10} = g_8/g_4$; $g_{11} = g_9 + g_{10}$; $g_{12} = g_{10}/g_4$; $g_{13} = g_{12}g_6$; $g_{14} = g_9 - g_{13}$; $g_{15} = g_{11} + g_9$; $g_{16} = g_{11} + g_9$ along with $p_0 = 7$, $p_4 = 14$, $p_3 = 16$, $p_1 = 18$, $p_2 = 20$. Note p_5 is no longer useful because x_5 no longer exists, and there are no other p_k. This computes the vector

$$\nabla f_0 = (\frac{x_2}{x_4},\frac{x_1}{x_4},\frac{1}{x_4},\frac{-x_1x_2 - x_3}{x_4^2})$$

which thankfully is correct, along with the value of f_0.

We started with 3 field operations, 2 of which were "hard" (multiplies and divides), plus 4 non-field operations (7 total steps). We finished with 10 field operations, 5 of which were "hard" (multiplies and divides), plus 6 non-field operations (16 total steps).

8.3.2.4 Dividing by Zero

The behavior during a division by zero is left undefined in the above model. If we are given a program for f that computes f on all points of its domain, then this implies there are no superfluous divisions by zero possible. Of course, f does not exist at its poles, and the program may divide by zero there.

Luckily, under this assumption, the generated program for ∇f will never divide by zero either (at any point in the domain of f). This is because a division operation is only appended in cases 12, 13 and 14. In cases 12 and 13, we consider x_i/x_j and c/x_j, and the newly introduced divisions are by x_j. Thus, at any value of x_j in the domain of f, we know $x_j \neq 0$, because otherwise f would be dividing by zero, which we assumed was not the case in the domain.

Likewise, in case 14, we consider x_i/c, where $c \in \mathbb{F}$. If $c = 0$ then f would always divide by zero, and its domain would be empty. This would trivially satisfy the requirement that ∇f never divide by zero on a point in the domain of f.

Therefore we can thankfully conclude that division by zero is not of concern to this model.

8.3.2.5 Impossibility of the Hessian

Given that we can find the gradient of a rational function in roughly 5 times as many operations as evaluating the function, the next natural question is if we can calculate the Hessian. Recall, the Hessian of an n variable function is the $n \times n$ matrix A such that $A_{ij} = (\partial^2 f)/(\partial x_i \partial x_j)$. Sadly, the answer is decidedly negative.

The following is from [180], in which Morgenstern accredits the proof to Stoss.

Lemma 54 (Stoss and Morgenstern). *Let* $\mathbf{u} = (u_1, u_2, \ldots, u_n)$ *and* $\mathbf{v} = (v_1, v_2, \ldots, v_n)$. *Let X and Y be two $n \times n$ matrices, and let $Z = XY$. Let*

$$f(u_1, u_2, \ldots, u_n, v_1, v_2, \ldots, v_n, x_{11}, \ldots, x_{nn}, y_{11}, \ldots, y_{nn}) = \mathbf{u}^T X Y \mathbf{v}$$

then

$$\frac{\partial^2 f}{\partial u_i \partial v_j} = Z_{ij}$$

and furthermore f is a polynomial of degree 4, and f can be computed using at most $4n^2 - 1$ field operations, and $2n^2 + n$ field multiplications (with 0 divisions).

Proof. Proof omitted. See [180] or calculate by hand. But note that $\mathbf{u}^T X Y \mathbf{v} = (X^T \mathbf{u})^T (Y \mathbf{v})$, and so two matrix-vector-products, followed by a dot-product, are all that are required. A matrix-vector product requires n^2 multiplies and $n^2 - n$ additions. Then the dot product requires n multiplies and $n - 1$ additions. The total is as stated. $\qquad\square$

That then implies

Lemma 55. *Let m and b be any positive real numbers. If it is the case that all rational functions which can be evaluated using at most t field operations, can each have all of its second partial derivatives calculated in at most mt + b field operations, then multiplying two n × n matrices will require ∼ 4mn² field operations.*

Proof. Proof is obvious.   □

Finally, we conclude with

Theorem 56. *Let m and b be any positive real numbers, and t any positive integer. There is some rational function f which can be evaluated using at most t field operations, but it is impossible to calculate all of its second partial derivatives using at most mt + b field operations.*

Proof. Ran Raz and Amir Shpilka proved that matrix multiplication of two $n \times n$ matrices is $\omega(n^2)$ [199, 198]. Applying this to the contrapositive of Lemma 55 gives the desired result.   □

8.3.3 Consequences for the Determinant and Inverse

Now finally we can draw the last edge in our diagram.

Theorem 57. *For any $b \geq 2$, if finding the determinant of an $n \times n$ matrix is $O(n^b)$ over the field $\mathbb{F}$, then matrix inversion is $O(n^b)$ over the field $\mathbb{F}$.*

Proof. Suppose one could calculate the determinant of an $n \times n$ matrix in $O(n^b)$ steps, and one is given an invertible $n \times n$ matrix A to invert.

Then one can find the cofactor matrix in $O(n^b)$ steps, using Theorem 53 on Page 121. This is because the set of partial derivatives of the determinant are the entries of the cofactor matrix, by Definition 50 on Page 120. With n^2 field operations, one can divide each entry of the cofactor matrix by the determinant, and thus obtain the matrix inverse via Fact 51 on Page 120.   □

Chapter 9
The Method of Four Russians

As we've seen throughout this text, solving a linear system of $\mathbb{GF}(2)$ equations lies at the heart of many cryptanalytic techniques. Some examples include stream cipher cryptanalysis via the XL algorithm and its many variants [22, 23, 24, 25, 66, 67, 69, 79, 134, 82]; the algebraic attacks on the HFE public-key cryptosystem and Quartz [65, 111, 78, 72, 77, 68, 70]; cryptanalysis of QUAD [42]; and solving the matrix square root (provably NP-Hard for matrices over the boolean semiring) with the XL algorithm [155].

Gaussian Elimination is a natural choice of algorithm for these problems. However, for dense systems, its cubic-time complexity makes it far too slow in practice. The algorithm in this chapter achieves a speed-up of 3.36 times for a 32000×32000 $\mathbb{GF}(2)$-matrix that is generated by random fair coins. The theoretical complexity of the algorithm is $O(n^3/\log n)$, but it should be remembered that frequently n is the cube or higher power of a parameter of the system being attacked, and so frequently is in the millions.

At first it may seem surprising that so much attention is given to an algorithm of complexity $O(n^3/\log n)$, since Strassen's Algorithm for Matrix Multiplication has complexity $O(n^{\log_2 7})$. But, in the end, we will combine the two algorithms, for further improvement.

The algorithms in this chapter form the backbone of a linear algebra suite coded by the author, and are now part of SAGE [7], an open source competitor to MAGMA [2]. Some of the experiments cited in this chapter were performed by SAGE volunteers and staff, as noted in each case.

Performance is formally modeled, and some experiments are given to confirm/explain that reasoning. These times are obsolete however, as the work has progressed immensely, and three years have gone by so the computers are faster. Rather than provide new timings, which would be obsolete in another three years, we refer the reader to http://m4ri.sagemath.org/, where the very latest data will be posted. The ratios of timings to each other should change much more slowly of course, and so are informative.

The algorithm is named the Method of Four Russians for Inversion or M4RI, to be pronounced "Mary", in honor of the matrix multiplication algorithm from which

G.V. Bard, *Algebraic Cryptanalysis*, DOI: 10.1007/978-0-387-88757-9_9
© Springer Science + Business Media, LLC 2009

it emerged, the Method of Four Russians for Multiplication (M4RM). The "Four" are Arlazarov, Dinic, Kronrod, and Faradzev [21], but later information showed that not all are Russian[1] .

9.0.4 The Fair Coin Assumption

Here we explore the effects of the algorithm on random matrices that are filled with fair coins. Of course, one could also consider a weighted coin. But the fair coin model is particularly appropriate to cryptography where the key is generated by having an equal probability of one or zero. In the event that zero is very common in the matrix, the matrix is called sparse, and this is discussed in Appendix D on Page 323.

9.1 Origins and Previous Work

A paper published by Arlazarov, Dinic, Kronrod, and Faradzev [21] in 1970 on graph theory contained an $O((\log d)(v^3/\log v))$ algorithm for finding the transitive closure of a directed graph of v vertices and diameter d. This problem is of course equivalent to exponentiation of a boolean matrix (the adjacency matrix) and the community quickly realized that it was useful not only as a matrix squaring algorithm, but also a matrix multiplication algorithm, because

$$\begin{bmatrix} A & B \\ 0 & 0 \end{bmatrix}^2 = \begin{bmatrix} A^2 & AB \\ 0 & 0 \end{bmatrix}$$

and therefore squaring a matrix and matrix multiplication are equivalent (See also Theorem 31 in Section 8.2.2). The running time of the algorithm so produced (given in Section 9.3 on Page 137 below), is $O(n^3/\log n)$ for an $n \times n$ matrix. This equivalence is not as inefficient as it might seem, as one can trivially calculate the upper-right quadrant of the answer matrix without calculating the other three-fourths of it. The matrix multiplication algorithm arising from this appears in Aho, Hopcroft, and Ullman's book, which gives the name "the Method of Four Russians... after the cardinality and the nationality of its inventors" [13, Ch. 6]. While that text states this algorithm is for boolean matrices, one can easily see how to adapt it to $\mathbb{GF}(2)$ or even to $GF(q)$ for very small q. We will discuss the $\mathbb{GF}(2)$ case in Section 9.3 on Page 137 and the $\mathbb{GF}(q)$ case in Section 9.9 on Page 152.

---

[1] Dan Bernstien has suggested the name "Kronrod's Algorithm", as the central theorem in the [21] paper was attributed to Kronrod alone. However, many in the community already know this algorithm as the Method of Four Russians and so we imagine that such a name change might create confusion.

A similarly inspired matrix inversion algorithm was known anecdotally among some cryptanalysts. The author would like to express gratitude to Nicholas Courtois who explained the following algorithm to him after Eurocrypt 2005 in Århus, Denmark. It appears that this algorithm has not been published, either in the literature or on the Internet. We call this newer algorithm the "Method of 4 Russians for Inversion" (M4RI) and the original as the "Method of 4 Russians for Multiplication" (M4RM).

9.1.1 Strassen's Algorithm

Strassen's famous paper [212] has three algorithms—one for matrix multiplication, one for inversion, and one for the calculation of determinants. The last two are for use with any matrix multiplication algorithm taken as a black box, and run in time big-Theta of matrix multiplication. However, substantial modification is needed to make these work over $\mathbb{GF}(2)$. Details can be found in Section 7.6.1 on Page 100 but for now, recall the running time,

$$\sim \left(\frac{n}{n_0}\right)^{\log_2 7} M(n_0)$$

where $M(n_0)$ is the time required to multiply an $n_0 \times n_0$ matrix in the "fall-back" algorithm. Strassen's algorithm will repeatedly cut a matrix in half until the pieces are smaller than n_0. After this point, the tiny pieces are resolved with the fall-back algorithm, and the answer is constructed. For this reason, if $M(n_0)$ is smaller with M4RM rather than the naïve algorithm, or likewise M4RI versus Gaussian Elimination, then Strassen's Algorithm will be proportionally improved for all sufficiently large matrices. Since n_0 might be large, a speed-up of $\log n_0$ is not trivial.

This, in the end, is the reason we are interested in both M4RM and M4RI, that they will be new "fall back" algorithms for the small parts during a Strassen recursive call, instead of the naïve algorithms.

9.2 Rapid Subspace Enumeration

The following step is crucial in the Method of Four Russians family of algorithms. An n-dimensional subspace of a vector-space over $\mathbb{GF}(2)$ has 2^n vectors in it, including the all-zero vector. Given n basis vectors for that subspace, how can we rapidly enumerate these 2^n vectors?

Obviously, any vector in the subspace can be written as a linear combination of the basis vectors. In $\mathbb{GF}(2)$, a linear combination is just a sum of a subset. There will be 1 vector with 0 basis vectors in that subset, n vectors will be written as a "sum" of one basis vector alone, $\binom{n}{2}$ will be written as a sum of two basis vectors, ..., $\binom{n}{n} = 1$

will be written as a sum of all the basis vectors. Thus the expected number of basis vectors in the sum for any particular subspace vector is given by

$$\frac{\sum_{i=0}^{i=n} i \binom{n}{i}}{2^n} = \frac{\frac{n}{2} 2^n}{2^n} = n/2$$

Thus computing an arbitrary vector in the subspace requires $n/2 = \Theta(n)$ vector adds, or $\Theta(n^2)$ field additions. The entire subspace would require $\Theta(n^2 2^n)$ additions.

Instead, [21] contains an indirect description of a faster way. A k-bit Gray Code is all 2^k binary strings of length k, ordered so that each differs by exactly one bit in one position from each of its neighbors. For example, one 3-bit Gray Code is $\{000, 001, 011, 010, 110, 111, 101, 100\}$ [129]. Note that the Gray code is named for Frank Gray of Bell Labs, not the color gray. Now consider the ith bit of this code to represent the ith basis vector. This means that the all-zero string represents the all-zero vector, and the all-ones string represents the sum of all the basis vectors. The Gray Code will cycle through all 2^n vectors in the subspace. Furthermore, each sum can be obtained from the previous sum by only one vector addition.

The reason for this is that each codeword differs in exactly one bit from its predecessor. Thus, given a codeword, suppose bit i is flipped to produce the next codeword. If it was a $0 \rightarrow 1$ transition, adding the ith basis vector to the sum will produce the correct new sum. But, if it was a $1 \rightarrow 0$ transition, adding the ith basis vector to the sum will also produce the correct sum because $\mathbf{x} + \mathbf{x} = 0$ in any vector space whose base field is of characteristic two. Thus, starting with the all-zero string, and cycling through all 2^n codewords, we can start with the all-zero vector, and cycle through all 2^n basis vectors, using only one vector-addition at each step.

This requires $2^n - 1$ vector additions instead of $(n/2)2^n$, and is a speed-up of $\Theta(n)$. Since a vector addition is a $\Theta(n)$ operation, this rapid subspace enumeration method requires $\Theta(n2^n)$ instead of $\Theta(n^2 2^n)$ bit-operations. Since there are $n2^n$ bits in the output of the algorithm, we can see that this method is optimal in the sense of Big-Θ. (In other words, an algorithm that found the basis vectors instantly would still need $\Theta(n2^n)$ bit operations to output all of them. For exact matrix memory operation counts, observing that $\sim 3n$ matrix-memory operations are needed for a vector addition (one read-read-write cycle for each position in the vector), or a total of $\sim 3n(2^n - 1)$ operations, is the requirement to enumerate an n-dimensional subspace. This is $\Theta(n2^n)$ instead of $\Theta(n^2 2^n)$ for comparison purposes.

Note, the Gray Codes should be computed only once and stored for long term use in a library. To enumerate the Gray Code can be done recursively. To output a Gray Code of length $\ell + 1$, take a Gray Code of length ℓ, and place a 0 in front of each word. Then relist the codewords in backwards order, this time prepending a 1 instead of a 0. Each word differs by its neighbor only in one slot and no possible word is omitted or repeated.

9.3 The Four Russians Matrix Multiplication Algorithm

This matrix multiplication algorithm is derivable from the original algorithm published by Arlazarov, Dinic, Kronrod, and Faradzev [21], but does not appear there. It has, however, appeared in books including [13, Ch. 6]—which is out of print. Consider a product of two matrices $AB = C$ where A is an $a \times b$ matrix and B is a $b \times c$ matrix, yielding an $a \times c$ for C. In this case, one could divide A into b/k vertical "stripes" $A_1 \ldots A_{b/k}$ of k columns each, and B into b/k horizontal stripes $B_1 \ldots B_{b/k}$ of k rows each. (For simplicity assume k divides b). The product of two stripes, A_iB_i is an $a \times b/k$ by $b/k \times c$ matrix multiplication, and yields an $a \times c$ matrix C_i. The sum of all k of these C_i equals C.

$$C = AB = \sum_{i=0}^{i=k} A_iB_i$$

The algorithm itself proceeds as given in Algorithm 13 on Page 137.

INPUT: A parameter k, a matrix A of size $a \times b$, and a matrix B of size $b \times c$, both divided each into k stripes (see Section 9.3 on Page 137).
OUTPUT: A matrix C of size $a \times c$, with $C = AB$.
1: Initialize the $a \times c$ matrix C with all zeroes.
2: For $i = 1, 2, \ldots, b/k$ do

    1: Make a Gray Code table of all the 2^k linear combinations of the k rows of the "stripe" denoted B_i, via rapid subspace enumeration (see Section 9.2 on Page 135). Denote the xth row T_x.

    (Costs $(3 \cdot 2^k - 4)c$ reads/writes, see Stage 2, in Section 9.5 on Page 145).
    2: For $j = 1, 2, \ldots, a$ do
        1: Read the entries $a_{j,(i-1)k+1}, a_{j,(i-1)k+2}, \ldots, a_{j,(i-1)k+k}$.
        2: Let x be the k bit binary number formed by the concatenation of $a_{j,(i-1)k+1}, \ldots, a_{j,ik}$.
        3: Add T_x to row j of C. (Costs 3c reads/writes).

Algorithm 13: Method of Four Russians, for Matrix Multiplication [after Arlazarov, Dinic, Kronrod, and Faradzev]

9.3.1 Role of the Gray Code

The Gray Code step is useful for two reasons. First, if any particular linear combination of the rows is required more than once, it is only computed once. That is an important savings alone. Second, even if each linear combination of rows is required exactly once, the Gray Code works $\sim n/2$ times faster than the naïve way of calculating those linear combinations (i.e. calculating them as needed, by direct addition, never remembering them). But, for matrices of various sizes and various values of

k, the expected number of times any particular linear combination is required will vary. Thus it is better to calculate the running time directly to see the effects of the algorithm.

The innermost loop out requires $k + 3c$ steps, and then the next requires $(3 \cdot 2^k - 4)c + a(k + 3c)$ steps. If the $(3 \cdot 2^k - 4)c$ is puzzling, note that $(2^k - 1)$ vector additions are required. This would normally require $(3 \cdot 2^k - 3)c$ matrix-memory read/writes. In the first iteration, we save an additional c by noting the first gray code vector is always all zeroes and so we do not have to actually read it.

Finally the entire algorithm requires

$$\frac{b((3 \cdot 2^k - 4)c + a(k + 3c))}{k} = \frac{3b2^k c - 4cb + abk + 3abc}{k}$$

matrix memory operations. Choose $k = \log b$, so that $2^k = b$, and observe

$$\frac{3b^2 c - 4cb + ab\log b + 3abc}{\log b} \sim \frac{3b^2 c + 3abc}{\log b} + ab$$

For square matrices this becomes $\sim (6n^3)/(\log n)$.

9.3.2 Transposing the Matrix Product

Since $AB = C$ implies that $B^T A^T = C^T$, one can transpose A and B, and transpose the product afterward. The transpose is a quadratic, and therefore cheap, operation. This has running time $(3b^2 a + 3abc)/(\log b) + cb$ (obtained by swapping a and c in the earlier expression) and some algebraic manipulations show that this more efficient when $c < a$, for any $b > 1$. Therefore the final complexity is $\sim (3b^2 \min(a,c) + 3abc)/(\log b) + b\max(a,c)$. To see that the last term is not optional, substitute $c = 1$, in which case the last term becomes the dominant term.

Thus we should add the step if $c < a$ then return $Prod(B^T, A^T)^T$ instead of $Prod(A, B)$.

9.3.3 Improvements

In the years since initial publication of M4RM, several improvements have been made, in particular, in reducing the memory requirements [28, 205], and the base fields upon which the algorithm can work [204].

9.3.4 A Quick Computation

Suppose we start again with the complexity expression,

$$\frac{b((3 \cdot 2^k - 4)c + a(k + 3c))}{k}$$

but substitute $a = b = c = n$ (i.e. a square matrix times a square matrix of the same size). One obtains, after some algebraic manipulations,

$$\sim \frac{3n^2 2^k + 3n^3}{k}$$

Then substitute $k = \gamma \log n$ and observe,

$$\frac{3n^{(2+\gamma)} + 3n^3}{\gamma \log n}$$

Immediately, we see that $\gamma > 1$ would cause the numerator to have a higher-than-cubic term. That would make it inferior to even the naïve algorithm. Further observation shows that $\gamma < 1$ is inferior to $\gamma = 1$ because of the coefficient γ in the denominator. Thus this quick analysis predicts $k = \log n$ is optimal, in theory. We will perform experiments to obtain a precise optimization, in practice.

9.3.5 M4RM Experiments Performed by SAGE Staff

Martin Albrecht, a member of the SAGE Project, evaluated the author's library for inclusion in SAGE [7] back in 2006. The library is a Level 1, 2, and 3, BLAS (Basic Linear Algebra System), and includes matrix multiplication and inversion via the algorithms in this chapter, as well as LUP-factorization, and any matrix-vector and vector-vector operations to support them. The tests were primarily for M4RI, but also included tests for M4RM. The crossover between M4RM and naïve appears to be slightly larger than 6000×6000. The results are in Table 9.1 on Page 140.

A log-log plot showed that MAGMA is using Strassen's Algorithm (or one with an equal exponent to it). It should be noted that MAGMA is hand optimized in assembly language, for several processors, including the Opteron, the processor used by the tests. This was later confirmed by e-mails with a MAGMA developer.

Now since Strassen's Algorithm has running-time $\Theta(n^{2.807})$ and we have running-time $\Theta(n^3/\log n)$, we know that there must be some point at which MAGMA, which presumably uses Strassen paired with the naïve algorithm, must cross over (have equal running time to) our code. And it does, because the plots cross each other.

But, we know that the running time of Strassen's algorithm is

$$M(n) \sim (n/n_0)^{\log_2 7} M(n_0)$$

as shown in Section 7.6.1 on Page 101. Thus, for any given n_0, such as for example 2000×2000, if we are 3.825 times faster than naïve, we then expect to be 3.825 times faster in general, for matrices larger than $n_0 \times n_0$. The 3.825 figure comes from Table 9.9 on Page 157.

The numerically-intensive computer used in the tests was provided by the NSF to SAGE. The following quotation can be found on the machine's website.

> *This is computer [sic] for very open collaboration among SAGE developers and testing of intense SAGE calculations. It is a special-purpose 64-bit computer built by Western Scientific that has 64GB of RAM and 16 AMD Opteron cores. You can browse everybody's home directories. It was purchased for SAGE development using my NSF Grant (No. 0555776).*

Also, due to the remarks in Section 9.5.1 on Page 147, an experiment was performed to try $k = 0.75 \log_2 n$ instead of $k = \log_2 n$. The results show that this change is not for the better.

Table 9.1 M4RM Running Times versus MAGMA

Matrix Size	M4RM (SAGE)	Strassen (MAGMA)
1000×1000	0.01 sec	0.02 sec
2000×2000	0.03 sec	0.16 sec
3000×3000	0.11 sec	0.24 sec
4000×4000	0.26 sec	0.48 sec
5000×5000	0.70 sec	1.03 sec
6000×6000	1.64 sec	1.67 sec
7000×7000	3.32 sec	2.61 sec
8000×8000	5.39 sec	3.34 sec
9000×9000	8.09 sec	5.45 sec
10000×10000	11.29 sec	7.28 sec

Table 9.2 Confirmation that $k = 0.75 \log_2 n$ is not a good idea.

	$k = \log n$		$k = 0.75 \log n$	
Matrix Size	k	time	k	time
8000	10	5.404	13	8.742
16000	10	46.310	14	64.846
32000	11	362.066	15	472.384

9.3.6 Multiple Gray-Code Tables and Cache Management

Additional work on this topic has been performed by Martin Albrecht, with William Hart and the author of this work. One trick was invented by Martin Albrecht and is very worth mentioning.

For the moment, assume k is divisible by 4. Instead of using 1 Gray Code table for k rows (size 2^k) we could use 2 Gray Code tables of $k/2$ rows (size $2^{k/2} \times 2$) or we could use 4 Gray Code tables of $k/4$ rows (size $2^{k/4} \times 4$). The point of this is that the memory usage is shrinking, and that raises the probability that everything can fit in the L2 cache. In turn, that makes an enormous speed improvement. The penalty is that instead of running one add operation in the very last step, we must make 2 or 4 additions. This tradeoff can only be established experimentally, and we have omitted a few details. More can be found in [15].

9.4 The Four Russians Matrix Inversion Algorithm

While the title of this section contains the words "matrix inversion", the algorithm which follows can be used either for matrix inversion or for triangulation and back-substitution, by the same mechanism that this is also true for Gaussian Elimination. As stated earlier, even if one has several $b_1, b_2, b_3, \ldots, b_c$, where $c = \min(m, n)$ it is far more efficient to solve $Ax_i = b_i$ by appending the b_i as columns to the end of the $m \times n$ matrix A, and putting matrix A in unit upper triangular form (UUTF). Then, one can solve for each x_i by back substitution to obtain the x_i. (The back-solve is quadratic, thus cheap, step). The alternative is to invert A, and Section 9.4.5 on Page 145 contains changes for that approach, by adjoining A with an identity matrix and processing it into row reduced echelon form (RREF).

In Gaussian Elimination to UUTF of an $m \times n$ matrix, each iteration i operates on the submatrix $a_{ii} \ldots a_{mn}$, with the objective of placing a one at a_{ii} and a zero at every other entry of the column i below row i, and leaving all above untouched. In the Method of Four-Russians Inversion (M4RI) algorithm, k columns are processed at once, producing a $k \times k$ identity matrix in the correct spot ($a_{ii} \ldots a_{(i+k-1),(i+k-1)}$), with all zeros below it, and leaving the region above the submatrix untouched.

Each stage will now be described in detail.

9.4.1 Stage 1:

Denote the first column to be processed in a given iteration as a_i. Then, perform Gaussian elimination on the first $3k$ rows after and including the ith row to produce an identity matrix in $a_{i,i} \ldots a_{(i+k-1),(i+k-1)}$, and zeroes in $a_{(i+k),i} \ldots a_{(i+3k-1),(i+k-1)}$ (To know why it is reasonable to expect this to succeed, see Lemma 1 in Section 9.5.3 on Page 148).

INPUT: A parameter k, and a matrix A of size $m \times n$.
OUTPUT: The row-echelon form A.
1: For $i = 1, k+1, 2k+1, 3k+1, \ldots \min(m,n)$ do

    1: Perform Gaussian Elimination on rows $i, i+1, \ldots, i+3k-1$, to establish a $k \times k$ identity matrix in cells $a_{ii} \ldots a_{i+k-1, i+k-1}$.
    2: Construct a gray-code table to enumerate the $2^k - 1$ non-zero vectors in the subspace generated by rows $i \ldots i+k-1$.
    3: For each row $j = i + 3k \ldots m$ do
        1: Read the entries in the k columns $i, i+1, \ldots, i+k-1$ of row j, and treat them as a k-bit binary number x.
        2: Add "the entry in the Gray Code table that has x as a prefix," to row j.

Algorithm 14: Method of Four Russians, for Inversion [Unknown]

9.4.2 Stage 2:

Construct a table consisting of the 2^k elements of the vector subspace generated by the first k rows among the $3k$ generated in the previous step, via the Gray Code and rapid subspace enumeration. Because of the submatrix mentioned above, which is equal to the identity matrix, we know these k rows are linearly independent as vectors. Thus with only 2^k vector additions (as explained in Section 9.2 on Page 136), all possible linear combinations of these k rows have been precomputed.

9.4.3 Stage 3:

One can rapidly process the remaining rows from $i + 3k$ until row m (the last row) by using the table. For example, suppose the jth row has entries $a_{ji} \ldots a_{j,(i+k-1)}$ in the columns being processed. Selecting the row of the table that starts with this k-bit string, and adding it to row j, will force the k columns to zero, and adjust the remaining columns from $i+k$ to n in the appropriate way, as if Gaussian Elimination had been performed.

No searching of the table is required. Each prefix will always appear in the same location, and so a dereferencing array can trivially render the act of finding the correct table row into a simple lookup action.

The process is then repeated $\min(m,n)/k$ times. As each iteration resolves k columns, instead of one column, one could expect that this algorithm is k times faster. The trade-off for large k is that Stage 2 can be very expensive. It turns out (see Section 9.6 on Page 150) that selecting the right value of k is critical.

9.4.4 A Curious Note on Stage 1 of M4RI

We have shown (See Section 9.5.3 on Page 148) that a $3k \times k$ submatrix, beginning at $a_{i,i}$ and extending to $a_{i+3k-1,i+k-1}$ is very unlikely to be singular. Therefore, the Gaussian Elimination (which is done on the rows $i, \ldots, i + 3k - 1$) will be successful and will produce a $k \times k$ identity matrix. But, this is not the whole story. With probability around 28.8%, the $3k \times 3k$ matrix will be full-rank, and so actually an identity matrix of size $3k \times 3k$ will be available. (Recall this is the probability that a sufficiently large random $\mathbb{GF}(2)$-matrix will be invertible). With probability 57.6%, the matrix will have nullity one (proof given as Theorem 58 on Page 143 below) and so a $(3k - 1) \times (3k - 1)$ identity matrix (with one row of zeroes under it) will be present. This means that the next *two* iterations of the algorithm will have essentially no work to do at all in Stage 1, with probability around 86.6% or so. The cases nullity 2, nullity 3, and nullity 4 absorb nearly all remaining probability (proved in Theorem 59 on Page 144 and shown in Table 9.3 on Page 145), and the probability that "only" $3k \times 2k$ will be in the form of an identity matrix (with k rows of zeroes underneath) is already approaching zero as ℓ gets large, with a probability that can be calculated using the aforementioned theorem. Therefore, Stage 1's cost is actually near to one-third its listed value. Since Stage 1 is not significant in the final complexity, we do not carry this analysis further.

The reader may wish to review Theorem 25 on Page 84 before proceeding.

Theorem 58. *The probability that an $n \times n$ $\mathbb{GF}(2)$-matrix, filled with the output of independent fair coins, is nullity 1 equals $(1 - 2^{-n})(1 - 2^{-n+1}) \cdots (1 - 2^{-n+n-2})(1 - 2^{-n})$. Also for large n, the ratio of the number of nullity one $n \times n$ matrices to the number of nullity zero matrices is ~ 2.*

Proof. Let A be a matrix that is $n \times n$ and nullity one. The null space contains $2^1 = 2$ vectors. Since the null space always contains the zero vector, it therefore contains one other vector $\mathbf{v}$.

There is a change-of-basis matrix B such that $B\mathbf{v} = \mathbf{e_1} = \{1, 0, \ldots, 0\}$, or $\mathbf{v} = B^{-1}\mathbf{e_1}$. Since $A\mathbf{v} = \mathbf{0}$ then $AB^{-1}\mathbf{e_1} = \mathbf{0}$ also and therefore $BAB^{-1}\mathbf{e_1} = \mathbf{0}$. Note that B, by virtue of being an $n \times n$ change-of-basis matrix, is non-singular, and so B^{-1} exists and is square, with the "correct" size.

The fact that $BAB^{-1}\mathbf{e_1} = 0$ means that the first column of BAB^{-1} is all zeroes. Note that BAB^{-1} and A have the same characteristic polynomial, nullity, rank, determinant, etc. . . .

The first column is all zeroes, but the rest of the matrix has to be full-rank for the nullity to be exactly one, and so the second column can be anything but all zeroes, the third column cannot be the second column nor all-zeroes, the fourth column cannot be the third, the second, nor their sum, and so on. For the ith column, we have ruled out the span of the $i - 2$ dimensional subspace generated by the previous $i - 1$ columns.

The original $\mathbf{v}$ in the null-space could be any non-zero vector, or $2^n - 1$ choices. We have therefore,

$$Pr[\text{nullity} = 1] = \frac{(1)(2^n - 1)(2^n - 2) \cdots (2^n - 2^{n-2})}{2^{n^2}} (2^n - 1)$$

$$= (1 - 2^{-n})(1 - 2^{-n+1}) \cdots (1 - 2^{-2})(1 - 2^{-n})$$

As one can see, compared to the nullity zero case, we have removed a $(1 - 2^{-1})$ term and replaced it with an extra $1 - 2^{-n}$ term, which asymptotically doubles the whole product. $\qquad\square$

Theorem 59. *If A is an $n \times n$ matrix filled with fair coins, the probability that it has nullity k is given by*

$$\frac{(1 - 2^{-n})(1 - 2^{-n+1}) \cdots (1 - 2^{-n+k-1}) \times (1 - 2^{-n})(1 - 2^{-n+1}) \cdots (1 - 2^{-k-1})}{(2^k - 1)(2^k - 2) \cdots (2^k - 2^{k-1})}$$

Proof. Suppose the nullity of A is k and thus the nullspace of A has $2^k - 1$ non-zero vectors in it. Choose k of them, $\mathbf{v}_1, \ldots, \mathbf{v}_k$ such that they are linearly independent.

There is a change-of-basis matrix B that maps the vectors so that $B\mathbf{v_i} = \mathbf{e_i}$, or $\mathbf{v_i} = B^{-1}\mathbf{e_i}$, for $i \in \{1, \ldots, k\}$. This further implies that $\mathbf{0} = A\mathbf{v_i} = AB^{-1}\mathbf{e_i}$ for $i \in \{1, \ldots, k\}$ and thus $BAB^{-1}\mathbf{e_i} = \mathbf{0}$. This means that the first k columns of BAB^{-1} are all zero.

The remaining $n - k$ columns have the following properties, because the remainder of the matrix must be full rank. The first remaining column cannot be all zeroes, the next cannot be the first nor all zeroes, and so forth. For $i > k + 1$, the ith column cannot be in the $(i - k - 1)$-dimensional subspace generated by the previous $i - 1$ columns, of which k are all-zero and $i - k - 1$ are non-zero.

Obviously for the non-zero columns we have $(2^n - 1)(2^n - 2) \cdots (2^n - 2^{n-k-1})$ choices, since they need to be non-zero and linearly independent. For the vectors in the null space, we have $(2^n - 1)(2^n - 2) \cdots (2^n - 2^{k-1})$ choices, but a permutation of those vectors produces the same final matrix for a different value of B, so a correction factor is needed.

Basically, the $\mathbf{v}_1, \ldots, \mathbf{v}_k$ was a basis for the nullspace, and nothing more. So, the correction factor to prevent overcounting of the same A generated by different B is just the number of bases of an n-dimensional space. The first vector could be any one of the $2^k - 1$ non-zero vectors in the space. The second vector can be anything but the first or zero, and the third can be anything except zero, the first, the second, or their sum. The ith can be anything not in the $i - 1$ dimensional subspace of the previous $i - 1$ vectors, which is $2^k - 2^{i-1}$. Essentially, there are $|GL_k(\mathbb{GF}(2))|$ ways to choose a basis.

Finally, we have:

$$Pr[A \in M_n(\mathbb{GF}(2)); \text{nullity}(A) = k] =$$

$$= \frac{(1 - 2^{-n})(1 - 2^{-n+1}) \cdots (1 - 2^{-n+k-1})(1 - 2^{-n})(1 - 2^{-n+1}) \cdots (1 - 2^{-k-1})}{(2^k - 1)(2^k - 2) \cdots (2^k - 2^{k-1})}$$

$$= \frac{\left(\prod_{i=1}^{i=k} 1 - 2^{-n+i-1}\right) \left(\prod_{i=1}^{i=n-k} 1 - 2^{-n+i-1}\right)}{\left(\prod_{i=1}^{i=k} 2^k - 2^{i-1}\right)}$$

and that is what we wanted to prove.   □

Table 9.3 Probabilities of a Fair-Coin Generated $n \times n$ matrix over $\mathbb{GF}(2)$, having given Nullity

nullity $n = 1000$	$n = 8$	$n = 3$	$n = 2$
0	0.28879	0.28992	168/512	6/16
1	0.57758	0.57757	294/512	9/16
2	0.12835	0.12735	49/512	1/16
3	5.2388×10^{-3}	5.1167×10^{-3}	1/512	0
4	4.6567×10^{-5}	4.4060×10^{-5}	0	0
5	9.6914×10^{-8}	8.5965×10^{-8}	0	0
6	4.8835×10^{-11}	3.7903×10^{-11}	0	0
7	6.0556×10^{-15}	3.5250×10^{-15}	0	0
8	1.8625×10^{-19}	5.4210×10^{-19}	0	0

9.4.5 Triangulation or Inversion?

While the above form of the algorithm will reduce a system of linear equations over $\mathbb{GF}(2)$ to unit upper triangular form, and thus permit a system to be solved with back substitution, the M4RI algorithm can also be used to invert a matrix, or put the system into reduced row echelon form (RREF). Simply run Stage 3 on rows $0 \cdots i - 1$ as well as on rows $i + 3k \cdots m$. This only affects the complexity slightly, changing the 2.5 coefficient to 3 (calculation done in Section 9.5.2 on Page 147). To use RREF to invert a matrix, simply concatenate an identity matrix (of size $n \times n$) to the right of the original matrix (of size $n \times n$), producing a $n \times 2n$ matrix. Using M4RI to reduce the matrix to RREF will result in an $n \times n$ identity matrix appearing on the left, and the inverse matrix on the right.

9.5 Exact Analysis of Complexity

Assume the matrix is $m \times n$ and for simplicity that k divides n and m. To calculate the cost of the algorithm one need only tabulate the cost of each of the three stages, which will be repeated $\min(m, n)/k$ times. Let these stages be numbered $i = 1 \ldots \min(m, n)/k$.

The first stage is a $3k \times (n - ik)$ underdefined Gaussian Elimination (RREF), which requires $\sim 1.5(3k)(n - ik)^2 - 0.75(3k)^3$ matrix memory operations (See Section 7.5.3 on Page 96). This will be negligible.

The second stage, constructing the table, requires $3(n - ik - k)$ steps per row. The first row is all zeroes and can be hard-coded, and the second row is a copy of the appropriate row of the matrix, and requires $(n - ik - k)$ reads followed by writes. Thus one obtains $2(n - ik - k) + (2^k - 2)(3)(n - ik - k) = (3 \cdot 2^k - 4)(n - ik - k)$ steps.

The third stage, executed upon $(m - ik - 3k)$ rows (if positive) requires $2k + 3(n - ik - k)$ reads/writes per row. This becomes $(m - ik - 3k)(3n - 3ik - k)$ matrix memory operations in total, when that total is positive. For example, in a square matrix the last 2 iterations of stage 1 will take care of all of these rows and so there may be no work to perform in Stage 3 of those iterations. To denote this, let $pos(x) = x$ if $x > 0$ and $pos(x) = 0$ otherwise.

Adding steps one, two, and three yields

$$\sum_{i=0}^{i=\ell/k-1} 1.5(3k)^2(n - ik) - 0.75((3k)^3) + (3 \cdot 2^k - 4)(n - ik - k) + $$
$$(pos(m - ik - 3k))(3n - 3ik - k)$$

$$= \left[\sum_{i=0}^{i=\ell/k-3} 1.5(3k)^2(n - ik) - 0.75((3k)^3)(3 \cdot 2^k - 4)(n - ik - k) + \right.$$
$$\left. (m - ik - 3k)(3n - 3ik - k) \right]$$
$$+1.5(3k)^2(n - \ell + 2k) - 0.75((3k)^3)(3 \cdot 2^k - 4)(n - \ell + k)$$
$$+1.5(3k)^2(n - \ell + k) - 0.75((3k)^3)(3 \cdot 2^k - 4)(n - \ell)$$
$$\leq \sim \frac{1}{4k} \left[2^k(-6k\ell + 12n\ell - 6\ell^2) - 6m\ell^2 - 6n\ell^2 + 4\ell^3 + 12mn\ell \right]$$

Recalling $\ell = \min(m, n)$ and substituting $k = \log \ell$ and thus $2^k = \ell$, we obtain,

$$\sim \frac{1}{4 \log \ell} \left(6n\ell^2 - 2\ell^3 - 6m\ell^2 + 12mn\ell \right)$$

Thus for the over-defined case $(\ell = n)$ this is $(4n^3 + 6n^2 m)/(4 \log n)$, and for the under-defined case $(\ell = m)$ this is $(18nm^2 - 8m^3)/(4 \log m)$, and for square $(5n^3)/(2 \log n)$.

9.5.1 An Alternative Computation

If we let $n = 2m$, which would be the case for inverting a square matrix, we obtain:

$$\sim \frac{1}{4k} \left[2^k(-6km + 18m^2) + 10m^3 \right]$$

Now substitute

$$k = \gamma \log m + \delta$$

and observe,

$$\sim \frac{2^{\delta}}{\gamma \log m + \delta} \left[18m^{2+\gamma} \right] + \frac{10m^3}{\gamma \log m + \delta}$$

from which it is clear that $\gamma > 1$ would result in a higher-than-cubic complexity. Also, $\gamma < 1$ is suboptimal because of the gamma in the denominator of the first term. As for δ, the picture is less clear. But what is interesting is that experimentation shows $\gamma \approx 0.75$ is around best in practice. The net result is that the computational cost model which I propose is approximate at best, due, perhaps, to the cache consequences which the model cannot consider.

9.5.2 Full Elimination, not Triangular

Like Gaussian Elimination, the M4RI algorithm can be used not only to reduce a matrix to Row-Echelon Form (making it upper triangular if it were full-rank), but to Reduced Row-Echelon Form (making the left-hand part of the matrix the $m \times m$ identity matrix if it were full-rank). The only change is that in Step 3, we process all rows other than the $3k$ rows processed by Gaussian Elimination. Before, we only processed the rows that were below the $3k$ rows, not those above. Thus instead of $m - 3k - ik$ row additions in stage 3, we will require $m - 3k$. Otherwise the calculation proceeds exactly as in Section 9.5 on Page 146.

$$\sum_{i=0}^{i=\ell/k-1} 1.5(3k)^2(n - ik) - 0.75((3k)^3) + (3 \cdot 2^k - 4)(n - ik - k) +$$

$$(pos(m - 3k))(3n - 3ik - k)$$

$$= \frac{\ell}{8k} \left(-14n + 24mn + 4mk - 72kn - 12k^2 - 162k^3 + 2^k(24n - 12k - 12\ell) \right.$$

$$\left. +25k + 7\ell - 12m\ell + 36\ell k \right)$$

$$\sim \frac{\ell}{8k} \left(24mn - 12m\ell + 2^k(24n - 12\ell) \right)$$

As before, if $k = \log_2 \ell$, then

$$\sim \frac{1}{2\log_2 \ell} \left[6mn\ell - 3m\ell^2 + 6n\ell^2 - 3\ell^3 \right]$$

and thus if $\ell = n$ (the over-defined case), we have $\frac{3mn^2 + 3n^3}{2\log_2 n}$ and if $\ell = m$ (the under-defined case), we have $\frac{6m^2n - 3m^3}{\log_2 m}$. In the case that $m = n$ (the square case), we have

$3n^3 / \log_2 n$. As specified in Section 9.4.5 on Page 145, this is just the same formula with 3 taking the place of $5/2$.

9.5.3 The Rank of $3k$ Rows, or Why $k + \varepsilon$ is not Enough

The reader may be curious why $3k$ rows are selected instead of k rows at the small Gaussian Elimination step (Stage 1 of each iteration). Normally to guarantee non-singularity, a linear system with k variables and an abundance of equations is solved with k plus some small integer number of equations, the others being discarded as redundant. Of course, all equations must be checked after the solution is found, to verify that it is indeed a solution. In fact, Theorem 59 on Page 144 encourages this by saying that nullity greater than 8 is quite rare.

However, this does not work in the M4RI algorithm, because $\ell / \log \ell$ submatrices must be reduced by Gaussian Elimination, and the algorithm fails if any one of these submatrices is singular.

The answer is that the probability of k vectors of length $3k$ having rank k is very high, as proved below. The small Gaussian Elimination will fail to produce the identity matrix followed by rows of zeroes if and only if this submatrix is not of full rank.

Lemma 60. *A random* $\mathbb{GF}(2)$ *matrix of dimension* $3k \times k$, *filled by fair coins, has full rank with probability* $\approx 1 - 2^{-2k}$.

Proof. Consider the columns of the matrix as vectors. One can attempt to count the number of possible full rank matrices. The first vector can be any one of $2^{3k} - 1$ non-zero length $3k$ vectors. The second one can be any non-zero vector distinct from the first, or $2^{3k} - 2$ choices. The third one can be any non-zero vector not equal to the first, the second, or their sum, or $2^{3k} - 4$. The ith vector can be any vector not in the space spanned by the previous $i - 1$ vectors (which are linearly independent by construction). Thus $2^{3k} - 2^{i-1}$ choices are available. Therefore, the probability of any k vectors of length $3k$ being linearly independent is

$$\frac{\prod_{i=1}^{i=k} (2^{3k} - 2^{i-1})}{(2^{3k})^k} = \prod_{i=1}^{i=k} (1 - 2^{i-1} 2^{-3k})$$

$$\approx 1 - \sum_{i=1}^{i=k} 2^{i-1} 2^{-3k}$$

$$\approx 1 - 2^{-3k} (2^k - 1)$$

$$\approx 1 - 2^{-2k}$$

and this is the desired result. $\square$

Even in the case $k = 5$, the actual probability of less than full rank is 9.46×10^{-4}, and the above formula has a relative error of 3.08×10^{-6}, and is even more accurate for higher k. Also, note when $k = c \log \ell$ then the probability of full rank is $1 - \ell^{-2c}$. Since there will be $(\ell)/(\log \ell) - 1$ iterations, the probability of even one failure during all passes is approximately $1/(\ell^{2c-1} \log \ell)$, which is very low, considering that ℓ may approach the millions.

Note that even if $2k \times k$ were chosen, then the probability of failure over the whole algorithm would be $1/\log \ell$, which is non-trivial. In practice, when k was significantly lower than $\log \ell$, the algorithm would abort very frequently, whereas it never aborted in any of the experiments when k was set near $\log \ell$. (Aborts are marked with a star in Table 9.6 on Page 155).

9.5.4 Using Bulk Logical Operations

The above algorithm can be improved upon if the microprocessor has instructions for 32-bit (or even 64-bit) logical operations. Stages 2 and 3 essentially consist of repeated row additions. The matrix can be stored in an 8-bits per byte format instead of the 1-bit per byte format, and long XOR operations can perform these vector additions. Stage 1 is unaffected. However, stages 2 and 3 can proceed 32 or 64 times as fast as normal if single-instruction logical operators are available in those sizes, as they are on all modern PCs. Additional details can be found in [221, Ch. 5], in the context of a 32-bit parity check done in parallel. Since only stages 2 and 3 were non-negligible, it is safe to say that the algorithm would proceed 32 or 64 times faster, for sufficiently large matrices.

Experimentally the author found that the speed-up varied between 80% to 95% of this figure, depending on the optimization settings of the compiler chosen. However, there is absolutely no reason not to do this all the time, thus by using unsigned long long in the C language, the vector additions were always performed 64 entries at one time in the final library.

9.6 Experimental and Numerical Results

Five experiments were performed. The first was to determine the correct value of k for M4RI. The second was to determine the running time of both M4RI and Gaussian Elimination. In doing these experiments, we noted that the optimization level of the compiler heavily influenced the output. Therefore, the third experiment attempted to calculate the magnitude of this influence. The fourth was to determine if a fixed k or flexible k was superior for performance. The fifth was a spreadsheet calculation to find an optimal $k = c_1 + c_2 \log n$, given the equations for the running time of a square matrix for the algorithm, found in Section 9.5.1 on Page 146.

The specifications of the computer on which the experiments were run is given in Section 9.3.5 on Page 140. Except as noted, all were compiled under gcc with the highest optimization setting (level three). The experiments consisted of generating a matrix filled with fair coins, and then checking the matrix for invertibility by attempting to calculate the inverse using M4RI to RREF. If the matrix was singular, a new matrix was generated. If the matrix was invertible, the inverse was calculated again using Gaussian Elimination to RREF. These two inverses were then checked for equality, and finally one was multiplied by the original to obtain a product matrix which was compared with the identity matrix. The times were calculated using `clock()` from `time.h` built into the basic C language. The functions were all timed independently, so extraneous operations like verifying the correctness of the inverse would not affect running time (except possibly via cache coherency but this is both unlikely and hard to detect). No other major tasks were being run on the machine during the experiments, but `clock()` measures user-time and not time in the sense of a "wall clock."

In the first experiment (to determine the best value of k), the range of k was permitted to change. The specific k which resulted in the lowest running time was reported for 30 matrices. Except when two values of k were tied for fastest (recall that `clock()` on Linux has a granularity of 0.01 sec), the thirty matrices were unanimous in their preferred value of k in all cases. A linear regression on this data shows that $k = c_1(\log n) + c_2$ has minimum error in the mean-squared sense at $k = (3/4)(\log n) + 0$. For the next two experiments, k was fixed at eight to simplify addressing. Another observed feature of the first experiment was that the running time was trivially perturbed if the value of k was off by one, and by a few percent if off by two. The results are in Table 9.6 on Page 155.

Each trial of the second experiment consisted of the same code compiled under all four optimization settings. Since k was fixed at eight, addressing was vastly simplified and so the program was rewritten to take advantage of this. The third experiment simply used the code from the second experiment, with the compilation set to optimization level 3. The results are in Table 9.8 on Page 156 and Table 9.7 on Page 156.

In the fourth experiment, k was permitted to vary. This resulted in the best running times, which was a surprise, because the addressing difficulties were nontrivial, and varying k slightly has a small effect on running time. Yet in practice, letting k vary did vastly improve the running time of the algorithm. Therefore $k = \log n$ was chosen. See Table 9.6 on Page 155 for the affect of relatively adjusting k upward or downward.

A fifth mini-experiment was to take the computational cost expression for M4RI (as found in Section 9.5.1 on Page 146), and place it into a spreadsheet, to seek optimal values of k for very large values of n, for which experimentation would not be feasible. The expression $1 + \log n - \log \log n$ was a better fit than any $c_1 + c_2 \log n$. On the other hand, it would be very hard to determine the coefficient of the $\log \log n$ term in that expression, since a double logarithm differs only slightly from a constant, to any reasonable degree of accuracy.

9.7 M4RI Experiments Performed by SAGE Staff

Martin Albrecht also performed some experiments for M4RI, just as he did for M4RM. See also, Section 9.3.5 on Page 139.

9.7.1 Determination of k

In order to independently determine if fixed or flexible k is better, some SAGE experiments were performed on matrices of size $1000, 2000, \ldots, 14000, 15000$. The k attempted were $6, 8, 10$ for the flexible method, and $k = 8$ for fixed addressing (see Section 9.6 on Page 150). The results are summarized in Table 9.9 on Page 157. The lowest of the three options for k in the flexible column is listed in the column "least". The column "Gaussian" is the author's implementation of Gaussian Elimination, and one ratio is the ratio of the least costly flexible k and Gaussian Elimination. The other ratio is that of the fixed to the flexible M4RI. This shows that while the fixed addressing has an advantage if $k \approx 8$. On the other hand, when k "should" be far from 8, there is a penalty for picking the "wrong" k that overrules the advantage of more simplified addressing.

9.7.2 The Transpose Experiment

One experiment was to multiply a 200×1000 matrix with a $1000 \times 100,000$ matrix. Clearly, it would be faster to do $100,000 \times 2000$ times 2000×2000 using the "transpose trick" described in Section 9.3.2 on Page 138. The effects are given in Table 9.10 on Page 157. In particular, 180 msecs without a transpose and 190 msecs with one. However, this is likely because we are using a naïve approach for calculating the transpose, rather than butterfly shuffles or some other fast technique. This is an area for improvement.

9.8 Pairing With Strassen's Algorithm for Matrix Multiplication

As stated earlier, by observing the slope and shape of a log-log plot, comparing matrix dimensions and running times, we are able to ascertain that MAGMA uses Strassen's Algorithm for matrix multiplication for large $\mathbb{GF}(2)$-matrices, not the naïve approach. This is probably for two reasons. First, in finite fields, there is no rounding error. Second, the exponent is lower ($\log_2 7 \approx 2.807$ vs 3). Thus the normal running time versus accuracy tradeoff is absent and there is no reason not to use Strassen.

Since Strassen's Algorithm multiplies an $n \times n$ matrix in 7 calls to an $n/2 \times n/2$ algorithm, versus 8 for the naïve method, one can estimate the cross-over easily. The "break-even" point occurs when the time spent on the extra overhead of Strassen's algorithm (the 18 matrix additions, for example) equals the time saved by the one fewer matrix multiplication. Table 9.1 on Page 140 shows that this occurs at about slightly below 4000×4000. This is because a 2000×2000 requires 0.03 sec, and a 4000×4000 requires 0.26 sec, slightly more than eight times as much.

On the other hand, at 6000×6000 the M4RM algorithm is roughly equal to the Strassen-naïve combo that MAGMA is using (despite the fact that MAGMA is famously hand-optimized). Considering that 0.03 sec are required for M4RM and 0.16 for MAGMA in the 2000×2000 case, we can expect a roughly $16/3$ speed-up by combining M4RM with Strassen over MAGMA.

The implementation of this was carried out by Martin Albrecht, with William Hart and this author and is now a part of SAGE. The details can be found in [15].

9.8.1 Pairing M4RI with Strassen

Naturally, one might want to pair M4RI with Strassen's Matrix Inversion Formula, to get matrix inversion in even faster time. However, this is far more complex than one might first imagine, and is discussed in Section 7.7 on Page 101.

9.9 Higher Values of q

For fields $\mathbb{GF}(q)$ having $q > 2$ the M4RM and M4RI algorithms can still be used. We must first discuss how to make a Gray Code suitable for these fields. In particular, we want a sequence of n-symbol strings, each composed of elements of $\mathbb{GF}(q)$. This sequence must have two properties. First, all possible q^n strings of length n made up of the alphabet of q field elements must be present exactly once. Second, each string must differ from its successor and predecessor in exactly one spot.

Shortly before the printing of this book, this project was carried out by Tomas Boothby and Robert Bradshaw [53], see that paper for details.

9.9.1 Building the Gray Code over $\mathbb{GF}(q)$

Suppose one has such a Gray code for length n and $\mathbb{GF}(q)$, yet one wishes to write one for length $n + 1$. Suppose further that the original sequence starts with all zeroes. Observe, the original sequence would have q^n strings of length n.

First, write the n-length version forwards, then backwards, then forwards, then backwards, *et cetera*, until one has written the entire sequence q times. In this new sequence, of length $qq^n = q^{n+1}$ strings of length n, each string differs from its neighbors in exactly one spot, except for those points when one "reverses direction." Now, prepend each string in the first run with 0, then the second run with 1, and the third run with the next field element, and so forth, until the last run has been prepended with the last field element. At this point, each member of the entire new, long sequence differs in exactly one spot from each of its neighbors. Furthermore, all possible q^{n+1} strings have been constructed, and appear exaclty once in the sequence. Finally, the string begins with all zeroes.

Thus it is easy to use induction-style recursion, and the above algorithm, to produce a Gray code of any particular length. The base case, of course, is just a list of the field elements, in any order, repeating none, and begining with 0.

Faster but more complex methods exist, and are covered by Knuth in his treatise [149, Fascicle 2a].

9.9.2 Other Modifications

For finite fields bigger than $\mathbb{GF}(2)$, the chance of a random matrix being invertible is much higher than $\mathbb{GF}(2)$. This means that chosing $3k$ rows might be excessive. But it is not a dominant term in the complexity analysis and so this is probably not important.

While enumerating the subspace, by aid of the Gray Code, one might have to do some scalar multiples of the vectors during the vector addition process. For example, going from 001 to 021 would require adding 2 times the second vector. But this would have to be done in Gaussian Elimination anyway, and at least in this method, the scalar multiples are calculated only in Stage 2, and not for each row of the matrix.

9.9.3 Running Time

The change in the formulas for the running time is that the logarithm in $n^3/\log n$ is to be taken "base q," and likewise $k = \log_q n$.

The details are omitted, since only $\mathbb{GF}(2)$ is commonly used in large scale cryptanalysis. Note, other fields of characteristic two are used in cryptography, but these can be written as vector spaces over $\mathbb{GF}(2)$, for matrix multiplication.

9.9.4 Implementation

Immediately before the printing of this book, the author received word that this research has been carried out. Tomas Boothby and Robert Bradshaw have implemented M4RM over small finite fields, in the paper [53]. The software has been made a part of SAGE.

Table 9.4 Experiment 1— Optimal Choices of k, and running time in seconds.

Size	128	256	362	512	724	1020	1448	2048
Best k	5 or 6	6	7	7 or 8	8	8 or 9	9	9
M4RI	0.09571	0.650	1.68	4.276	11.37	29.12	77.58	204.1
Gauss	0.1871	1.547	4.405	12.34	35.41	97.99	279.7	811.0
Ratio	1.954	2.380	2.622	2.886	3.114	3.365	3.605	3.974

Table 9.5 Running times, in msec, Optimization Level 0

k	1,024	1,536	2,048	3,072	4,096	6,144	8,192	12,288	16384
5	870	2,750	6,290	20,510	47,590	—*	—*	—*	–
6	760	2,340	5,420	17,540	40,630	132,950	—*	1,033,420	–
7	710	2,130	4,850	15,480	35,540	116,300	—*	903,200	–
8	680	2,040	4,550	14,320	32,620	104,960	242,990	798,470	–
9	740	2,100	4,550	13,860	30,990	97,830	223,270	737,990	1,703,290
10	880	2,360	4,980	14,330	31,130	95,850	215,080	690,580	1,595,340
11	1,170	2,970	5,940	16,260	34,020	99,980	218,320	680,310	1,528,900
12	1,740	4,170	7,970	20,470	41,020	113,270	238,160	708,640	1,557,020
13	2,750	6,410	11,890	29,210	55,970	147,190	295,120	817,950	1,716,990
14	4,780	10,790	19,390	45,610	84,580	208,300	399,810	1,045,430	–
15	8,390	18,760	33,690	77,460	140,640	335,710	623,450	1,529,740	–
16	15,290	34,340	60,570	137,360	246,010	569,740	1,034,690	2,440,410	–

*Indicates that too many aborts occurred due to singular submatrices.
See Section 9.5.3 on Page 148.

Table 9.6 Percentage Error for Offset of K, From Experiment 1

error of k	1,024	1,536	2,048	4,096	6,144	8,192	12,288	16384
-4	—	—	48.0%	53.6%	38.7%	–	32.8%	—
-3	27.9%	34.8%	26.6%	31.1%	21.3%	–	17.4%	—
-2	11.8%	14.7%	11.7%	14.7%	9.5%	13.0%	8.5%	11.4%
-1	4.4%	4.4%	3.3%	5.3%	2.1%	3.8%	1.5%	4.3%
Exact	0.0%	0.0%	0.0%	0.0%	0.0%	0.0%	0.0%	0.0%
+1	8.8%	2.9%	3.4%	0.5%	4.3%	1.5%	4.2%	1.8%
+2	29.4%	15.7%	17.3%	9.8%	18.2%	10.7%	20.2%	12.3%
+3	72.1%	45.6%	47.7%	32.4%	53.6%	37.2%	53.7%	—
+4	155.9%	104.4%	110.8%	80.6%	117.3%	85.9%	124.9%	—
+5	304.4%	214.2%	229.1%	172.9%	250.2%	189.9%	258.7%	—
+6	602.9%	428.9%	458.9%	353.8%	494.4%	381.1%	—	—

Table 9.7 Results of Experiment 3—Running Times, Fixed k=8

Size	M4RI	Gaussian	Ratio
4,000 rows	18.97 s	6.77 s	2.802
6,000 rows	59.40 s	22.21 s	2.674
8,000 rows	135.20 s	51.30 s	2.635
12,000 rows	167.28 s	450.24 s	2.692
16,000 rows	398.12 s	1023.99 s	2.572
20,000 rows	763.92 s	1999.34 s	2.617

Note: The fixed k=8 option was rejected because of these inefficiencies.

Table 9.8 Experiment 2—Running time under different Compiler Optimization Settings, k=8

	Opt 0	Opt 1	Opt 2	Opt 3
4000 x 4000				
Gauss	91.41	48.35	48.37	18.97
Russian	29.85	17.83	17.72	6.77
Ratio	3.062	2.712	2.730	2.802
6000 x 6000				
Gauss	300.27	159.83	159.74	59.40
Russian	97.02	58.43	58.38	22.21
Ratio	3.095	2.735	2.736	2.674
8000 x 8000				
Gauss	697.20	371.34	371.86	135.20
Russian	225.19	136.76	135.21	51.30
Ratio	3.096	2.715	2.750	2.635

Table 9.9 Trials between M4RI and Gaussian Elimination (msec)

Matrix Size	Fixed K=8	Flex K=6	Flex K=8	Flex K=10	Flex Least	Gaussian	Ratio G/Least	Ratio fixed/flex
1,000	0	10	10	10	10	20	2.0000	0.0000
2,000	20	40	40	40	40	133	3.3125	0.5000
3,000	70	110	100	110	100	383	3.8250	0.7000
4,000	150	230	210	210	210	873	4.1548	0.7143
5,000	350	790	430	470	430	1,875	4.3605	0.8140
6,000	940	1,180	990	1,060	990	4,178	4.2197	0.9495
7,000	1,970	5,320	2,120	1,980	1,980	8,730	4.4091	0.9949
8,000	3,360	4,450	3,480	3,280	3,280	14,525	4.4284	1.0244
9,000	4,940	6,830	5,240	4,970	4,970	22,233	4.4733	0.9940
10,000	7,110	9,820	7,240	6,890	6,890	31,180	4.5254	1.0319
11,000	9,340	13,010	9,510	9,090	9,090	41,355	4.5495	1.0275
12,000	12,330	46,470	12,640	12,010	12,010	54,055	4.5008	1.0266
13,000	15,830	20,630	16,040	15,260	15,260	67,920	4.4509	1.0374
14,000	19,280	62,180	19,640	18,690	18,690	83,898	4.4889	1.0316
15,000	23,600	45,840	24,080	22,690	22,690	101,795	4.4863	1.0401

*The "fixed $k = 8$" includes the streamlined addressing as described in Section 9.6 on Page 150, which the "flexible $k = 8$" and other flexible k's do not have.

Table 9.10 The Ineffectiveness of the Transpose Trick

k	$C = AB$	$C = (B^T A^T)^T$
1	0.79 s	0.37 s
2	0.35 s	0.25 s
3	0.23 s	0.22 s
4	0.20 s	0.20 s
5	0.18 s	0.21 s
6	0.25 s	0.20 s
7	0.33 s	0.19 s
8	0.54 s	0.19 s
9	0.82 s	0.19 s
10	1.31 s	0.19 s
11	2.10 s	0.19 s

$(200 \times 1000$ by $1000 \times 100{,}000)$

Table 9.11 Optimization Level 3, Flexible k

Dimension	4,000	8,000	12,000	16,000	20,000	24,000	28,000	32,000
Gaussian	19.00	138.34	444.53	1033.50	2022.29	3459.77	5366.62	8061.90
7	7.64	–	–	–	–	–	–	–
8	7.09	51.78	–	–	–	–	–	–
9	6.90	48.83	159.69	364.74	698.67	1195.78	–	–
10	7.05	47.31	151.65	342.75	651.63	1107.17	1740.58	2635.64
11	7.67	48.08	149.46	332.37	622.86	1051.25	1640.63	2476.58
12	–	52.55	155.51	336.11	620.35	1032.38	1597.98	2397.45
13	–	–	175.47	364.22	655.40	1073.45	1640.45	2432.18
14	–	–	–	–	–	–	1822.93	2657.26
Min	6.90	47.31	149.46	332.37	620.35	1032.38	1597.98	2397.45
Gauss/M4RI	2.75	2.92	2.97	3.11	3.26	3.35	3.36	3.36

Chapter 10
The Quadratic Sieve

This chapter will discuss the Linear Sieve and Quadratic Sieve, algorithms for factoring the product of two distinct prime integers, or any other composite number. The main purpose of the algorithm is to break the famous cryptosystem RSA. The algorithms use matrices over $\mathbb{GF}(2)$, but they will be sparse matrices rather than dense matrices.

This chapter is here for several reasons. First, we have written primarily of dense matrices over $\mathbb{GF}(2)$, and the exposition would be incomplete without discussing sparse matrices. The sparse matrix techniques described in Appendix D can be used anywhere that sparsity occurs in cryptanalysis, but many were designed for the Quadratic Sieve. Second, we have written of how to break block ciphers and stream ciphers, so it would be a pity not to discuss how to break public-key systems as well. Third, the Quadratic Sieve algorithm, when taken with all its variants and modifications, stands as one of the most sophisticated algorithms in all of computer science, and fourth, it uses some elegant number theory.

This is only the tip of a very large iceberg. There are many variations, improvements, and enhancements which are omitted here. Many of those are crucial in factoring larger numbers. Furthermore, we exclude many other important factoring algorithms, because they are unrelated to the Quadratic Sieve. While the NFS (Number Field Sieve) has eclipsed the Quadratic Sieve as an algorithm, understanding the NFS is much easier after studying the QS. We hope this section will inspire the reader to read further on this vital topic.

10.1 Motivation

In order to understand why factoring is important, and particularly the case of the product of two primes, we must briefly mention the RSA algorithm, named for its inventors Ron Rivest, Adi Shamir, and Leonard Aldeman. We need not inform the reader of the enormous influence that the RSA algorithm has had on communications security, cryptography, and network security. Obviously, E-commerce would be much harder without public-key cryptosystems, and so the economic boom of the late 1990s can be accredited, in part, to the existence of public-key cryptography—the most common example and application of which is RSA.

To define a public-key cryptosystem rigorously would lead us too far astray in our discussion. However, it works as follows. Each user has a public key and a private key. To encrypt a message requires only the public key, but to decrypt it requires the private key. Thus if Alice publishes her public key on her web page, then anyone can write encrypted messages to Alice using that key. And further, if the private key is kept secret, only the owner of the private key (Alice) can read those encrypted messages, provided that the system used is cryptographically secure.

In any case, it turns out that if you can factor the product of two distinct primes, you can convert public keys into private keys, and thus read everyone's messages. This is a major motivation in factoring research and so we will quickly present the rudiments of the RSA algorithm.

10.1.1 A View of RSA from 60,000 feet

A user will generate two primes p and q, subject to further restrictions which we will not mention here, and calculate $n = pq$. Then he will generate e, d such that $ed = 1 \bmod \phi(n)$, which will always be possible if e and d are coprime to $\phi(n)$.

Here $\phi(n)$ is the Euler Totient function. By definition, this is the number of positive integers that are both coprime to n and less than n. That sounds irrelevant, but it turns out that $x^a = x^b \bmod n$ when $a = b \bmod \phi(n)$. Furthermore, when n is the product of two primes p and q (as is the case here), then $\phi(n) = (p-1)(q-1)$.

The public key will consist of (n, e) and the private key will consist of d.

If Alice wants to send a message m to Bob, then Alice must get Bob's public key (n, e), and calculate $m^e \bmod n = c$. Bob will then transmit c, the ciphertext to Alice. Then Alice will calculate

$$(c)^d = (m^e)^d = m^{ed} = m^1 = m$$

since $ed = 1 \bmod \phi(n)$. Now Alice has Bob's message. Thus it is imperative that Alice keep the private key d secret, because anyone who has it can read her messages.

There are many details which we exclude. An excellent introductory reference is [216, Ch. 6]. However, we will now show that if one can factor n, one can construct the private key given only the public key.

10.1.2 Two Facts from Number Theory

We need two more facts from number theory. The first is that the Euclidean Algorithm can be used to easily find a and b, given some x and y, such that $ax + by = 1$, provided that x and y are coprime. If they are not coprime, let $g > 1$ be their gcd. Then the same method produces $ax + by = g$. The other fact is much deeper.

Theorem 61 (The Two-Modulus Chinese Remainder Theorem). *Let c and d be coprime positive integers. Suppose $ac + bd = 1$ and also that $z = x_1$ mod c and $z = x_2$ mod d. Then*

$$z = x_1 bd + x_2 ac \text{ mod } cd$$

Proof. We are given the equation

$$ac + bd = 1$$

From this equation we have

1. By reducing the equation mod c, we learn $bd = 1$ mod c.
2. Because d divides bd then $bd = 0$ mod d.
3. By reducing the equation mod d, we learn $ac = 1$ mod d.
4. Because c divides ac then $ac = 0$ mod c.
5. Combining these we learn that if $z = x_1$ mod c and $z = x_2$ mod d then

$$x_1 bd + x_2 ac = z \text{ mod } cd$$

which can be verified by reducing it mod c to get $x_1 bd = x_1 = z$ and reducing it mod d to get $x_2 ac = x_2 = z$. $\square$

It should be noted that the above theorem still works with multiple moduli, specifically with $c_1, c_2, \ldots, c_n$ which are pairwise coprime (i.e. the gcd of c_i and c_j is 1 if $i \neq j$), and given $z = x_i$ mod c_i for $i \in \{1, 2, \ldots, n\}$. The final answer is given mod $c_1 c_2 c_3 \cdots c_n$. The statement and proof are only slightly more complex, see [101, Ch. 7.6].

If instead c and d are not coprime, then $ac + bd = g$ instead of $ac + bd = 1$. The first and third bullet have $= g$ instead of $= 1$. Let f denote our formula $x_1 bd + x_2 ac$. We have $f = zg$ mod cd, but because $g | f$ in the ordinary integers, we can write $(f/g) = z$ mod cd/g. Note that since g is the gcd of c and d, it divides each of them, and thus surely divides their product. One need merely check $k \in \{0, cd/g, 2cd/g, 3cd/g, \ldots, cd - cd/g\}$ to find $k + (f/g) = z$ mod cd, a total of g checks at worst.

10.1.3 Reconstructing the Private Key from the Public Key

The algorithm is given in Algorithm 15 on Page 163. Suppose we have a public key (n, e) and somehow we manage to factor n into p and q. Using only three calls to the Euclidean Algorithm, we can find the following coefficients

- Let $g = \gcd(p-1, q-1)$ and find r_0 and s_0 such that $r_0(p-1) + s_0(q-1) = g$.
- Find r_1 and s_1 such that $r_1 e + s_1(p-1) = 1$.
- Find r_2 and s_2 such that $r_2 e + s_2(q-1) = 1$.

By reducing the last two bullets mod $(p-1)$ and mod $(q-1)$ respectively, we know that $r_1 e = 1$ mod $(p-1)$ and $r_2 e = 1$ mod $(q-1)$, and so we can use the Chinese Remainder Theorem with $z = 1$. In other words, we have found equations for 1 mod $q-1$ and for 1 mod $p-1$, and we will use them to make one for 1 mod $(p-1)(q-1)$. Here, $a = r_0$, $c = p-1$, $b = s_0$ and $d = q-1$. Furthermore, $x_1 = r_1 e$ and $x_2 = r_2 e$. Then we have, via Theorem 61 on Page 161,

$$(r_1 e)s_0(q-1) + (r_2 e)r_0(p-1) = g \text{ mod } (p-1)(q-1)$$

and we can factor out an e to obtain

$$e[r_1 s_0(q-1) + r_2 r_0(p-1)] = g \text{ mod } (p-1)(q-1)$$

Notice furthermore, that since g is the gcd of $p-1$ and $q-1$, it divides both of them, and thus divides that which is between the brackets. We obtain

$$e\left[\frac{r_1 s_0(q-1) + r_2 r_0(p-1)}{g}\right] = 1 \text{ mod } \frac{(p-1)(q-1)}{g}$$

and since that which is in the brackets (call it w) multiplied by e is 1 mod $(p-1)(q-1)/g$, then w is the unique multiplicative inverse of e in the integers mod $(p-1)(q-1)/g$.

This is not quite what we need. We need the multiplicative inverse of e in the integers mod $(p-1)(q-1)$. Let $k \in \{0, (p-1)(q-1)/g, 2(p-1)(q-1)/g, 3(p-1)(q-1)/g, \ldots\}$. Then for some value of k, we will have $e(k+w) = 1$ mod $(p-1)(q-1)$, because only numbers of the form $k + w$ mod $(p-1)(q-1)$ are equal to w mod $(p-1)(q-1)/g$. Then $k + w = d$ is the private key. There are only g choices for k.

Note in practice, p and q are often restricted to being "strong primes" or "safe primes." The usual definition of this is that $(p-1)/2$ and $(q-1)/2$ are prime, but authors vary on this term. Because p and q are both odd, then $p-1$ and $q-1$ must have a gcd of at least 2. But if p and q are strong/safe primes, then the gcd is exactly 2. This is important in preventing certain attacks related to Pollard's "rho" Factoring Method [63, Ch. 31.9], but has the convenient effect of allowing us to only check at worst two candidates for d.

An example is in order. Suppose Alice's public key is $(n = 9797, e = 7)$. Then we start with factoring and get $p = 101$ and $q = 97$ as the prime factorization of $n = pq$. We can easily calculate $\phi(n) = (96)(100) = 9600$ and $\gcd(96, 100) = 4$. Following Algorithm 15, we obtain:

$$(1)(101) + (-1)(97) = 4 = g$$
$$(43)(7) + (-3)(100) = 1$$

INPUT: An RSA public key (n, e), and a factoring oracle F.
OUTPUT: The RSA private key d such that $ed = 1 \bmod \phi(n)$.
1: Use F to factor $n = pq$, a product of two distinct primes.
2: Using the Euclidean Algorithm, find r_0 and s_0 such that $r_0(p-1) + s_0(q-1) = g$.
3: Using the Euclidean Algorithm, find r_1 and s_1 such that $r_1 e + s_1(p-1) = 1$.
4: Using the Euclidean Algorithm, find r_2 and s_2 such that $r_2 e + s_2(q-1) = 1$.
   Note: Since we know $r_1 e = 1 \bmod (p-1)$ and $r_2 e = 1 \bmod (q-1)$, then we can use the Chinese Remainder Theorem $(r_1 e)s_0(q-1) + (r_2 e)r_0(p-1) = g \bmod (p-1)(q-1)$.
   Note: Which further implies $e\left[\frac{(r_1)s_0(q-1)+(r_2)r_0(p-1)}{g}\right] = 1 \bmod (p-1)(q-1)/g$
5: Let $w \leftarrow (r_1 s_0(q-1) + r_2 r_0(p-1))/g$.
6: For $i \leftarrow 0$ to $g-1$ do

   - Let $d \leftarrow w + i\frac{(p-1)(q-1)}{g}$.
   - Check if $ed = 1 \bmod (p-1)(q-1)$.
   - If YES: output d is the private key.

Algorithm 15: Reconstruction of an RSA Private Key from the Public Key [Rivest, Shamir, Aldeman]

$$(-41)(7) + (3)(96) = 1$$
$$7 \times [(43)(-1)(96) + (-41)(1)(100)] = 4 \bmod (100)(96)$$
$$7 \times [-8228] = 4 \bmod (100)(96)$$
$$7 \times [-2057] = 1 \bmod (100)(96)/4$$

We have found the multiplicative inverse of 7 mod 2400, but want the multiplicative inverse mod 9600. We consider adding $k \in \{0, 2400, 4800, \ldots,\}$ and discover that $7(4800 - 2057) = 7(2743) = 1 \bmod 9600$. Therefore the private key is $d = 2743$.

Thus anyone who can factor the product of two distinct primes can turn RSA public keys into private keys.

10.2 Trial Division

This section reviews some operations which are likely familiar or obvious to the reader. But when we wish to count operations or explore running time, we must be carefully precise even with obvious tasks.

Imagine that we want to know if n is prime. Since we know all positive integers are a product of finitely many primes, we can just simply divide by all the primes in $[2, n-1]$, and if none of them divide n, then we know n is prime. This is less foolish than dividing by all integers less than n, (we only need the primes, because if 2 and 3 fail to divide n then surely 6 will fail to divide n) but it is still very naïve because there is no need to check a prime bigger than $\sqrt{n}$.

Suppose p is a factor of n, and $p > \sqrt{n}$. Then $1/p < 1/\sqrt{n}$ and $n/p < n/\sqrt{n}$. But $n/\sqrt{n} = \sqrt{n}$ and so $n/p < \sqrt{n}$. Surely n/p is a factor of n, and so n has a factor less than $\sqrt{n}$. We do not know if n/p is prime, but its prime factors are smaller than it, and so therefore smaller than $\sqrt{n}$, as well. Therefore we have proven

Lemma 62. *All non-prime positive integers n have a prime factor* $\leq \sqrt{n}$.

This helps not only with primality tests, but also with factoring the product of two primes. Once you find one of the primes p, then the other is n/p.

This is not as good news as it sounds. The n we are interested in are in the range of 100-1000 digits, and so $\sqrt{n}$ has between 50–500 digits. There are roughly $x/\ln x$ primes less than x [226]. Accordingly, we would expect $2\sqrt{n}/\log n$ primes less than $\sqrt{n}$, and so there would be no hope of factoring such an n in a lifetime, by using repeated division.

Thus we might imagine that factoring integers is very difficult. However, with numbers that have a property called "smoothness" the story is quite different.

Definition 63. A number n is called B-smooth if all primes dividing n are less than or equal to B.

The following is a factoring strategy for smooth numbers n. If we knew n was square-free and B-smooth, then it is the product of some of the primes less than B. At most $\pi(B)$ trial divisions would be required. We divide n by each prime. If it divides, we work with the quotient from then on, and if not, we ignore the fractional result. If this process ends with 1, then we know the number was B-smooth, and if not, then whatever is left is a product of only primes greater than B.

For the non-square-free case, we have to deal with the case that p might divide n several times. Naturally, we simply divide n by 2 repeatedly until it no longer divides n, and take the last integer result. Then we repeatedly divide by 3, 5, and so on. This operation of removing the p-part of x is quite useful.

This algorithm is given here as "Process(x,p)" (see Algorithm 16 on Page 165), which gives the value of m in $x = p^k m$ such that p does not divide m, or equivalently k is the largest possible. In other words, take all of the prime factorization of n, except any instances of p. Another way to put it is $\lim_{n\to\infty} \gcd(x,p^n)$ This function is called "remove" in the GNU Multiple-Precision Library for large integer arithmetic, GMP.

Then, to find out if n is smooth, we can execute $n_1 =$Process$(n,2)$, $n_2 =$Process$(n_1,3)$, $n_3 =$Process$(n_2,5)$, ..., $n_{i+1} =$Process(n_i, p_{i+1}), ... where p_i is the ith prime number. We definitely stop after reaching any prime larger than B. If we ever have $n_i = 1$, at any time, then we know the number is B-smooth, and moreover, is p_{i-1}-smooth. If the process terminates and the last n_i is not 1, then that number is a product of primes greater than B. Also, we get the entire factorization of the B-smooth original input for free with this method, assuming we retain how many times we had to divide by each prime.

It is easy to see that if $n = 2^{e_1}3^{e_2}5^{e_3}7^{e_4}\cdots p_i^{e_i}\cdots p_\ell^{e_\ell}$ then

$$(e_1 + 1) + (e_2 + 1) + (e_3 + 1) + \cdots + (e_\ell + 1) = \ell + \sum_{i=1}^{i=\ell} e_i$$

integer divisions would be required.

INPUT: A number x, and a prime p.
OUTPUT: The number x after removing the p-part. This can be thought of in three ways:

- The value of m where $x = p^k m$ and $k > 0$, but p does not divide m.
- The value of $\lim_{n \to \infty} \gcd(x, p^n)$.
- The product of all the primes in the factorization of x not equal to p, including multiple primes multiple times.

1: $t1 \leftarrow x$
2: $t2 \leftarrow x/p$
3: $i \leftarrow 0$
4: While $t2$ is an integer do

    a. $t1 \leftarrow t2$
    b. $t2 \leftarrow t2/p$
    c. $i \leftarrow i+1$

5: Return i
    Note: The number of integer divisions used is $k+1$.

Algorithm 16: Process(x,p): Removing all factors of p from x. [Classic]

10.2.1 Other Ideas

10.2.1.1 Classification by Difficulty

Let us consider the positive integers less than a trillion. For prime numbers, there are special primality tests, which we do not have need of here. For numbers that are the product of two primes, we know that at least one is less than a million—there are 78,498 such primes. By similar argument, for numbers that are the product of three primes, we know that at least one is less than 10,000—there are 1229 such primes. For all others, they surely have 4 or more primes dividing them, and so they all have a prime less than 1000 dividing them—there are only 168 such primes.

How many numbers less than N are a product of two primes? (Some authors call these numbers a semiprime, but that sounds funny because an integer is either prime or not). Let us classify the semiprimes less than N by their lowest prime divisor. If it is 2, the other prime can be up to N/2. If it is 3, the other prime can be N/3, and so forth. And note that these sets are mutually exclusive. The lower prime cannot simultaneously be 2 and 3.

Thus we have $\pi(N/2) + \pi(N/3) + \pi(N/5) + \cdots + \pi(N/p)$ where p is the largest prime less than $\sqrt{N}$. For $N = 10^{12}$, it turns out $p = 999,983$, which is the 78,498th prime. Summing the approximation $\pi(x) \sim x/\ln x$, which is only partially accurate, over $\pi(N/2) + \cdots + \pi(N/999983)$ yields 127.7 billion. The approximation $10^{12}/\ln 10^{12}$ approximates that there are 36.2 billion primes less than a trillion. The product of three or more primes is the classification, therefore, of the remaining 836.1 billion positive integers less than a trillion.

This is a fantastically bizarre distribution of difficulty. Of the trillion numbers under consideration, 83.61% of them can be factored by considering only 1229 primes—and for many of them, far fewer than that. But for 12.77% of them, we must consider 78,498 primes. The remaining 3.62% are primes, and so are easily found with primality tests. We have also neglected to mention that perfect squares and perfect cubes are very easy to identify, since a square-root or cube-root is a cheap operation and identifies them immediately. There are only a million perfect squares in our domain, so they are not significant overall as a percentage.

Therefore, we can conclude that the difficulty in factoring strongly rests with the product of two distinct primes case. Thus, it seems natural as the underlying hard problem for a public-key cryptosystem.

10.2.1.2 Easy Factorization

We can turn the above idea on its head. First we use some primality test on n. Then given n is composite, we can engage in trial division, testing only those primes less than B. If n is B-smooth, then surely we succeed in factoring. If n is the product of a large prime and a number which is B-smooth, then the trial divisions end with a number that is a large prime, which can then be detected by a primality test. Baring these two cases, we end with a product of two or more primes, all of which are bigger than B—call this the "hard part" of n.

Such an algorithm is called "easy factorization". If we are searching for B-smooth numbers, we can simply use this method. Once we know that the number is not B-smooth, we need not find its large prime components. This requires very few integer divisions, and will be a step in the Linear Sieve. It turns out that Carl Pomerance found a much better way of finding smooth numbers, and this was the origin of the Quadratic Sieve.

10.2.1.3 Testing Divisibility with GCDs

Suppose we wish to know if a number n is 101-smooth. We can first do trial division on 2, 3, 5, 7, 11, 13, and 17. Then we take the remainder, m, and evaluate

$$\gcd(m, 19 \times 23 \times 29 \times 31 \times 37 \times 41 \times 43)$$

as well as

$$\gcd(m, 47 \times 53 \times 57 \times 59 \times 61 \times 67 \times 71)$$

and finally

$$\gcd(m, 73 \times 79 \times 83 \times 89 \times 91 \times 97 \times 101)$$

and only if one of those gcds is not 1 do we need to bother with any of those "medium" primes. For the size of numbers we consider in this chapter, "medium primes" might have five digits each.

10.2.2 Sieve of Eratosthenes

To generate primes in bulk, we can use a method invented[1] by 'Ερατοσθενης ο Κυρηνης. It is named the Sieve of Eratosthenes in his honor. Basically, one writes all the integers from 2 to n, inclusive. At each iteration, one circles the first unmarked number. This will be 2 on the first round, and so one crosses out every second number (4, 6, 8, ...). On the next round, it will be 3, and one crosses out every third number (6, 9, 12, ...). And on the ith round, it will be the ith prime number, p_i, and one crosses out $2p_i, 3p_i, 4p_i, \ldots$

The crossed out numbers are composite and the circled numbers are prime. One property is that after the greatest prime less than $\sqrt{n}$, no new composite numbers are crossed out. Thus at that point, any numbers remaining unmarked are primes and can be circled.

INPUT: An integer $n > 2$.
OUTPUT: The set of primes inside $[2, n]$.
1: For $i = 2$ to n do Marked$[i] =$"unknown"
2: $P \leftarrow \{\}$
3: While there is an i marked "unknown" do

    a. Let i be the lowest integer marked "unknown".
    b. if $i > \sqrt{n}$ then see note below.
    c. Mark i as "prime" and insert i into P.
    d. $j \leftarrow 2i$
    e. While $j \leq n$ do
        i. Mark j as "composite".
        ii. $j \leftarrow j + i$

4: Return P.
Note: in the if statement after the outermost while loop, if $i > \sqrt{n}$, no more numbers will ever be newly marked as composite. Therefore, you can mark all remaining unmarked numbers as "prime" and insert them into P. This is what would happen anyway, but it is a speed-up.

Algorithm 17: To generate a list of all primes in $[2, n]$. [Eratosthenes of Cyrene]

Of course, this requires $n - 1$ integers to be stored in memory, or if coded more efficiently, at least n bits of memory. To identify all the primes less than 10^{50} (in order to factor a number roughly 10^{100}) would require roughly $10^{50} \approx 2^{159.25\cdots}$ bits of memory, at absolute minimum. At the time this was written, the most expensive PCs had 128 gigabytes of RAM or 2^{50} bits. So there is absolutely no hope of generating (or for that matter, storing) these primes.

Nonetheless, it is useful because sometimes one merely needs a list of the lowest 10,000 prime numbers for some computational purpose.

---

[1] Eratosthenes of Cyrene, who also had some noted results in astronomy and geography, including the famous experiment which measured the curvature of the earth.

10.2.2.1 Smooth Version

To find all the smooth numbers, the Sieve of Eratosthenes only needs the following tiny change. This is a simple version of the Quadratic Sieve, which also generates smooth numbers in bulk.

INPUT: An integer $n > 2$.
OUTPUT: The set of all p_B-smooth numbers inside $[2,n]$.
1: For $i = 2$ to n do Sandbox$[i] \leftarrow i$
2: For $i = 1$ to B do

    a. Let p be the lowest integer such that Sandbox$[p] \neq 1$
    b. $j \leftarrow p$
    c. While $j \leq n$ do
        i. Sandbox$[j] \leftarrow$ Process(Sandbox$[j]$, p)
        ii. $j \leftarrow j + p$

3: $L \leftarrow \{\}$
4: For $i = 2$ to n do if Sandbox$[i] = 1$ then insert i into L.
5: Return L.
Note: Process is defined in Algorithm 16 on Page 165.

Algorithm 18: To generate a list of all p_B-smooth numbers in $[2,n]$. [Variant of Eratosthenes, from [192]]

We stated earlier that if $x = p^k m$, with p not dividing m, then the number of divisions used by "process" is $k + 1$. Since we only call process when x is divisible by p, then every number will experience 2 or more divisions by p. Of these, $1/p$ will experience 3 or more divisions by p, and $1/p^2$ will experience 4 or more, and $1/p^i$ will experience $i + 2$ or more. This means the expected number of divisions is

$$2 + 1\left(\frac{1}{p} - \frac{1}{p^2}\right) + 2\left(\frac{1}{p^2} + \frac{1}{p^3}\right) + 3\left(\frac{1}{p^3} + \frac{1}{p^4}\right) + \cdots = 2 + \frac{1}{p} + \frac{1}{p^2} + \frac{1}{p^3} + \cdots = \frac{2p - 1}{p - 1}$$

An Interesting Trick

Incidentally, another trick is to simply take $\gcd(n, p^{20})$ if the prime is not "too large". Then one can call Process(x, p). Only if n is divisible by p^{21} or larger will Process use more than one integer division. And taking the gcd of two positive integers is not all that much more of an operation than multiplying them.

10.3 Theoretical Foundations

The entire structure and method of the Linear Sieve and Quadratic Sieve are based on the following result. First, we present the most general case, and that shows that both algorithms are useful for factoring any composite number. However, it turns out that the proof is far easier to understand in the special case that concerns us here, namely that $n = pq$, the product of two distinct prime numbers. We prove that corollary from scratch, so the reader should not be dismayed if the proof of the general case is too confusing.

The theorem has probably been known for a long time, but was first used in factoring by Maurice Kraitchik, according to [192].

Theorem 64. *If $x^2 = y^2 \bmod n$ and $x \neq y \bmod n$ and $x \neq -y \bmod n$ then $\gcd(x+y,n)$ and $\gcd(x-y,n)$ are non-trivial factors of n.*

Note, all positive integers n are divisible by 1 and n, and so therefore, we call those the trivial factors. Any other positive integer factor is a non-trivial factor.

Proof. Since $x^2 = y^2 \bmod n$ then $x^2 - y^2 = 0 \bmod n$, or $(x+y)(x-y) = 0 \bmod n$. In the integers, this means that $(x+y)(x-y) = kn$.

Let p be a prime dividing n. Then p divides kn and since p divides the right-hand side surely it must also divide the left-hand side. Thus p divides $(x+y)(x-y)$. It is a basic result in number theory that if p divides ab then either p divides a or p divides b, provided that p is prime. Thus either p divides $(x+y)$ or p divides $(x-y)$, or both.

Thus we have proven that for all primes dividing n, either that prime divides $(x+y)$ or that prime divides $(x-y)$.

Let the factorization of $n = p_1^{e_1} p_2^{e_2} \cdots p_\ell^{e_\ell}$. Suppose for all $i \in \{1, 2, \ldots, \ell\}$ we have $p_i^{e_i}$ divides $(x+y)$. Then surely n divides $(x+y)$ but that would require $(x+y) = 0 \bmod n$ or $x = -y \bmod n$, which we forbade.

Thus for some p_i, there is an $f_i < e_i$ such that $p_i^{f_i}$ divides $(x+y)$ but $p_i^{e_i}$ does not. The same argument works for $(x-y)$ also, by reliance upon $x = y \bmod n$ being forbidden.

On the one hand $f_i = 0$ is allowed. But we know p_i must divide either $(x+y)$ or $(x-y)$, and so $f_i > 0$ for either one or the other, or both.

Of course, $(x+y)$ and $(x-y)$ can have factors that are not of the form p_i to some power. But since n is composed only of p_i to various powers, then $\gcd(x+y,n)$ and $\gcd(x-y,n)$ cannot have any primes other than p_i to various powers in their factorizations. Also, the exponent of p_i in either of the gcds cannot exceed the exponent in the factorization of n. Finally, because of the $f_i < e_i$ argument, the factors are non-trivial. □

The argument is much simpler if n is square free. Then all the exponents in the factorization are one, and this means that some of the primes dividing n divide $(x+y)$, but not all, and some of the primes dividing n divide $(x-y)$, but not all.

In the case of RSA moduli, n is the product of two primes, and so this theorem gives us much more than just a pair of non-trivial factors. Since there are only two primes in the factorization, then one gcd is one of the primes and the other gcd is the other prime.

Corollary 65. *If n is the product of two primes, and $x^2 = y^2 \bmod n$ but $x \neq y \bmod n$ and $x \neq -y \bmod n$ then $\gcd(x+y, n)$ and $\gcd(x-y, n)$ are those primes.*

Because the proof of the previous theorem is somewhat difficult, we will prove the corollary from scratch.

Proof. Since $x^2 = y^2 \bmod n$ then $x^2 - y^2 = 0 \bmod n$, or $(x+y)(x-y) = 0 \bmod n$. In the integers, this means that $(x+y)(x-y) = kn$.

We know that p divides n and q divides n, so clearly both must also divide kn, and since they divide the right-hand side, they must divide the left-hand side as well. Thus we have both p and q dividing $(x+y)(x-y)$.

It is a basic result in number theory that if p divides ab then either p divides a or p divides b. Thus either p divides $(x+y)$ or p divides $(x-y)$, or both, and likewise q. We now claim that it cannot be the case that both p and q divide $(x+y)$. If this were the case, surely pq would divide $(x+y)$ and then n divides $(x+y)$ which means that $x + y = 0 \bmod n$ or $x = -y \bmod n$, which we forbade.

Likewise, if both p and q divide $(x-y)$ then $x = y \bmod n$ which we forbade. So either p divides $(x+y)$ but not $(x-y)$ and q divides $(x-y)$ but not $(x+y)$, or alternatively, q divides $(x+y)$ but not $(x-y)$ and p divides $(x-y)$ but not $(x+y)$. Because the factorization of $n = pq$, then either $\gcd(x+y, n) = p$ and $\gcd(x-y, n) = q$ or alternatively, $\gcd(x+y, n) = q$ and $\gcd(x-y, n) = p$. Either way, we have factored n. $\qquad\square$

10.4 The Naïve Sieve

We could imagine the following algorithm. Generate a random number x, and find $y = x^2 \bmod n$. Add this pair (x, y) to a list. If at any time, two pairs in the list have the same y value, then check to see that the x values are not equal mod n, nor are they additive inverses mod n. If two pairs have matching y values without these two caveats, then Corollary 65 on Page 170 gives the correct values of p and q. This is summarized as Algorithm 19 on Page 171.

This algorithm is very similar to that of Pierre de Fermat's factoring method. Check to see if $n + y^2$ is a perfect square, for $y \in \{0, 1, 2, \ldots\}$. Surely if $n + y^2 = x^2$ then $n = (x+y)(x-y)$. While Fermat requires $n = x^2 - y^2$, we require $kn = x^2 - y^2$.

It turns out this is very inefficient. Suppose $x_1^2 = 2 \bmod n$ and $x_2^2 = 3 \bmod n$, but also $x_3^2 = 6 \bmod n$. Clearly $(x_1 x_2)^2 = x_1^2 x_2^2 = 2 \times 3 = 6$, and $x_3^2 = 6$. So we should check that $(x_1 x_2) \neq x_3$ and also $(x_1 x_2) \neq -x_3 \bmod n$. But after checking for those two caveats, we can use Corollary 65 on Page 170 to obtain the factors. The naïve version of the algorithm would never find this, because $2 \neq 6$ and $3 \neq 6$.

INPUT: The product of two primes, n.
OUTPUT: The factorization of n.
1: $i \leftarrow 0, L \leftarrow \{\}$
2: For "a long time" do

    a. $i \leftarrow i + 1$.
    b. Generate a random x_i in the set $\{2, \ldots, n-1\}$.
    c. Calculate $y_i = x_i^2 \bmod n$.
    d. If there is any $(x_j, y_j) \in L$ for $j \in \{1, 2, 3, \ldots, i-1\}$ such that $y_i = y_j$ then
        • if $x_i \neq x_j$ and $x_i \neq n - x_j$ then output

$$\gcd(x_i + x_j, n) \times \gcd(x_i - x_j, n) = n$$

       is the factorization, and halt.
    e. Add (x_i, y_i) to the set L.

Algorithm 19: A Naïve Version of the Quadratic Sieve [Unknown]

10.4.1 An Extended Example

Suppose we wish to factor 9797. We could pick a few numbers in the range $(\sqrt{9797}, 9797)$ and square them. Then we could see if any square to the same answer.

$x_1^2 = 235^2 = 6240$ $x_7^2 = 1275^2 = 9120$

$x_2^2 = 296^2 = 9240$ $x_8^2 = 1670^2 = 6552$

$x_3^2 = 541^2 = 8568$ $x_9^2 = 1763^2 = 2520$

$x_4^2 = 568^2 = 9120$ $x_{10}^2 = 1765^2 = 9576$

$x_5^2 = 753^2 = 8580$ $x_{11}^2 = 2163^2 = 5400$

$x_6^2 = 1124^2 = 9360$

At this point, you might believe we have failed. But actually, this is sufficient data to factor the number, as we will see in Section 10.7 on Page 178. As it comes to pass, precisely 15 combinations[2] of these squares (multiplied together) result in perfect squares in the integers, and so can be used to factor n provided that they do not fall into the $x = \pm y \bmod n$ trap.

10.5 The Gödel Vectors

In number theory, it is extraordinarily common to deal with numbers in their factored form. The prime factorization yields a great deal of information about the number. Many techniques arise from this notation.

As it turns out, if we restrict ourselves to numbers that are 20-smooth, for example, then because there are only 8 primes less than 20, all the factorizations can be written as $2^a 3^b 5^c 7^d 11^e 13^f 17^g 19^h$ or an 8-dimensional vector (a, b, c, d, e, f, g, h).

---

[2] Actually, there are 16 combinations if you count the zero vector, but that is not useful in factoring.

Thus by limiting ourselves to numbers that are p_B-smooth, where p_B is the Bth prime number, we can use B-dimensional vectors to represent the factorization. Note that 1 is represented by the zero vector.

This may not seem to be all that much of a gain, and we loose the ability to write numbers that are not p_B-smooth, which is a serious loss. But as it comes to pass, many operations which are otherwise highly non-trivial on the integers become extremely simple and fast in this notation. Because the author needs to call these vectors something, we shall call them the Gödel vectors, for reasons which will become clear momentarily.

10.5.1 Benefits of the Notation

Let us first examine the numerous number-theoretical operations that are speeded by using positive integers encoded in this form. This is in no small part aided by the fact that k-smooth numbers are closed under multiplication.

Suppose we have two numbers, a and b, with Gödel vectors **a** and **b**. Denote the ith element of the vector **a** as a_i and similarly b_i.

1. **Multiplication:** To find **c** such that $c = ab$, then just add the vectors **a** and **b**.
2. **Squaring:** Accordingly, since $a^2 = aa$ then squaring a requires simply doubling each entry of the vector **a**.
3. **Is Perfect Square?** If **b** is a perfect square, then all its vector entries must be even.
4. **Square Root:** If **b** is a perfect square, because all its vector entries are even, one can divide each by two to get the integer square root.
5. **nth Powers:** To raise a^n, with n being a positive integer, we need merely multiply the vector **a** by n.
6. **Is Perfect nth Root?** Again, if and only if **b** is a perfect nth power, for any positive integer n, will all of its entries be multiples of n.
7. **Extracting the nth Root:** Accordingly, if **b** has all its entries being multiples of n, then dividing each by n will yield the integer $\sqrt[n]{b}$.
8. **Divisibility:** If a divides b, then $a_i \leq b_i$ for all entries i of the vectors.
9. **Division:** Accordingly if $a_i \leq b_i$ for all entries i, then a divides b and $c = b/a$ can be calculated by $\mathbf{c} = \mathbf{b} - \mathbf{a}$.
10. **Least Common Multiple:** To find the least common multiple c of a and b, note that $c_i = \max(a_i, b_i)$ will yield **c**.
11. **Greatest Common Factor:** Likewise, to find the greatest common divisor g of a and b, note that $g_i = \min(a_i, b_i)$ will yield **g**.
12. **Tau:** To find $\tau(a)$, the number of positive integers dividing a (see Theorem 15 on Page 46), then simply increment all the non-zero entries of **a** and multiply them.

Only the last one really requires examples. We will work with 12-smooth numbers, so our vectors have 5 entries, one each for 2, 3, 5, 7, and 11.

1. $\tau(12) = \tau((2,1,0,0,0)) = 3 \times 2 = 6$
2. $\tau(5) = \tau((0,0,1,0,0)) = 2$
3. $\tau(25) = \tau((0,0,2,0,0)) = 3$
4. $\tau(30) = \tau((1,1,1,0,0)) = 8$

Using this notation, it is much easier to prove that $\tau(x) = 5$ implies that x is the fourth power of a prime number—a standard exercise in an elementary number theory course, trivial in this notation but otherwise somewhat longer.

10.5.2 Unlimited-Dimension Vectors

We will cap the length of the vector at dimension B in order to be able to have matrices of a predictable and bounded dimension. Therefore, we are restricted to p_B-smooth positive integers. However, one need not do so. Particularly, if one uses a linked list to store the non-zero entries of the vector, there is no need to cap the length. One can even include negative integers by introducing one additional slot for the "prime" -1, but taking care not to include the exponent for -1 in operations like testing for divisibility or finding lcms and gcds.

10.5.3 The Master Stratagem

We have shown that a perfect square has a Gödel vector which is all even. And thus, we seek to find a linear combination of the Gödel vectors of the y_is, so that the sum is all-even. Recall $y_i = x_i^2 \bmod n$. More precisely, some subset of the y_i, such as $y_{a_1}, y_{a_2}, \ldots, y_{a_\ell}$ will have $\mathbf{y}_{a_1} + \mathbf{y}_{a_2} + \cdots + \mathbf{y}_{a_\ell} = \mathbf{z}$ and $\mathbf{z}$ is an all-even Gödel vector. This will yield a perfect square in the integers $s^2 = z$, and so we can take the integer square root s by dividing by two. Then we will have a product

$$y_{a_1} y_{a_2} \cdots y_{a_\ell} = x_{a_1}^2 x_{a_2}^2 \cdots x_{a_\ell}^2 = (x_{a_1} x_{a_2} \cdots x_{a_\ell})^2 = z = s^2 \bmod n$$

and finally $\gcd(x_{a_1} x_{a_2} \cdots x_{a_\ell} + s, n)$ and $\gcd(x_{a_1} x_{a_2} \cdots x_{a_\ell} - s, n)$ will give us the factors we desire, provided of course $\pm x \neq s \bmod n$.

How can we find that linear combination? This will become the essence of the Linear and Quadratic Sieve algorithms. But first, the author will justify the terminology "Gödel Vectors."

10.5.4 Historical Interlude

As we noted, there are many advantages to writing numbers in their factored form. But this is only occasionally seen as a vector of exponents.

Kurt Gödel showed how to write any finite-length logical sentence in any (countable) alphabet of symbols as a positive integer. Because the alphabet is countable, each symbol can be identified with a positive integer. For example, in first order logic, we could assign the five operators "and"=1, "or"=2, "not"=3, "implies"=4, "iff"=5), and the two parenthesis to 6 and 7, and then have countably many variables $x_0 = 8, x_1 = 9, x_2 = 10, \ldots$ *ad infinitum*. Then we could number the sentence $x_1 \wedge (x_2 \vee x_3)$, as $\{x_1, \wedge, (, x_2, \vee, x_3,)\}$ and construct

$$2^9 \times 3^1 \times 5^6 \times 7^{10} \times 11^2 \times 13^{11} \times 17^7$$

which is a 39-digit integer. But, every logical sentence, regardless of its length, is identified by a positive integer this way, and two distinct sentences cannot map to the same integer. This was a useful step in his proof of the famous Incompleteness Theorem [124, Ch. III]. Furthermore, if you cap the length of the logical sentences used (which Gödel did not), then you get vectors similar to ours here. Gödel considered arbitrary-length vectors.

In honor of this very deep trick, and its profound consequences, we call these vectors Gödel vectors. Pomerance calls them "exponent vectors" in [192].

10.5.5 Review of Null Spaces

If $c_1, c_2, \ldots, c_n$ are m-dimensional vectors then we can make a matrix M, whose ith column is c_i. In fact, M will have dimension $m \times n$.

Let $\mathbf{k} = (k_1, k_2, \ldots, k_n)$ be some n-dimensional vector. By the definitions of matrix multiplication, we would have

$$M\mathbf{k} = k_1 c_1 + k_2 c_2 + \cdots + k_n c_n$$

which the reader is urged to verify. For example, if all the ks are 0, except $k_n = 1$, then $M\mathbf{k}$ is the last column of M. Likewise, if all the ks are 0, except $k_1 = k_2 = 1$, then $M\mathbf{k}$ is the sum of the first two columns of M.

Now suppose that there were some linear combination of the vectors $c_1, c_2, \ldots, c_n$ that made zero. For example $a_1 c_1 + a_2 c_2 + \cdots + a_n c_n = 0$. Surely taking the coefficients of that linear combination, and making them into a vector (e.g. $k_i = a_i$) would result in $M\mathbf{k} = 0$. Likewise, if there were some vector $\mathbf{v}$ such that $M\mathbf{v} = 0$ then $v_1 c_1 + v_2 c_2 + \cdots + v_n c_n = 0$.

Recall, the set of vectors $\mathbf{v}$ such that $M\mathbf{v} = 0$ is called the null space of M. In this sense, the null space of the matrix M is the set of coefficients of any linear combinations of the column vectors of M that will make the zero vector.

Thus, we have established an isomorphism between the null space of M and the set of linear combinations of its columns that sum to the zero vector. The coordinates of the null space vector are precisely the coefficients of the linear combinations.

If the columns of M are linearly independent, then only the zero vector will be in the null space. However, if the columns of M are linearly dependent, there will be several vectors (at least 2 over $\mathbb{GF}(2)$) in the null space, and each of these is a linear combination of the columns to make the zero vector. See Section 27 on Page 87 for additional review of null spaces, and Algorithm 5 on Page 87 describes how to find the null space efficiently, from the RREF of M.

10.5.6 Constructing a Vector in the Even-Space

Suppose we now have M, the matrix where every column is a Gödel vector. Let M_2 be the matrix M reduced mod 2, and we shall use some algorithm (Gaussian Elimination, Block Wiedemann, etc...) to find the null space of M_2 in the field $\mathbb{GF}(2)$. This is some set of vectors $\mathbf{v}$ such that $M_2\mathbf{v} = \mathbf{0}$ in $\mathbb{GF}(2)$, or

$$v_1\mathbf{c_1'} + v_2\mathbf{c_2'} + \cdots + v_n\mathbf{c_n'} = \mathbf{0} \bmod 2$$

where c_i' is the ith column of M_2. Because this is the case, then

$$v_1\mathbf{c_1} + v_2\mathbf{c_2} + \cdots + v_n\mathbf{c_n} = \mathbf{s}$$

when calculated in the ordinary integers, and with $\mathbf{c_i}$ being the ith column of M, will produce $\mathbf{s}$ such that $\mathbf{s}$ is an integer vector, and $\mathbf{s} = \mathbf{0} \bmod 2$. Of course, this means that $\mathbf{s}$ is all even. And because the v_i are all 0 or 1, and the entries of M are all non-negative, then all the entries of $\mathbf{s}$ are non-negative.

Note that the v_i are either 0 or 1, because they came from a linear algebraic operation mod 2. We can simplify our notation a lot by letting $v_{a_1}, v_{a_2}, \ldots, v_{a_\ell}$ be precisely those entries of $\mathbf{v}$ that are 1. In other words, ℓ is the number of ones in $\mathbf{v}$, sometimes called the weight of $\mathbf{v}$. All the other entries are zero, so

$$v_1\mathbf{c_1} + v_2\mathbf{c_2} + \cdots + v_n\mathbf{c_n} = v_{a_1}\mathbf{c_{a_1}} + v_{a_2}\mathbf{c_{a_2}} + \cdots + v_{a_n}\mathbf{c_{a_n}} = \mathbf{c_{a_1}} + \mathbf{c_{a_2}} + \cdots + \mathbf{c_{a_n}} = \mathbf{s}$$

As noted, every vector $\mathbf{v}$ in the null space of M_2 is a vector which produces an all-non-negative-integer all-even $\mathbf{s}$. Interpreted as a Gödel vector, then $\mathbf{s}$ is a perfect square (because it has no odd exponents) and multiplying it by $1/2$ produces an all integer vector $\mathbf{r}$. We can easily convert $\mathbf{s}$ and $\mathbf{r}$ back to integers s and r. Then we have that $r^2 = s$ in the integers, so certainly it will be the case that $r^2 = s \bmod n$.

Let the columns of M be the Gödel vectors of the squares we found in our search, mod n. In other words, column $\mathbf{c_i}$ is the Gödel vector of $x_i^2 \bmod n$. We know that

$$\mathbf{c_{a_1}} + \mathbf{c_{a_2}} + \cdots + \mathbf{c_{a_n}} = \mathbf{s}$$

but that converts out of Gödel vector notation into

$$x_{a_1}^2 x_{a_2}^2 \cdots x_{a_\ell}^2 = s \bmod n = \left(x_{a_1} x_{a_2} \cdots x_{a_\ell}\right)^2$$

and also $r^2 = s \bmod n$, and so we need merely check that $\pm r \neq x_{a_1} x_{a_2} \cdots x_{a_\ell} \bmod n$ and baring those two disappointments, we know that $\gcd(x_{a_1} x_{a_2} \cdots x_{a_\ell} - r, n) = p$ and $\gcd(x_{a_1} x_{a_2} \cdots x_{a_\ell} + r, n) = q$ or vice-versa.

10.6 The Linear Sieve Algorithm

And so the algorithm can be thought of as in two stages. First, we generate lots of squares mod N, search for those that are smooth, in the sense that their prime factorization comes only from the first B primes. Here, B is a parameter of the algorithm set in advance. If the square is smooth, we will store the square's Gödel vector, as well as the number itself. We will do this until we get some pre-set number of vectors, n_{max}. Note, we reject squares that are insufficiently smooth in order to bound the length of our Gödel vectors.

In the second stage, we will take those vectors as the columns of a matrix, reduce the matrix mod 2, and find the matrix's null space. Each vector in the null space represents a linear combination of the columns that makes zero mod 2, or a linear combination of the original columns that makes an all-even vector. Of course, an all-even vector is a perfect square, call that s, and its square root r.

Then we reconstruct the number t by multiplying the x_i that produced all the vectors that are used with non-zero coefficient in the linear combination. We know $t^2 = s \bmod N$, and since $r^2 = s$ in the integers, then $r^2 = s \bmod N$. Now we have $t^2 = r^2 \bmod N$. We need only check that $\pm r \neq t \bmod N$, and baring those two disappointments, we are done. This algorithm is called the Linear Sieve because of its dependance on linear algebra.

An important note is that if the product of the x_{a_i}s is less than $\sqrt{n}$, that the square root in the ordinary integers r might actually be the product of the x_{a_i}, namely t, because the operations to multiply the x_{a_i}s and square the answer might not "wrap around" the modulus n. Therefore, we should confine our xs to $[\lfloor \sqrt{n} \rfloor + 1, n - 1]$, to make it more likely that $\pm t \neq r$. This point can be summarized by saying $2^2 = 4$ mod any 100-digit prime, just as $2^2 = 4$ in the integers.

Variants of the Quadratic Sieve which consider -1 to be "a prime" in the factor base resolve this issue, and further, allow one to search both below and above $\sqrt{n}$.

10.6.1 Matrix Dimensions in the Linear & Quadratic Sieve

Note, the matrix will have B rows and n_{max} columns. Recall that a matrix is likely to have a large null space if it has many more columns than rows. The dimension of the null space is called the nullity of the matrix, and over $\mathbb{GF}(2)$, a nullity of k means 2^k vectors in the null space. A matrix of dimension $B \times (B + 10)$ would be expected to have nullity between 10–13, (see Table 9.3 on Page 145) and so have between 1024 and 8196 vectors in the null space. Roughly one half of these would

INPUT: A number N to be factored, and two parameters, B and n_{max}.

OUTPUT: The factors p and q so that $N = pq$.

1: $L \leftarrow \{\}$ and $n \leftarrow 0$.

2: Let p_B be the Bth prime number.

3: While $n < n_{max}$ do

    a. Generate a random x in $\{2, 3, \ldots, N-1\}$.

    b. Calculate $y = x^2 \bmod n$.

    c. Do an easy factorization of y, stopping after B primes.

      • If y is not p_B-smooth, reject it, and start a new iteration of the while loop.

      • If y is p_B-smooth

        i. Calculate its Gödel vector $\mathbf{g}$

        ii. Insert $(x, \mathbf{g})$ into the list L.

        iii. $n \leftarrow n+1$

Note: There are now exactly n_{max} entries in the list L.

Note: Above this point is "Stage One" and below this point is "Stage Two".

4: Construct a matrix M, with each Gödel vector as a column.

Note: The matrix M will have dimension $B \times n_{max}$.

5: Reduce M mod 2 and call that M_2.

6: Find the null space of M_2 and call it $\mathcal{N}$.

7: For each $\mathbf{v} \in \mathcal{N}$ do

    a. $\mathbf{g} \leftarrow \mathbf{0}, r \leftarrow 1$

    b. For $i = 1, 2, \ldots n_{max}$ do

      • if $v_i = 1$ then

        i. Fetch $(x_i, \mathbf{g_i})$ from the list L.

        ii. $\mathbf{g} \leftarrow \mathbf{g} + \mathbf{c_i}$

        iii. $t \leftarrow tx_i$

    c. $\mathbf{h} \leftarrow \frac{1}{2}\mathbf{g}$

    Note: Since $\mathbf{g}$ is composed of only all-even non-negative integers, $\mathbf{h}$ is composed of only non-negative integers also.

    d. Convert $\mathbf{h}$ into an integer r. Note r^2 has Gödel vector $\mathbf{g}$.

    e. If $r = \pm t \bmod N$ then reject, go to the next $\mathbf{v}$.

    f. Otherwise output $p = \gcd(r-t, N)$ and $q = \gcd(r+t, N)$.

8: Fail.

Algorithm 20: The Linear Sieve [Richard Schroeppel]

be expected to have $\pm r \neq t \bmod n$ and so we are very likely to find at least one, which is all that we require.

Let us assume we set $n_{max} = B + 10$. What if B is too small? Then it may be very hard to find x such that $x^2 \bmod n$ has no primes in its factorization above p_B. What if we make B too large? Finding the null space of a $10,000 \times 10,000$ matrix is extremely fast with the Method of Four Russians (see Chapter 9). Even for $100,000 \times 100,000$ this is a relatively simple computation. But at $1,000,000 \times 1,000,000$ we run into the problem that that the matrix has one trillion entries. Even storing 8 entries per byte this is 125 gigabytes.

Luckily, the matrix will be extremely sparse. But we are required to use a sparse matrix method, because if we try the Method of Four Russians or a Gaussian Elimination method, the matrix will rapidly become dense (a process called "fill-in"). Once the matrix is dense, we will surely run out of memory on the computer.

Thus the Linear Sieve and Quadratic Sieve algorithms gave new impetus to an entire area of research on sparse matrices over finite fields, particularly $\mathbb{GF}(2)$, which then found other uses. Incidentally, the Number Field Sieve can also make use of very sparse finite field matrices—in fact the structure of the matrices is essentially the same. Carl Pomerance (the inventor of the Quadratic Sieve) invented a null-space finding algorithm specifically for these sorts of matrices, and it is described in Section D.6 on Page 327.

10.6.2 The Running Time

The dilemma is that it is not easy to establish the number of x which must be squared to get n_{max} smooth values of $x^2 \bmod n$. Therefore, let us say R values of x are attempted, and write the running time in terms of R.

We must test divide each x by all the primes less than B. There are roughly $B/\log B$ of those. If a number is of the form $p_1^{e_1} p_2^{e_2} p_3^{e_3}$ we must divide by p_1 exactly $e_1 + 1$ times, and by p_2 we must divide $e_2 + 1$ times, and so forth. However, of the numbers divisible by p_i, roughly $1/p_i$ of those are divisible by p_i^2 and of those, $1/p_i$ of them are divisible by p_i^3, and so forth.

We calculated in Section 10.2.2.1 on Page 168 that roughly 2 divisions of each prime would be the expected value in general. Thus 2 divisions per prime, and $B/\log B$ primes available means that $2RB/\log B$ integer divisions must be made.

Of course, there are many ways to speed this up and we have been a bit naïve in our approach, but the Quadratic Sieve will be categorically much faster at generating these squares, and so we will not spend time discussing how to improve the Linear Sieve.

10.7 The Example, Revisited

The numbers we have selected in Section 10.4.1 on Page 171 produce the following squares and smooth factorizations.

$$x_1^2 = 235^2 = 6240 = 2^5 \times 3 \times 5 \times 13$$
$$x_2^2 = 296^2 = 9240 = 2^3 \times 3 \times 5 \times 7 \times 11$$
$$x_3^2 = 541^2 = 8568 = 2^3 \times 3^2 \times 7 \times 17$$
$$x_4^2 = 568^2 = 9120 = 2^5 \times 3 \times 5 \times 19$$
$$x_5^2 = 753^2 = 8580 = 2^2 \times 3 \times 5 \times 11 \times 13$$
$$x_6^2 = 1124^2 = 9360 = 2^4 \times 3^2 \times 5 \times 13$$
$$x_7^2 = 1275^2 = 9120 = 2^5 \times 3 \times 5 \times 19$$
$$x_8^2 = 1670^2 = 6552 = 2^3 \times 3^2 \times 7 \times 13$$

$$x_9^2 = 1763^2 = 2520 = 2^3 \times 3^2 \times 5 \times 7$$
$$x_{10} = 1765^2 = 9576 = 2^3 \times 3^2 \times 7 \times 19$$
$$x_{11} = 2163^2 = 5400 = 2^3 \times 3^3 \times 5^2$$

Taking the Gödel vectors, and writing them vertically, we get the following matrix (mod 2)

$$M_2 = \begin{bmatrix} 1 & 1 & 1 & 1 & 0 & 0 & 1 & 1 & 1 & 1 & 1 \\ 1 & 1 & 0 & 1 & 1 & 0 & 1 & 0 & 0 & 0 & 1 \\ 1 & 1 & 0 & 1 & 1 & 1 & 1 & 0 & 1 & 0 & 0 \\ 0 & 1 & 1 & 0 & 0 & 0 & 0 & 1 & 1 & 1 & 0 \\ 0 & 1 & 0 & 0 & 1 & 0 & 0 & 0 & 0 & 0 & 0 \\ 1 & 0 & 0 & 0 & 1 & 1 & 0 & 1 & 0 & 0 & 0 \\ 0 & 0 & 1 & 0 & 0 & 0 & 0 & 0 & 0 & 0 & 0 \\ 0 & 0 & 0 & 1 & 0 & 0 & 1 & 0 & 0 & 1 & 0 \end{bmatrix}$$

After reducing it with Gaussian Elimination, we find it has nullity 4, and the null-space basis vectors are:

$$n_1 = (0,0,0,1,0,0,1,0,0,0,0)$$
$$n_2 = (0,1,0,0,1,0,0,1,0,0,0)$$
$$n_3 = (0,1,0,0,1,1,0,0,1,0,0)$$
$$n_4 = (1,1,0,1,1,0,0,0,0,1,0)$$

and so we will find 16 vectors in the null space, one of which is the zero vector.

Just taking n_1 on its own suggests x_4 and x_7 should be multiplied. We get

$$x_4^2 x_7^2 = 568^2 \times 1275^2 = 9019 \times 9019$$

but $\gcd(568 \times 1275 + 9019, 9797) = 1$. Bad luck!

Just taking n_2 on its own suggests x_2, x_5, and x_8 should be multiplied. We get

$$x_2^2 x_5^2 x_8^2 = 296^2 \times 753^2 \times 1670^2$$
$$= 9240 \times 8580 \times 6552$$
$$= 519437318400 = 720720^2$$

and $\gcd(296 \times 753 \times 1670 + 720720, 9797) = 1$. Again, bad luck.

Just taking n_3 on its own suggests x_2, x_5, x_6, and x_9 should be multiplied. We get

$$x_2^2 x_5^2 x_6^2 x_9^2 = 296^2 \times 753^2 \times 1124^2 \times 1763^2$$
$$= 9240 \times 8580 \times 9360 \times 2520$$
$$= 1869974346240000 = 43243200^2$$

and the correct factorization is produced:

$$\gcd(296 \times 753 \times 1124 \times 1763 + 43243200, 9797) = 101$$
$$\gcd(296 \times 753 \times 1124 \times 1763 + 43243200, 9797) = 97$$

10.8 Rapidly Generating Smooth Squares

First off, finding x^2 mod n is equivalent to finding $x^2 - n$ in the range $n < x^2 < 2n$ or more plainly $\sqrt{n} < x < \sqrt{2n}$. For the type of n we are discussing in RSA (roughly 100 to 1000 digits), we can't even search a fraction of that range, so it is safe to restrict ourselves to it.

For any particular prime p, either n is a quadratic residue mod p or it is not. When one says "n is a quadratic residue mod p", it means there is some s such that $s^2 = n$ mod p. There are three cases.

- First, no such s exists. In that case, we can remove p from consideration entirely. This p will not divide *any* values of $x^2 - n$ and so it certainly will not divide the smooth values. To see this, suppose p divides $x^2 - n$. Then $x^2 - n = 0$ mod p and $x^2 = n$ mod p. Thus we have produced a value x so we know n is a quadratic residue mod p. The contrapositive is that if n is not a quadratic residue mod p, then p does not divide $x^2 - n$, and $x^2 - n \neq 0$ mod p.
- Second, the s exists and is zero. Then this means $0^2 = n$ mod p or p divides n. In this very lucky outcome, the factors are p and n/p and we are done.
- In the third case, $0 < s < p$. Now suppose p divides $x^2 - n$. Again $x^2 - n = 0$ mod p or $x^2 = n$ mod p which implies $x^2 = s^2$ mod p. This further results in $x^2 - s^2 = 0$ mod p or $(x+s)(x-s) = 0$ mod p. Since operating mod p means we are operating in a field, then we know either $x - s = 0$ mod p or $x + s = 0$ mod p, which implies $x = \pm s$ mod p. It is noteworthy that in this case the only two solutions will be $s^2 = (-s)^2 = n$ mod p.

Therefore, when handling divisibility by p, we can merely check twice every p when searching for available x. This is much like the sieve of Eratosthenes. The first integer in the range $[a,b]$ to be 0 mod p will be $\lceil a/p \rceil p$. Therefore we can check

$$x \in \{ \lceil a/p \rceil p - s, \lceil a/p \rceil p + s, \lceil a/p \rceil p + p - s, \lceil a/p \rceil p + p + s, \lceil a/p \rceil p + 2p - s,$$
$$\lceil a/p \rceil p + 2p + s, \lceil a/p \rceil p + 3p - s, \lceil a/p \rceil p + 3p + s, \ldots \}$$

which is a tremendous savings over checking every possible x.

We must define what we mean by checking. We know these values will be divisible by p. Therefore, we can (in the spirit of the smooth version of the sieve of Eratosthenes given in Algorithm 18 on Page 168) make a list of all the integers in the range $[a,b]$, and divide through by p those selected by the above sequence.

Also like the sieve of Eratosthenes, we must consider numbers that are divisible by $p^2, p^3, \ldots$ if we are to be searching for smooth numbers. Therefore, after the first division, we divide again. If it is an integer still, we continue to divide until it is not an integer. We keep the last value that is an integer, and store that. This is precisely the algorithm "Process(x,p)" given in Algorithm 16 on Page 165.

Quadratic Residues

All that remains is we must specify how to detect if n is a quadratic residue mod p. The following lemma can be found in [101, Ch. 17.3].

Lemma 66. *If $n \neq 0$ mod p then $n^{(p-1)/2} = 1$ if and only if n is a quadratic residue mod p.*

10.8.1 New Strategy

Therefore, we have the following strategy. Assemble a "factor base" of primes, all less than p_B, such that n is a quadratic residue mod p. Make a list of all the integers x from $[a, b]$ along with $x^2 - n$. Note that $\sqrt{n} < a < b < \sqrt{2n}$, so $x^2 - n$ is equivalent to x^2 mod n. For each prime in the factor base, search the sequence mentioned above. For each x in the sequence, divide by p until the result is no longer an integer, and then store the last integer result.

After all the primes in the "factor base" have been used, any 1s in the list correspond to smooth values of $x^2 - n$. We collect these smooth numbers, compute their Gödel vectors, and proceed with the linear algebra "Stage Two" as before.

For any specific p, we must recall that $(2p-1)/(p-1) \approx 2$ divisions will be used for each call of process. There will be roughly $(b-a)/p$ iterations of the loop, and so $2(b-a)/p$ calls to Process(x, p) will be made, and this comes to $(b-a)(4p-2)/(p^2-p) \approx 4(b-a)/p$ integer divisions. See Section 10.2.2.1 on Page 168.

We will do this for all the primes less than p_B that have n as a quadratic residue mod p. Here, it is handy to use the amazing result that the sum of the reciprocals of the primes less than p_B is equal to roughly $\log \log p_B$, where log indicates the natural logarithm. Note that summing up $4(b-a)/p$ for a subset of all primes less than p_B is surely less than $4(b-a)$ times summing up $1/p$ for all primes less than p_B. Thus we have proven.

Theorem 67 (Carl Pomerance). *At most $\sim 4(b-a) \log \log p_B$ integer divisions will be used by Stage 1 of the Quadratic Sieve to identify the values of $x \in [a, b]$ such that $x^2 - n$ is p_B-smooth.*

When compared to $2RB/\log B$ found in Section 10.6.2 on Page 178, one can see that $4R \log \log B$ is a great improvement. The coefficients can be improved. There are many optimizations, including

- Allowing one or two "medium sized primes", in addition to smooth numbers.
- Allow -1 to be a "prime" in the factor base.
- Using polynomials other than $x^2 - n$.
- One can take

$$gcd(n, 2 \times 3 \times 5 \times 7 \times 11 \times 13 \times 17 \times 19 \times 23 \times 29)$$

INPUT: The integer n to be factored, a largest acceptable prime p_B, a long range $[a,b]$ to be searched.

OUTPUT: The set of positive integers x in $[a,b]$ such that $x^2 - n$ is p_B-smooth

Note: Stage 0, Determine a Factor Base.

1: $F \leftarrow \{\}$
2: For each prime $p \le p_B$ do

    a. Calculate $z = n^{(p-1)/2} \bmod p$.
      • If $z = -1$, skip to next prime.
      • If $z = 1$, insert p into F.

3: Create a sufficiently set of large intervals $[a,b]$ that are non-overlapping, and are contained in $(\sqrt{n}, \sqrt{2n})$.

    Note: Stage 1, Generate smooth x.

4: For each interval $[a,b]$ (often given to separate CPUs) do

    a. For $i = a$ to b do
      i. Residue$[i - a] \leftarrow i^2 - n$
    b. For each prime $p \in F$ do
      i. Find s such that $s^2 = n \bmod p$.
      ii. $i \leftarrow \lfloor a/p \rfloor p$
      iii. While $i - s \le b$ do
        • if $(a < i - s < b)$ then Residue$[i - a] \leftarrow$ Process$(i - s, \text{p})$
        • if $(a < i + s < b)$ then Residue$[i - a] \leftarrow$ Process$(i + s, \text{p})$
        Note: Process is defined in Algorithm 16 on Page 165.
        • $i \leftarrow i + p$
    c. $L \leftarrow \{\}$
    d. For $i = a$ to b do
      • if Residue$[i - a] = 1$ then insert i into the set L.
    e. Output the list L.

5: Collect and union together all the lists L.
6: For each $x \in L$, calculate the Gödel vector of $x^2 - n$.

    Note: Stage 2, Find the Nullspace mod 2, take gcds.
    Note: Identical from here onward to the Linear Sieve.

7: Construct a matrix M, with each Gödel vector as a column.

    Note: The matrix M will have dimension $|F| \times |L|$.

8: Reduce M mod 2 and call that M_2.
9: Find the null space of M_2 and call it $\mathcal{N}$.
10: For each $\mathbf{v} \in \mathcal{N}$ do

    a. $\mathbf{g} \leftarrow \mathbf{0}, r \leftarrow 1$
    b. For $i = 1, 2, \ldots |L|$ do
      • if $v_i = 1$ then
        i. Fetch $(x_i, \mathbf{g_i})$ from the list L.
        ii. $\mathbf{g} \leftarrow \mathbf{g} + \mathbf{c_i}$
        iii. $t \leftarrow t x_i$
    c. $\mathbf{h} \leftarrow \frac{1}{2}\mathbf{g}$
    Note: Since $\mathbf{g}$ is all-even non-negative integers, $\mathbf{h}$ is all non-negative integers also.
    d. Convert $\mathbf{h}$ into an integer r. Note r^2 has Gödel vector $\mathbf{g}$.
    e. If $r^2 = \pm t^2 \bmod N$ then too bad, go to the next $\mathbf{v}$.
    f. Otherwise output $p = \gcd(r - t, N)$ and $q = \gcd(r + t, N)$.

Algorithm 21: The Quadratic Sieve [Carl Pomerance]

several times to "batch process" the first 10 primes, which after all, are the most common. The next prime is 31, and so it is better to loop through 2 out of every 31 entries in the range $[a, b]$ than 2 out of every 3 or 5.

- Instead of using "process" one can merely divide by p. This means that non-square-free smooth numbers are not detected, but $1/\zeta(2) = 6/\pi^2 \approx 0.6079$ of integers are square-free. Or one can merely divide by p and p^2, failing to detect smooth numbers that are not cube-free. Then $1/\zeta(3) \approx 0.8319$ of integers are cube-free [223], which is more tolerable.
- ...and many other variations.

10.9 Further Reading

The interested reader may find the following useful

- "A Tale of Two Sieves", by Carl Pomerance, published in the December 1996 issue of the Notices of the American Mathematical Society [192]. This is an expository introduction.
- "The Magic Words are Squeamish Ossifrage" by Derek Atkins, Michael Graff, Arjen Lenstra and Paul Leyland, published in ASIACRYPT in 1994 [27].
- For the application to RSA, Chapter 6 "The RSA Algorithm" in *Introduction to Cryptography with Coding Theory* by Wade Trappe and Lawrence Washington [216]. This includes a detailed description of the linear sieve, which is denoted there as the quadratic sieve.
- *Factorization and Primality Testing* by David Bressoud, a book published in the "Undergraduate Texts in Mathematics" series by Springer-Verlag.
- *The Development of the Number Field Sieve*, a book of several authors edited by the Lenstra brothers [159], published in 1993.

10.10 Historical Notes

Maurice Kraitchik first used differences of the form $x^2 - y^2 = kn$ for factoring [192], but factoring using the difference of two squares goes back to Pierre de Fermat in the 17th century (see Section 10.4 on Page 170), with an intermediate method due to John Dixon [99]. The algorithm presented here was designed by Carl Pomerance [192], but first implemented by Joseph Gerver [122]. There is also the Pollard "rho" Method [63, Ch. 31.9]. Other examples of factoring algorithms include the continued fraction method [181], the Elliptic Curve Factoring Method (which is better for numbers that are not the product of two primes, but are not very smooth either) [216, Ch. 16.3], and the Number Field Sieve (the current optimal method) [192]. At Eurocrypt 2009, it seems that the Elliptic Curve Factoring Method might be making a comeback, as it can benefit from the ultrahigh-speed massively-parallel GPUs (Graphical Processing Units) in modern graphics cards (see [45]).

Part III
Polynomial Systems and Satisfiability

Chapter 11
Strategies for Polynomial Systems

11.1 Why Solve Polynomial Systems of Equations over Finite Fields?

Before we devote numerous pages to the topic of polynomial systems of equations over finite fields, we should pause and ask why one would want to do this. This will give us an opportunity to highlight the several applications of this interesting area.

Before that, however, an important distinction must be made. In this book, when one has a system of equations over the finite field $GF(p^n)$, we assume that one is interested *only in those solutions which are also elements of* $GF(p^n)$. If one is also interested in solutions in some extension field, (e.g. $GF(p^m)$ with $n|m$), then since $GF(p^n)$ is a subfield of $GF(p^m)$, it is safe to consider the system of equations as if it were over $GF(p^m)$, or at worse the splitting field.

Algebraic Cryptanalysis   The previous few chapters of this book demonstrate the crucial role of polynomial systems of equations in block cipher design and analysis, due to the necessity of engaging in or thwarting algebraic attacks.

Pure Mathematics   Polynomials and finite fields are both simple, elegant objects. Relationships among them are interesting in their own right. But the integers are a still more basic and fundamental domain, and one can inquire if a polynomial system has any solutions over the integers. These are called "diophantine" equations, for[1] $\Delta\iota o\phi\alpha\nu\tau o\varsigma\ o\ \text{'}A\lambda\epsilon\xi\alpha\nu\delta\rho\epsilon\upsilon\varsigma$, the Greek mathematician who first studied these problems. In particular, if a set of polynomial equations with integer coefficients has a solution over $\mathbb{Z}$, then the polynomial system formed by taking the image of the coefficients "mod p" has a solution in the field of p elements, $GF(p)$. One solution is, in fact, the image of the original solution "mod p". Therefore, by using the contrapositive, if the system has no solutions in $GF(p)$, then it has no solutions in $\mathbb{Z}$. For small numbers of variables, it might be much easier to solve in $GF(p)$ than over $\mathbb{Z}$.

---

[1] Diophantos of Alexandria, 3rd century AD.

G.V. Bard, *Algebraic Cryptanalysis*, DOI: 10.1007/978-0-387-88757-9_11   187
© Springer Science + Business Media, LLC 2009

Multivariate Cryptography   It is believed that solving an arbitrary system of polynomial equations is hard (in fact, if a polynomial time algorithm for general polynomial systems of equations is found, then $P = NP$, which would be a surprise). However, evaluating a polynomial at a point (or a vector, if the system is multivariate) is easy. This creates a one-way function, and one-way functions are important building blocks in the construction of cryptographic primitives. So while polynomials are more associated in cryptology with *breaking* codes, the topic of multivariate cryptography wishes to use facts from this discipline in *making* codes. There have been several examples, such as QUAD: a stream cipher [42], SFLASH: a signature scheme [72, 78, 70], and several public-key systems like HFE [65, 187], as summarized in [97, Chs. 2,3,4,6].

The Universal Map   It turns out that any map, from any finite set to any other finite set, can be written as a polynomial system of equations over $\mathbb{GF}(p)$, for any prime p. This extremely general result will be proven as Theorem 72 on Page 190. Being able to solve such systems means that one can compute pre-images under these arbitrary maps, no matter how strange or intricate they might be.

NP-Completeness Oracle   All NP-Complete problems are polynomially reducible to each other. This means that if one can solve any particular NP-Complete problem in time upper-bounded by $f(n)$, for input size n, then for any other NP-Complete problem there exist polynomials $p_1(n), p_2(n), p_3(n)$ such that the solution time is upper-bounded by $p_1(n) + p_2(f(p_3(n)))$. Of course, if f is a polynomial then $P = NP$, which would be a surprise, but there may be other efficient approximation algorithms, probabilistic algorithms, heuristic methods, or techniques that work in special cases.

Graph Coloring   The problem of finding a coloring of a graph is searching for an assignment of one of c colors to each vertex. It is required that if an edge connects v_i and v_j, that the color of v_i be distinct from the color of v_j. This problem can be solved as a sparse system of polynomial equations over $\mathbb{GF}(q)$ for any $q \geq c$. If $c > 2$ the problem is also NP-hard, and even detecting if such a coloring exists is NP-Complete. However, there are special cases which are very useful in applications, which can be colored rapidly, see Section C.4 on Page 318.

Radio Channel Assignments   The Graph Coloring problem is useful not only in drawing colorful atlases, but also for radio channel assignments. Regions, irregularly shaped due to mountains and hills, are vertices, and edges are drawn between regions where broadcasts upon the same frequency would result in non-trivial interference. The graph is colored, and each collection of vertices with the same color is assigned the same set of frequencies.

Compiler Optimization   The optimization of code to a specific microprocessor during compiling is usually viewed as an empirical process—a series of shortcuts which collectively improve running time. But register allocation, just as an example, can be done via graph coloring. More detail will be given in Section C.3.2 on Page 318.

11.2 Universal Maps

Like most material in this chapter, the author believes this has been known for a very long time.

Theorem 68. *Let F be a finite field of prime order p. Any map from a finite-dimensional F-vector space V to F can be written as a polynomial with coefficients in F.*

Proof. Consider the set Φ of all maps from V to F, such that all values map to zero, except for one privileged element $\mathbf{x} \in V$, which is mapped to one. This map will be called $\phi_{\mathbf{x}}$, and there is one for each $\mathbf{x} \in V$.

Obviously, Φ is a basis for the set of all maps from V to F, in the sense that every map from V to F can be written as a linear combination of $\phi_{\mathbf{x}}$'s. In fact, for any $f : V \to F$, the coefficient of $\phi_{\mathbf{x}}$ in the linear combination is merely the value of $f(\mathbf{x})$. Furthermore, one cannot construct $\phi_{\mathbf{x}}$ as a linear combination of all the other maps in $\Phi - \{\phi_{\mathbf{x}}\}$ because all of those maps evaluate to 0 at $\mathbf{x}$, and so therefore any linear combination of them will also.

Therefore, in order to prove the theorem, it suffices to show that $\phi_{\mathbf{x}}$ is a polynomial, because a linear combination of polynomials is a polynomial. Let $\mathbf{x} \in V$ be equal to $(x_1, \ldots, x_n)$. Using Corollary 94 on Page 315, we can write $(y_i - x_i)^{p-1} - 1 = 0$, which is non-zero if and only if $y_i = x_i$. The product

$$\prod_{i=1}^{n} \left((y_i - x_i)^{p-1} - 1\right)$$

is non-zero if and only if all n terms are each non-zero. This happens if and only if $y_i = x_i$ for all $i \in \{1, 2, \ldots, n\}$.

Thus we can write:

$$\psi_{\mathbf{x}}(y_1, y_2, \ldots, y_n) = \prod_{i=1}^{n} \left((y_i - x_i)^{p-1} - 1\right)$$

which is zero everywhere but $\mathbf{x}$, where its value is non-zero and finally,

$$\phi_{\mathbf{x}}(y_1, y_2, \ldots, y_n) = \frac{\psi_{\mathbf{x}}(y_1, y_2, \ldots, y_n)}{\psi_{\mathbf{x}}(x_1, x_2, \ldots, x_n)}$$

which is clearly a polynomial, since the demonenator is a constant. $\square$

These $\phi_{\mathbf{x}}$ functions are the finite-field analog of Lagrange polynomials. They are sometimes called needle functions or haystack functions.

Corollary 69. *Let F be a finite field of characteristic p. Any map from a finite-dimensional F-vector space V to $\mathbb{GF}(p)$ can be written as a polynomial with coefficients in $\mathbb{GF}(p)$.*

Proof. Here we have changed the requirement on the field F. In the previous theorem, it had to be of order p. Now it is merely of characteristic p, but still finite.

The reason for this is that every finite field of characteristic p is a $\mathrm{GF}(p)$-vector space. And so a map from $\mathrm{GF}(p^n)^m \to \mathrm{GF}(p)$ can be rewritten as a map from $\mathrm{GF}(p)^{mn} \to \mathrm{GF}(p)$. Then previous work yields the required condition. $\qquad\square$

Corollary 70. *Let F be a finite field of characteristic p. Any map f from a finite-dimensional F-vector space V to a finite-dimensional $\mathrm{GF}(p)$-vector space U, can be written as a polynomial system of equations with coefficients in $\mathrm{GF}(p)$.*

Proof. Suppose U is dimension m over $\mathrm{GF}(p)$. Then for any $\mathbf{x} \in F$, $f(\mathbf{x}) = (u_1, u_2, \ldots, u_m)$ with each $u_i \in \mathrm{GF}(p)$. Therefore we can write m functions, $f_1(\mathbf{x}), f_2(\mathbf{x}), \ldots, f_m(\mathbf{x})$, with $f_i(\mathbf{x})$ equal to the ith element of the vector $f(\mathbf{x})$.

Each of these $f_i(\mathbf{x})$ is a map from V to $\mathrm{GF}(p)$, and so is a polynomial with coefficients in $\mathrm{GF}(p)$. Therefore, taken together, one has m polynomials with coefficients in $\mathrm{GF}(p)$, or a polynomial system of equations. $\qquad\square$

Corollary 71. *Let F, G be finite fields of characteristic p. Any map f from a finite-dimensional F-vector space V to a finite-dimensional G-vector space U, can be written as a polynomial system of equations with coefficients in $\mathrm{GF}(p)$.*

Proof. Suppose $G = \mathrm{GF}(p^m)$, and $U = \mathrm{GF}(p^m)^n$. Then since G can be written as an m-dimensional vector space over $\mathrm{GF}(p)$, surely U can be written as an mn-dimensional vector space over $\mathrm{GF}(p)$. Then we apply the previous corollary. $\qquad\square$

Up to this point, if the map had certain properties pertaining to addition (e.g. if $f(\mathbf{x} + \mathbf{y}) = f(\mathbf{x}) + f(\mathbf{y})$) then all the polynomials representations would continue to respect this property, for some choice of basis during the conversion from the finite field to a vector space over the base field. That is because the addition operation of $\mathrm{GF}(p^n)$ is isomorphic to that of $\mathrm{GF}(p)^n$. But any multiplication properties, such as $f(k\mathbf{x}) = kf(\mathbf{x})$, might not be preserved.

This final theorem, which includes as special cases all the previous corollaries following from Theorem 68 on Page 189, could have been proven earlier, but drops the addition preservation. Yet it allows one to pick *any prime number* for the characteristic of the field.

Theorem 72 (Universal Mapping Theorem). *Any map from a finite set S to a finite set T can be written as a polynomial system of equations over $\mathrm{GF}(p)$, for any prime p.*

Proof. Select a prime p. Let n be a positive integer such that $p^n \geq |S|$ and likewise m such that $p^m \geq |T|$. Label the elements of S with elements from $\mathrm{GF}(p^n)$ and the elements of T with elements from $\mathrm{GF}(p^m)$. For the remaining elements of $\mathrm{GF}(p^n)$, select any values from $\mathrm{GF}(p^m)$ that you like as outputs. Proceed as in the previous corollary. $\qquad\square$

And thus, any map, from any finite set to any other finite set, can be written as a polynomial system of equations over $\mathrm{GF}(p)$, for any prime p. This is in stark contast to $\mathbb{Z} \to \mathbb{Z}$, where for example

$$f(x) = \begin{cases} 1 \ x \text{ if is a perfect square} \\ 0 \ \text{ otherwise} \end{cases}$$

which cannot be a polynomial, because it has infinitely many zeros. In $\mathbb{Z}[x]$, only the zero polynomial has infinitely many zeros.

11.3 Polynomials over $\mathbb{GF}(2)$

It is interesting to reflect on what polynomials over $\mathbb{GF}(2)$ look like, and what simple notions like equality signify.

11.3.1 Exponents: $x^2 = x$

First, since $1 \times 1 = 1$ and $0 \times 0 = 0$ then $x^2 = x$ for all elements of $\mathbb{GF}(2)$. Likewise, $x^k = x$ for all $k > 0$. Therefore, for any given polynomial in n variables, if we change all the non-zero exponents to one, we will not change its value on any of the 2^n possible inputs.

11.3.2 Equivalent versus Identical Polynomials

Normally, when cryptographers discuss $\mathbb{GF}(2)$-polynomials, they would consider any two polynomials with n variables to be equal if they agree on all 2^n possible inputs. (They might not match on inputs from extension fields, but that is another matter entirely). Mathematicians sometimes point out that x^2y^2 and xy, for example, are in fact distinct polynomials. There are two ways around this dilema.

First, define a relation $\approx$ on the set of $\mathbb{GF}(2)$-polynomials, with $f \approx g$ if and only if they have the same variables, and they agree on all inputs (i.e. $f(x) = g(x)$ for all x in the domain). This map is obviously reflexive and symmetric. The transitive property follows from the transitive property of equality. Therefore this is an equivalence relation. Throughout this book, we will use $=$ to mean precisely this equivalence, denoted $\approx$ here. Thus over $\mathbb{GF}(2)$, in this book, $x^2y^2 = xy$.

The other way is to simply "mod out" by the non-zero polynomials which are zero on all their inputs. Call these polynomials "zero-like." These are $x^2 - x$, $y^2 - y$, et cetera.... Thus, strictly speaking, when working with n-dimensional $\mathbb{GF}(2)$-polynomials in this book, we are not working in $\mathbb{GF}(2)[x_1, \ldots, x_n]$ but rather in

$$\mathbb{GF}(2)[x_1, \ldots, x_n] / \left(x_1^2 - x_1, x_2^2 - x_2, \ldots, x_n^2 - x_n \right)$$

The set that we are "mod"ing out by is the set of all polynomials which evaluate to zero on all their inputs—the zero-like polynomials. It is clear that the sum of two of zero-like polynomials also evaluates to zero on all inputs, and the product of a zero-like polynomial with any other polynomial will also evaluate to zero on all inputs. Thus the set of zero-like polynomials is an ideal, and we are permitted to "mod" out by it, and still have a ring.

The ideal is not prime, because $x - 1$ is not zero-like, and x is not zero-like, but $x(x-1) = x^2 - x$ is zero-like. Thus our ring has zero-divisors, including x and $x - 1$.

11.3.3 Coefficients

It should be noted that since the only non-zero element of $\mathbb{GF}(2)$ is 1, there is never a need to write a coefficient. If the coefficient of a monomial is zero, we omit the monomial, as always. If it is one, we need not write one, of course.

Thus a polynomial is either a sum of monomials, where each monomial is a product of variables, or zero, with the possibility of a "plus 1" added to either case. For example, $x + yz + wxz + 1$.

11.3.4 Linear Combinations

Since the only possible coefficients are 1 or 0, the concept of linear combination is a bit peculiar. The set of linear combinations of n polynomials will consists of assignments of coefficients (1 or 0) to these n polynomials, which are then added together. Some of the n polynomials will get 0, which we ignore, and others will get 1. Thus we can identify the linear combination itself with the subset of the n polynomials which happen to get coefficient 1.

Therefore, there are as many linear combinations of n polynomials as there are subsets of those n polynomials, which is, of course, a total of 2^n choices. It turns out that one can rapidly enumerate these $2^n - 1$ non-zero polynomials. In fact, this will be a crucial step in the Method of Four Russians, explained in Section 9.2 on Page 135.

11.4 Degree Reduction Techniques

Unlike most of this book, this section should be considered over any field whatsoever. This includes $\mathbb{R}$, $\overline{\mathbb{Q}}$, $\mathbb{C}$, $\mathbb{Q}$, or the field of rational functions over those fields, or any other field.

When solving polynomial systems of equations, over finite fields or over the rational numbers, the number of equations, number of variables, and the maximum degree are the crucial measures of difficulty. This will be discussed in more detail in Section 11.6 on Page 203. However, we prove below that one can write a second system of equations, with a solution set in bijection with the first, so that the degree of all equations in the second system is at most two. Furthermore, we give an algorithm which accomplishes this conversion. We also show that the number of new equations and new variables added to the system to accomplish this, is polynomial in the number of original variables, for any fixed maximum degree of the original system.

All this is accomplished by using the very old trick: $w = abcd$ if and only if $x_1 = ab$, $x_2 = x_1 c$, and $x_3 = x_2 d = w$. And so, if we add the variables x_1, x_2, x_3 to the system, and those three equations, we can substitute x_3 anywhere when we see w. Repeatedly doing this introduces no spurious solutions, destroys no solutions, and eventually drops the degree to two, which we will prove shortly.

This technique undoubtedly has been known for a long time. At the least, it is used in [33]. But we are not aware of any proof or complexity calculations. For an example of this algorithm "in action," see Section 2.7 on Page 14.

11.4.1 An Easy but Hard-to-State Condition

Consider an Algorithm A which takes a system of equations as an input, and outputs a new system of equations (with m' equations in n' unknowns instead of m equations and n unknowns). It will also output a function ϕ, such that $\phi : \mathbb{F}^{n'} \to \mathbb{F}^n$. It is required that if x' is a solution of the new system then $\phi(x')$ is a solution of the old system. Likewise the algorithm outputs a ψ such that if x is a solution to the old system of equations, then $\psi(x)$ is a solution to the new system of equations. Furthermore, both ϕ and ψ must be injective.

Now suppose Algorithm A converts the second system to a third system in m'' equations and n'' unknowns, with a function ϕ' such that for any solution x'' to the third system, $\phi'(x'')$ is a solution to the second system. And it outputs a ψ' such that if x' is a solution to the second system then $\psi'(x')$ is a solution to the third system.

It is clear that if x'' is a solution to the third system, then $\phi(\phi'(x''))$ is a solution to the first system. Likewise, if x is a solution to the first system that $\psi'(\psi(x))$ is a solution to the third.

So if we construct an Algorithm A that meets the specified conditions, then we can freely apply that Algorithm as many times as we like. We could say that "repeated evaluations of Algorithm A introduce no spurious solutions and destroy no solutions," because the solutions of the first are in bijection with the solutions of the last.

We have proven:

Theorem 73. *If Algorithm A takes as input a system of m polynomial equations in n unknowns, over $\mathbb{F}$, and outputs*

- *a system of m' equations in n' unknowns, also over $\mathbb{F}$, and*
- *an injective map $\phi : \mathbb{F}^n \to \mathbb{F}^{n'}$ such that if x solves the old system, then $\phi(x)$ solves the new system, and*

- *an injective map* $\psi : \mathbb{F}^{n'} \to \mathbb{F}^n$ *such that if* x' *solves the new system then* $\psi(x')$
  solves the old system,

then repeated evaluations of Algorithm A neither create spurious solutions nor destroy solutions. That is to say that the solution set, regardless of the number of evaluations of A, is in bijection with the original solution set.

11.4.2 An Algorithm that meets this Condition

Algorithm A works as follows. If no equation has a monomial of degree greater than 2, then terminate. Otherwise, call this monomial w. For all equations, in the system, we can write

$$f_i(x_1, \ldots, x_n) = g_i(x_1, \ldots, x_n) + c_i w$$

for some constant c_i, in such a way as the monomial w does not appear in g_i. If w already does not appear in f_i, we choose $c_i = 0$ otherwise we choose c_i to be the coefficient of w in f_i. Then $g_i = f_i - c_i w$ and so it will be the case that w does not appear in g_i

Let the degree of w be d, and recall $d \geq 3$, because that is how we chose w. Then $w = x_{a_1} x_{a_2} \cdots x_{a_d}$. Note the $a_1, \ldots, a_d$ are integer indexes, in the range $[1, n]$. We introduce the following new variables, $r_1, \ldots, r_{d-2}$ as dummy variables.

And we create the following new equations:

$$0 = x_{a_1} x_{a_2} - r_1$$
$$0 = r_1 x_{a_3} - r_2$$
$$0 = r_2 x_{a_4} - r_3$$
$$\vdots \quad \vdots \quad \vdots$$
$$0 = r_{d-4} x_{a_{d-2}} - r_{d-3}$$
$$0 = r_{d-3} x_{a_{d-1}} - r_{d-2}$$

and furthermore note that $w = r_{d-2} x_{a_d}$.

We will append the first $d - 2$ of these to the system of equations, and use $r_{d-2} x_{a_d}$ in place of w. This will introduce no terms of degree 3 or higher. We do not append the equation $w = r_{d-2} x_{a_d}$. Also, these equations are all satisfied if and only if $r_{i-1} x_{a_{i+1}} = r_i$. This means all can be satisfied if and only if

$$x_{a_1} x_{a_2} = r_1; \quad r_1 x_{a_3} = r_2; \quad r_2 x_{a_4} = r_3; \quad \cdots; \quad r_{d-2} x_{a_d} = w$$

By substitution, this means that the equations can only be satisfied if

$$w = x_{a_1} x_{a_2} x_{a_3} \cdots x_{a_d}$$

Therefore it is safe to substitute the $+c_{i,j}w$ with $+c_{i,j}r_{d-2}x_{a_d}$. This is because in any satisfying solution, $r_{d-2}x_{a_d} = w$, and the other equations are unchanged.

In the input we had m equations. We now have $m+d-2$ equations: The first m equations are the $f_1(x) = 0, f_2(x) = 0, \ldots, f_m(x) = 0$, and the next $d-2$ are those written above. Previously, we had n variables. We now have $n+d-2$ variables, the additional $d-2$ unknowns being the r's.

Suppose $x = (x_1, x_2, \ldots, x_n)$ is a solution to the old system. Then let $r_1 = x_{a_1}x_{a_2}$ and $r_i = r_{i-1}x_{a_{i+1}}$, for $i = 2, 3, \ldots, d-2$. Then define $\psi(x) = x' = (x_1, x_2, \ldots, x_n, r_1, r_2, \ldots, r_{d-2})$. Clearly, $\psi(x)$ is a solution to the new system of equations.

If $x' = (x_1', x_2', \ldots, x_{n+d-2}')$ is a solution to the new system, then simply lop off the last $d-2$ values and let $\phi(x') = (x_1', x_2', \ldots, x_n')$. Since x' satisfied all the new equations, including the first m of them, then it satisfies the old equations too, because *the old equations are* the first m of the new equations. Note this is only possible because the first m equations of the new system (i.e. the equations of the old system) do not contain the variables $r_1, r_2, \ldots, r_{d-2}$. Therefore those m equations "do not care" what values were assigned to those variables.

The injectivity of ϕ is clear, because if $\phi(x') = \phi(y')$, then the first n terms of x' must equal the first n terms of y', because that is what ϕ does (truncation). The other $d-2$ terms are calculated soley from these, and so they are also equal. Therefore $x' = y'$.

The injectivity of ψ is even simpler. If $\psi(x) = \psi(y)$, then in particular $\psi(x)$ and $\psi(y)$ agree on the first n terms. But the first n terms of $\psi(x)$ are all n terms of x, and the first n terms of $\psi(y)$ are all n terms of y. Thus $x = y$.

Finally, note that the number of cubic or higher degree monomials has dropped by one since w was removed.

Since there are finitely many equations, and finitely many monomials per equation, the entire system has finitely many cubic or higher degree monomials. Thus the algorithm terminates in finitely many steps.

11.4.3 Interpretation

So we have found such an Algorithm A that satisfies our theorem. If z distinct cubic and higher degree monomials occur in the system of equations, then observe that each iteration of the algorithm reduces by exactly 1 the number of cubic and higher degree monomials. So after z iterations, there will only be terms of degree 2 or 1, and constants. By the theorem, any solutions to the outputted system will give rise to solutions to the original system (via the ϕ's), and vice versa (via the ψ's). No solutions are created and destroyed.

The number of equations and variables added is $D - 2z$, where D is the sum of the degrees of all of the cubic and higher monomials that exist somewhere in the system, never counting the same monomial twice. Furthermore, z is the number of such unique cubic and higher monomials.

11.4.4 Summary

Surely, this algorithm will be useful to those who use Gröbner Basis algorithms to solve systems of polynomial equations, because many such methods have running time dependent on the degree of the system of equations. But more importantly, we have proven the theorem

Theorem 74. *If $f_1(x_1,\ldots,x_n) = 0, f_2(x_1,\ldots,x_n) = 0,\ldots,f_m(x_1,\ldots,x_n) = 0$ is a system of equations, then there is another system of equations $g_1(x_1,\ldots,x_{n'}) = 0, g_2(x_1,\ldots,x_{n'}) = 0,\ldots,g_{m'}(x_1,\ldots,x_{n'}) = 0$ with solutions in bijection with the system of the f's, but with degree at most 2.*

Let the number of monomials of degree 3 or higher which have non-zero coefficient somewhere in the system of the f's be z. Let the sum of their degrees be D. Then $m' \leq m + D - 2z$ and $n' \leq n + D - 2z$.

11.4.5 Detour: Asymptotics of the "Choose" Function

To properly calculate the complexity of this algorithm, we need the following lemma.

Lemma 75. *The value of $\binom{a}{b}$, for sufficiently large a is approximately*

$$\frac{1}{\sqrt{2\pi b}} \left(1 + \frac{b}{a-b}\right)^{a-b+1/2} \left(\frac{a}{b}\right)^b$$

Proof. Note, that Sterling's Approximation is $x! \approx x^x e^{-x} \sqrt{2\pi x}$. With that in mind

$$
\binom{a}{b} = \frac{a!}{(a-b)!\,b!}
$$

$$
= \frac{a^a}{(a-b)^{a-b} b^b} \frac{e^b e^{a-b}}{e^a} \frac{\sqrt{2\pi}}{\sqrt{2\pi}\sqrt{2\pi}} \sqrt{\frac{a}{(a-b)b}}
$$

$$
= \frac{a^{a-b} a^b}{(a-b)^{a-b} b^b} \frac{1}{\sqrt{2\pi}} \sqrt{\frac{a}{(a-b)b}}
$$

$$
= \left(\frac{a}{a-b}\right)^{a-b+1/2} \left(\frac{a}{b}\right)^b \frac{1}{\sqrt{2\pi b}}
$$

$$
= \frac{1}{\sqrt{2\pi b}} \left(1 + \frac{b}{a-b}\right)^{a-b+1/2} \left(\frac{a}{b}\right)^b
$$

$\square$

Corollary 76. *In the limit as $a \to \infty$,*

$$\binom{a}{b} = \frac{1}{\sqrt{2\pi b}} \left(1 + \frac{b}{a-b}\right)^{a-b+1/2} \left(\frac{a}{b}\right)^b = \frac{1}{\sqrt{2\pi b}} \left(\frac{a}{b}\right)^b \frac{e^b}{\sqrt{1-b/a}} = \frac{e^b}{\sqrt{2\pi b}} \left(\frac{a}{b}\right)^b$$

Proof. The result follows from

$$\left(1 + \frac{b}{a-b}\right)^{a-b} = e^b$$

in the limit as $a - b$ goes to infinity but b remains constant. However, this limit is not a good approximation except for extraordinarily large a, and so we recommend using the lemma instead of the corollary except when working asymptotically. □

Corollary 77. *For a fixed b,*

$$\binom{n}{b} \sim \frac{e^b}{b^b \sqrt{2\pi b}} n^b = \Theta(n^b)$$

Proof. Follows directly from previous corollary. □

11.4.6 Complexity Calculation

With this accomplished, note that if the maximum degree of the entire system of f's is Δ, then

$$\binom{n}{\Delta} + \binom{n}{\Delta - 1} + \binom{n}{\Delta - 2} + \cdots + \binom{n}{3} = Z$$

is the maximum value of z.

The proper estimation of that sum, keeping both n and Δ free, requires the use of hypergeometric series. Thus it is hard to estimate. However, we can consider the question of what values of i will make $\binom{n}{i}$ as large as possible.

If one writes Pascal's Triangle, one sees that this occurs for even n at $\binom{n}{n/2}$. For odd n, there is a tie at $\binom{n}{(n-1)/2}$ and $\binom{n}{(n+1)/2}$. Thus we can generalize for all n, that the optimal is at $\binom{n}{\lfloor (n+1)/2 \rfloor}$.

And the series surely contains $\Delta - 2$ values. Since we have identified the largest member, we can obviously (and this is a grotesque over-estimate) state that the sum is less than $(\Delta - 2)$ times the largest element.

Normally, if $\Delta \ll n$, which would almost certainly be the case in any rational example, we can estimate

$$D = \Delta Z = \Delta (\Delta - 2) \binom{n}{\Delta}$$

and be rather confident. However, if $\Delta > (n+1)/2$ then it is not $\binom{n}{\Delta}$ which is the largest member of the series.

And we split into two cases. Either Δ is fixed, or it is not. If it is fixed, then as n goes to infinity, then surely $\Delta \leq (n+1)/2$. With this in mind, $D = \Delta Z = \Delta(\Delta - 2)\binom{n}{\Delta} \in \Theta(n^\Delta)$ and the algorithm is polynomial time. If Δ is not fixed, then we have to consider the possibility that $\Delta = n/2$. This, in turn, will be $\Theta(e^{n/2} n^{n/2})$ which is most decidedly *not* polynomial time.

Thus the "simple degree dropper" algorithm is polynomial time, for any fixed Δ, and not polynomial time if Δ is not fixed. This is usually stated as "the algorithm is pseudo-polynomial time in Δ."

11.4.7 Efficiency Note

Suppose $w = x_2 x_3 x_5 x_7$ and another monomial $v = x_1 x_3 x_5$ appears somewhere in the system. If w is processed first, we will have new equations such that $r_1 = x_2 x_3$, $r_2 = r_1 x_5$, and $r_3 = r_2 x_7$. None of these can be substituted into v.

Instead, suppose we were to reinterpret w as $w = x_3 x_5 x_2 x_7$ which is obviously the same thing since fields are commutative in both operations. Then $r_1 = x_3 x_5$, $r_2 = r_1 x_2$, and $r_3 = r_2 x_7$. But now $v = x_1 r_1$. So we could remove two monomials of degree greater than 2 in one change, saving a step and several equations. Since this also reduces the number of dummy variables introduced as well, its very useful to take advantage of such opportunities if they arise.

Therefore, the algorithm we showed above, which we name the "simple degree-dropper algorithm" should not be considered optimal. However, we present a variant, using the Greedy Algorithm, which should work quite fine. For differentiation, we called it the "greedy degree-dropper algorithm."

11.4.8 The Greedy Degree-Dropper Algorithm

Make an array of counters for every possible pair of variables $x_i x_j$, with $i \leq j$. For every equation, and for every monomial of degree 2 or higher $x_{a_1} x_{a_2} \cdots x_{a_d}$, increment the counter for $x_{a_i} x_{a_j}$ for all $i < j$. Thus $(d-1) + (d-2) + \cdots + 2 + 1 = d(d-1)/2$ increments are made. Also, an array of flags, initially all set to false, should exist for all possible pairs $x_i x_j$, with $i \leq j$. The flag will be set to true if this pair ever appears in any monomial of degree 3 or higher.

Take the most common pair $x_i x_j$, such that the flag is true. Then introduce a new variable $y = x_i x_j$, and substitute it where ever $x_i x_j$ is found, either by itself, or inside any monomial.

The purpose of the flag is that $x_i x_j$ might be very common but only appear by itself and never in a cubic or higher degree monomial—in which case, we would not want to eliminate it at the cost of a new variable.

Note that if there is a quintic monomial, it will only drop to quartic in this step if it drops at all. Another step will take it to cubic, and a third to quadratic. But, this is at worse $d - 2$ times as much effort as before, as the example of a quintic monomial shows. Therefore, the "greedy degree-dropper" algorithm is also polynomial time, as it will take Δ times as many passes as the simple one, at worst. Nonetheless, it is obvious this will almost always do better than the simple algorithm.

11.4.9 Counter-Example for Linear Systems

So since we can reduce any polynomial system of equations to a larger one that is degree two, the natural question is if we can drop to degree 1.

If $P \neq NP$ the answer is clearly no, because solving a polynomial system of equations over a finite field is NP-Hard. Solving a linear system is possible in polynomial time—in fact, cubic time or better.

Of course, if there were exponentially many equations, or even super-polynomially many equations, then this argument fails.

11.5 NP-Completeness of MP

This section has been written is a slightly more rigorous tone than the rest of the book, and the author apologizes, but one must sometimes be very careful when making complexity claims.

The problem of solving a polynomial system of equations, over a finite field or over the rationals, is NP-Hard. For shorthand, when the degree is exactly two, we write this problem as MQ, and when the degree is at least two, as MP. Finally, if the degree is exactly 3, we write MC. We signify the associated decision problems, i.e. does this system have a solution or not, as MQD, MPD, and MCD, respectively.

For the decision problems, their membership in NP is obvious because given any particular solution, then one could simply plug the solution into all the variables, and see if it is indeed a solution, or not. This would be a very fast operation, and certainly polynomial time.

We will now prove some theorems and corollaries about MP, MC, and MQ, and their associated decision problems. But first, it is important to grasp that unless $P = NP$, which would be a surprise, this means that these problems will never have a polynomial time algorithm for solving them. Thus, as researchers, we must focus on special cases, randomized algorithms, moderately fast non-polynomial time algorithms, heuristics, and approximations.

Theorem 78. *The problem of MC, i.e. detecting whether a system of cubic polynomial equations over $\mathbb{GF}(2)$, has a solution in the base-field, is NP-Complete.*

Proof. The proof will proceed by assuming that we have a black-box that can detect if a cubic system of polynomial equations, over $\mathbb{GF}(2)$, has a solution. The black-box runs in polynomial time. We will write a converter to use this black-box to solve the 3-CNF SAT problem, which is known to be NP-Complete. The converter runs in polynomial time, and so MC is NP-Complete.

The SAT problem is as follows. One is given a logical expression in the five operators of predicate calculus $(\wedge, \vee, \sim, \Rightarrow, \Longleftrightarrow)$ but not the existential or universal quantifiers $(\exists, \forall)$. Then, one is asked if the logical expression has a setting for each of its variables, such that the entire expression evaluates to "true."

The 3-CNF SAT problem has the logical sentence written as a large conjunction (logical-AND). Each element of the conjunction is called a clause and consists of a disjunction (logical-OR) of three variables. The three variables can each be negated or not. Obviously, all clauses must evaluate to true for the conjunction (logical-AND) to be true. The reader may be interested to know that every SAT problem can be rewritten as a 3-CNF SAT problem [63, Ch. 34.4].

Given a 3-CNF problem, write the clauses in the form $(v_1 \vee v_2 \vee v_3)$ with v_i being either some variable x_j or its negation $\sim x_j$. Then we can write a cubic system of equations as follows.

We will write one equation for each clause by noting the following tautology:

$$
\begin{aligned}
(a \vee b \vee c) \quad &\Leftrightarrow \quad ((a \vee b) \wedge c) \oplus (a \vee b) \oplus c \\
&\Longleftrightarrow ((a \wedge c) \vee (b \wedge c)) \oplus (a \vee b) \oplus c \\
&\Longleftrightarrow ((a \wedge c \wedge b \wedge c) \oplus (a \wedge c) \oplus (b \wedge c)) \oplus ((a \wedge b) \oplus a \oplus b) \oplus c \\
&\Longleftrightarrow (a \wedge b \wedge c) \oplus (a \wedge c) \oplus (b \wedge c) \oplus (a \wedge b) \oplus a \oplus b \oplus c \\
&\Longleftrightarrow (abc + ac + bc + ab + a + b + c) = 1 \\
&\Longleftrightarrow abc + ac + bc + ab + a + b + c + 1 = 0
\end{aligned}
$$

Furthermore, if a were negated, substituting $1 + a$ for a would not change the degree of that polynomial, likewise for b and c. Thus each clause becomes one cubic polynomial equation. The number of variables is unchanged. And this is clearly a linear-time conversion, compared to the number of clauses. The increase in length is also linear, obviously (since it is written in linear-time).

The black-box now is given the cubic system of equations. If the system has a solution, then each polynomial is satisfied, and thus each clause is satisfied. Therefore, the original logical sentence is satisfied. Likewise, if there were a solution to the logical problem, then that would be a solution to the polynomial system. Thus the black-box always gives the right answer. □

Corollary 79. *The problem of MP, i.e. detecting whether a system of polynomial equations, over $\mathbb{GF}(2)$, has a solution in the base-field, is NP-Complete.*

Proof. The problem of MP contains the problem of MC. Any MP solving oracle solves MC. The problem of MC is NP-Complete. Thus any polynomial-time machine that could solve MP would solve an NP-Complete problem in polynomial time. Therefore, MP is NP-Complete. $\square$

The next one is non-trivial. It uses the result from Theorem 74 on Page 196, that every polynomial system of equations, for any fixed degree, can be rewritten as having degree two, with only polynomial time and polynomial growth. If the degree of the initial system is not fixed, then the conversion is not polynomial time.

Corollary 80. *The problem of MQ, i.e. detecting whether a system of quadratic equations over $\mathbb{GF}(2)$, has a solution in the base-field, is NP-Complete.*

Proof. Imagine one has a black-box that can solve MQ in polynomial time.

Suppose one has a problem of type MC. Use the technique Theorem 74 on Page 196 to rewrite it as an only polynomially-larger problem of type MQ. Since the original is of fixed degree (fixed at 3), this is a polynomial-time step. Then use the black-box to detect if the system of equations has a solution. This process will take, in its entirety, polynomial time. Thus, since MC is NP-Complete, likewise so is MQ. $\square$

Corollary 81. *The problem of finding a base-field solution to a polynomial system of equations over $\mathbb{GF}(2)$, i.e. MP, is NP-Hard.*

Proof. A black box that could do this could solve MP by a simple algorithm. Given a problem, ask the black box what a solution is. If it responds with the null-set, then there is no solution. If it responds with a solution, then a solution exists. Since a polynomial time algorithm for MP would solve an NP-Complete problem (MPD) in polynomial time then MP is NP-Complete. $\square$

Corollary 82. *The problem of finding a base-field solution to a cubic system of equations over $\mathbb{GF}(2)$, i.e. MC, is NP-Hard.*

Proof. Same proof as Corollary 81 on Page 201. $\square$

Corollary 83. *The problem of finding a base-field solution to a quadratic system of equations over $\mathbb{GF}(2)$, i.e. MQ, is NP-Hard.*

Proof. Same proof as the Corollary 81 on Page 201. $\square$

blarg

Lemma 84. *If there is an algorithm to solve a polynomial system of equations over the field $\mathbb{GF}(q)$, in polynomial time, then there algorithms to solve a polynomial systems of equations for every other finite field as well, and they all run in polynomial time.*

Proof. A system of equations over a finite field $\mathbb{F}$ with n variables and m equations can be thought of as a map from $\mathbb{F}^n$ to $\{1, 0\}$. If a particular assignment $\mathbf{x} \in \mathbb{F}^n$ of values to the n variables from will satisfy the system, then map that $\mathbf{x}$ to 1. If it will not satisfy, then map it to 0.

This is a map from a finite set to a finite set. Then by using Theorem 72 on Page 190, we can write it as a polynomial system over $\mathbb{GF}(p)$, for any p. We must be careful, however. The size of the new system must be upper-bounded by some polynomial $f(s)$, where s was the size of the old polynomial system. A careful reading of the proofs leading to that theorem will reveal that this is clearly the case.

Therefore, we merely choose $p = q$ and solve our system over $\mathbb{F}$ by converting to one over $\mathbb{GF}(p)$. $\square$

Theorem 85. *The problem of finding a base-field solution to a polynomial system of equations over any particular finite field, is NP-hard.*

Proof. This follows directly from Lemma 84 on Page 201 and Corollary 81 on Page 201. $\square$

Theorem 86. *The problem of detecting if a base-field solution exists for a polynomial system of equations, over any particular finite field, is NP-Complete.*

Proof. This follows directly from Lemma 84 on Page 201 and Corollary 79 on Page 200. $\square$

One Last Interesting Thought

Suppose that detecting if a quadratic system of equations over $\mathbb{GF}(2)$ has a solution or no solutions, could be done in time $f(n)$, where n is the number of variables. Then, if the system has a solution, one solution can be found in at most $(n+1)f(n)$ time. At start, we run the $f(n)$ time algorithm, and verify that a solution exists. Now, we proceed as follows.

First, assume that the first variable is one, and substitute accordingly. Then check for the existence of a solution. If the system has become unsolvable, then you know that the first variable must be 0 in any valid solution. If the system remains solvable, then you know that there is some solution with the first variable equal to a 1. In either case, substitute accordingly, and repeat this process, now with one fewer variable. Since there are n variables, then clearly a total of $n+1$ calls would be needed.

In practice, SAT-solvers as described in Chapter 14, could be this $f(n)$ time black-box, but instead they find a satisfying solution whenever a solution exists. So there is no use for this method at this time.

But it is possible to imagine a black-box that would simply report "solvable" or "unsolvable" for any quadratic system of equations over $\mathbb{GF}(2)(2)$, at least as a thought-experiment. It is important to realize that with only slightly more effort, such a black-box would solve the system of equations also.

For other finite fields, of size q, then $(q-1)nf(n)$ calls would be required to the $f(n)$ time algorithm, for reasons that are perhaps obvious.

11.6 Measures of Difficulty in MQ

Strictly speaking, the complexity of a problem is normally expressed as a function of the length of its inputs when encoded into binary. For example, this is the strict standard used in determining if an algorithm is polynomial time or not. However, this is not how most complexity expressions are written. For example, an $n \times n$ matrix has n^2 elements and its inversion via adjoining an identity matrix and performing Gaussian Elimination is $O(n^3)$. We do not say that an "n element square matrix" takes $O(n^{3/2})$ time to invert and $O(n)$ time to negate, for example, but these would be correct statements, in that world view. The metric of how many rows and columns a matrix has is much more useful to us, and so it is used as the measure for dense linear algebra

With that in mind, we can consider what metrics are appropriate for polynomial systems. The number of equations and the number of variables is key, as we will see in the description of Linearization in Section 12.3 on Page 211. Throughout this book we have been using m for the number of equations and n for the number of unknowns, as a parallel to the usage for a linear system induced by an $m \times n$ matrix, and an m dimensional vector of constants.

In the next section, we will discuss the role of m/n, often denoted c or γ.

11.6.1 The Role of Over-Definition

Normally, a system of equations is said to have m equations and n unknowns. However, sometimes it is useful to think of this as nc equations and n unknowns, where obviously, $c = m/n$. This is sometimes denoted γ, to contrast with β, the sparsity of the system.

Over-defined equations can have several consequences. First, if $c \approx n$, which means that $m \approx n^2$, then Linearization will likely solve the system. See Section 12.3 on Page 211. Also, if $c \approx 1/n$, or $m \approx 1$, other techniques might solve the equation [164, Ch. 4.3]. In general, a high c makes SAT-Solver based attack easier as well. See Chapter 13.

However, this is not the complete picture. To see this, we will do an extreme example.

11.6.2 Ultra-Sparse Quadratic Systems

Suppose a quadratic system of m equations, on n variables, has only $\alpha \log n$ unique degree 2 monomials with non-zero coefficients. Here α is a constant (or possibly upper-bounded by a constant in n). Let us call such a system hypersparse. Those $\alpha \log n$ unique monomials, if viewed as totally unrelated to each other, such as after a linearization, have $2^{\alpha \log n} = n^\alpha$ possible values. This is highly pessimistic of course, as there might be much smarter ways to iterate through all possibilities, but this is only an example. Also, observe for random n, each is 0 with probability $3/4$. So surely if one wishes to iterate, one should do it with the lowest-weight guesses first (i.e., guesses where most monomials are zero).

For any particular guess as to the values of the quadratic monomials, only linear monomials will remain, and a linear system in m equations and n unknowns can be solved in time $\Theta(mn\min(m,n))$ using only Gaussian Elimination.

The total running time is therefore at worst $\Theta(mn^{\alpha+1}\min(m,n))$. Since α is a constant, or upper-bounded by a constant, then this is polynomial time! Using the c notation from before, this is $\Theta(cn^{\alpha+2}\min(c,1))$.

To see that such a system is not unreasonable, note that $\alpha\log n$ is not the number of times a quadratic monomial appears in the system, but the number of unique monomials appearing. A monomial cannot be repeated in one equation but it can be repeated between equations. If all the monomials appeared in each equation, the sum of two equations would be linear (in the $\mathbb{GF}(2)$ case), so this is not a hard problem. But it would be difficult to solve if $3/4$ of them randomly appeared in each equation.

The key to this paradox is to realize that the system is very sparse. If there are n variables, then there are $(n^2+n)/2$ columns for quadratic and linear terms after linearization, plus one for constants. The sparsity β, which is the fraction of entries which are non-zero, is

$$\beta = \frac{4n+3\alpha\log n}{2n^2+2n+1} \approx \frac{1}{2n}$$

if we assume any particular 3/4ths of the unique monomials will appear in any given equation.

This technique was thought up by the author but it appears that it was independently thought up by Alexander Maximov and Alex Biryukov [175], in their paper "Two Trivial Attacks on Trivium." There are very few quadratic terms in that system of equations, and so this makes sense. See that paper for details.

An Interesting Observation

Amusingly if $\alpha = (\log n) - 2$, and $c = 1$ then the running time becomes $n^{\log n}$. This function is assymptotically greater than all polynomials, regardless of degree. It is assymptotically less than all exponential functions, $c_1(c_2)^{p(n)}$, where $p(n)$ is a polynomial (and $c_2 > 1$).

To see that

$$n^{c_0} < n^{\log n} < c_1(c_2)^{p(n)}$$

for all positive c_1, c_2, c_3, simply take the log of both sides and obtain

$$c_0\log n < (\log n)^2 < (\log c_1) + p(n)\log c_2$$

which is obviously true. There are many functions which are strictly assymptotically between polynomial and exponential. But many computer scientists speak of the polynomial-exponential divide as if there were nothing between the two classes, or as if no serious algorithm for a real-world problem has that running time.

11.6.3 Other Views of Sparsity

Connection To Linear Sparsity

Sparsity is important for other reasons. First, sparse linear systems are much easier to solve than dense systems of the same size. Since linear algebra is used at many stages of solving polynomial systems, including Linearization, XL, and also various steps when finding Gröbner Bases, we anticipate that sparse systems are easier to solve as a result. In part, this was shown, in part, in Section 11.6.2 on Page 203.

Memory Usage

Quite often, the limiting factor in solving a polynomial system of equations with a Gröbner Bases algorithm is not running time, which can be countered with human patience, but rather memory, especially with MAGMA. The algorithms of J. C. Faugère, called F4 and F5 [109, 110], use much memory, and this resulted in MAGMA frequently crashing in the experiments performed in Chapter 13, and Chapter 15.

SAT Solvers

For reasons not completely well understood, the sparsity of a system of equations was absolutely crucial in the ability of SAT-solvers to find a solution. This too will be detailed in Chapter 13.

11.6.4 Structure

This is perhaps the most under-appreciated factor. We show in Appendix C on Page 315, that a p-coloring can be found by solving a polynomial system of equations. This is no surprise, because MP is NP-Complete and all NP-Complete problems are reducible to each other, including graph coloring.

Graph coloring, as well as other graph theoretic problems, such as Maximum Clique, Maximum Independent Set, and Minimum Vertex Cover, are NP-Complete, but in the special case of bipartite they become polynomial time problems [63, Ch.

34.5]. Likewise the Maximum Cut problem is NP-Complete, but becomes polynomial time for planar graphs.

For example, any bipartite graph can be n-colored for all $n > 1$, unless it has no edges, in which case for $n \geq 1$. The maximum size of any clique in a bipartite graph is 2, unless it has no edges, in which case it is 1. Obviously these are trivial cases. The Minimum Vertex Cover case is not trivial, as it relates to König's Theorem, which states that in any bipartite graph, the number of edges in a maximum matching equals the number of vertices in a minimum vertex cover [168]. The matching can be found in polynomial time.

11.7 The Role of Guessing a Few Variables

While this idea is very simple, many researchers do not grasp all of its consequences. Sometimes one will "guess" a few variables of a system of equations. This occurs in at least three settings, and the general term for it is a "guess-and-determine" attack.

First, one might have a system of equations that one cannot break. Suppose there are 100 variables, and 100 equations. If you guess 10 variables, then 2^{10} guesses in worse case are required. It might be that solving the 90 variable system with 100 equations is 1024 times faster than solving the original. Perhaps not. But the "over-definition," earlier denoted by either γ or c, is changing from $c = 1$ to $c = 1.11\cdots$. In the special case of the XL algorithm (see Section 12.4 on Page 213), this is called Fix-XL (see Section 11.7.2 on Page 207 and Section 12.4.4 on Page 217).

Second, in testing algebraic cryptanalysis, one might simulate the above. Usually this is done as follows. First, one generates a random key. Second, one encrypts some plaintexts in that key. Third, one passes the ciphertext-plaintext pairs to an attack algorithm. One can also leak, perhaps, g of the bits of the key as "extra equations" of the form $k_i = 1, k_j = 0$, *et cetera....* Since the progammer knows the key, then he can always guess correctly. However, this also requires the progammer to verify that a false guess results in a rejection of the guess in time shorter than solving the system in the event of a good guess. In this case, if the average running time of a solution with a good guess is t, then an attacker who must truly guess would have worse case running time of $2^g t$, or in average, half that. An example of this is given in Section 2.8 on Page 15.

11.7.1 Measuring Infeasible Running Times

The final use is when a system of equations cannot be solved. For example, it is unlikely that one will find a system of equations to break the Advanced Encryption Standard in a reasonable amount of time. However, since the key can be 128 bits long, suppose one runs the attack and one leaks 112, 111, 110, ..., 96 bits to the attack algorithm. Then one gets t_{112}, t_{111}, t_{110}, ..., t_{96}, the average running times, in each case. Surely,

$$\min_{i \in [96,112]} \frac{2^i t_i}{2}$$

is a good lower bound for the average case running time of an attack on the system. It might be that t_{64} is actually much better, but that it takes far too long to run, and so we, as humans with a finite lifespan, will probably never know that.

The attack is better than brute force if

$$\frac{2^i t_i}{2} < \frac{2^{128} t_{ver}}{2}$$

where t_{ver} is the time for a brute-force attacker to simply verify a potential guess of the key. And this condition is met if $t_i < t_{ver} 2^{128-i}$. Therefore we have demonstrated

Theorem 87. *One can verify that an attack against a cryptosystem is faster than brute-force, even if one cannot determine how long the attack takes.*

For example, suppose $t_{ver} = 10^{-8} = 2^{-26.575}$ CPU seconds on some special machine, optimistic but not unreasonable at the time of the writing of this book (Fall of 2008). Suppose further that when 70 bits of the key are leaked to some sort of algebraic attack algorithm, the attack takes 300 CPU years, or $2^{33.139}$ CPU seconds. Since

$$2^{33.139} < 2^{-26.575} 2^{70} = 2^{43.425}$$

the attack is faster than brute-force. In fact, it is faster by a factor of

$$2^{43.425} / 2^{33.139} = 2^{10.286} \approx 1249$$

Assuming the algebraic attack is amenable to parallelization (which is a huge and possibly very unfair assumption), with a network of 1000 PC's, this algebraic attack could be verified in 109.5 days. This would be a large undertaking, but surely possible. However, the true running time of a real attacker would require 58 bits to be guessed in worse case, or 2^{57} runs of the algorithm on average. This would be $2^{57} \times 300$ CPU years, or 43.23 quintillion CPU years. Even with a trillion CPUs, this would take longer than the age of the universe.

11.7.2 Fix-XL

Nicolas Courtois has shown that guessing bits before solving a system of polynomial equations in many cases is a surprisingly effective way of accelerating the solution of polynomial systems of equations, either by XL or by Gröbner Bases. For example, see [79, 66]. This has sometimes been called Fix-XL, when used with the XL algorithm. This topic, a generalization of the "guess-and-determine" method, will be covered more in Section 12.4.4.

Chapter 12
Algorithms for Solving Polynomial Systems

Because of the previously mentioned NP-Complete status of MQ, the topic of how to actually solve these systems has the disadvantage that truly efficient (i.e. polynomial time) algorithms have not yet been found. In fact, they will never be found if $P \neq NP$. Nonetheless, there are algorithms available for experimentation, many of which we describe below.

12.1 A Philosophical Point on Complexity Theory

It is important to note that statements about assymptotic complexity are merely that: assymptotic. The absence of a polynomial time algorithm for a problem means less if the size of the problem can be bounded above. For example, all known versions of the "simplex" method [90] of solving linear systems of inequalities have worst-case exponential running time [130]. But for any typical medium-sized problem, usually the program simply terminates in a reasonable time (on modern machines, often too short to accurately measure) and outputs a result.

This occurs for three reasons. First, the number of variables in a typical problem is often not large, on the order of a few hundred. Second, the worst-case complexity and average-case complexity are quite distinct in the simplex method [130]. Third, over several decades, the code has become highly optimized.

The topic of solving polynomials over $\mathbb{GF}(2)$ has not had the benefit of the same length of time and number of researchers as linear programming over $\mathbb{R}$. But, several lessons remain. While our problems are, in fact, quite large sometimes reason alone can trim the problem to a smaller analog of itself, such as in Keeloq in the earlier chapters. Average-case analysis, rather than worse-case analysis, is very important. Finally, we should never lose focus of the fact that assymptotic results only matter in the limit, as the number of variables goes to infinity.

G.V. Bard, *Algebraic Cryptanalysis*, DOI: 10.1007/978-0-387-88757-9_12
© Springer Science + Business Media, LLC 2009

12.2 Gröbner Bases Algorithms

The Buchberger Algorithm, for calculating a Gröbner Basis, is an exact algorithm that can solve many problems in the area of polynomial systems. Sometimes it is used for triangulating a system of polynomial equations. This works identically to triangulating a linear system, in that the final equation is in terms of one variable, and each equation above it has one more variable than the equation immediately below it (naturally, some coefficients will be zero, and so some variables may be absent from some equations).

Many other uses of Gröbner Bases algorithms exist. An excellent reference on the topic is [86]. Since this topic has been so extensively covered by others, we will not go into detail here. We strongly encourage the reader to read at least the first half of [86].

12.2.1 Double-Exponential Running Time

Many cryptographers misunderstand the worse-case estimate, of double-exponential running time, for Gröbner Bases algorithms. There exists ideals [123], that have a basis made of polynomials of degree at most d, but whose Gröbner Basis has polynomials of degree

$$O(d^{(\sqrt{3})^n})$$

Those polynomials will have many monomials! Thus the algorithm which computes the Gröbner Basis of that ideal will run in double-exponential running time, or roughly $O(d^{2^{O(n)}})$, compared to the degree d of the input basis. This is Theorem B in [123].

However it is proven in [123], that with the exception of a set of measure zero, the ideal generated by m polynomials with random coefficients, in n variables of degree at most d, has a Gröbner Basis composed of polynomials of degree at most $(n+1)d - n$. This implies a running time of $O(2^{dn})$ at absolute worst, which is much better than double-exponential. Note, the author learned of this from [37], but the original paper goes even further to say that this set of measure zero is actually the complement of a non-empty open set in the Zariski topology—very small indeed. This is Theorem C in [123].

12.2.2 Remarks about Gröbner Bases

In practice, the author recommends that the polynomial system under consideration be written in the language of MAGMA [2] and SINGULAR [9], submitted to each, and then one should wait for the program to either crash due to a lack of memory, or output a result. In other words, the implementations of these algorithms have been done with such care (including hand-optimizations of the compiled code in some cases) that it is unlikely that a "home-grown" implementation will out-perform them. They can safely be viewed as black-boxes.

On the other hand, knowing precisely how an algorithm operates is very use-ful for determining how to prepare inputs to that algorithm. Many details of the implementation of the algorithm can dictate strategies for pre-conditioning. It is very unfortunate that much of the computer algebra software in the community is closed-source, which renders such pre-conditioning and research impossible. Luck-ily, the SAGE project [7], an open source competitor to MAGMA [2], MATLAB [5], MAPLE [3], and MATHEMATICA [4], is open-source and appears to equal or exceed MAGMA's performance in many cases. It should be noted that SINGULAR is also free software, like SAGE, and SAGE includes SINGULAR on shipment.

An important paper, entitled "Why You Cannot Even Hope to Use Gröbner Bases in Public-Key Cryptography—An Open Letter to a Scientist Who Failed and a Challenge to Those Who have Not Yet Failed" [37], published under several pseudonyms, caused many cryptographers to abandon Gröbner Bases approaches. But in reality, anyone who has actually read [37] knows that this paper deals with *constructing a cryptosystem* based on the hardness of Gröbner Bases problems, not *breaking a cryptosystem already constructed*, which is our focus in this book. There-fore, the utility of Gröbner Bases approaches in cryptanalysis is not to be dismissed because of the arguments of that article.

12.3 Linearization

Suppose one has a quadratic system of m equations in n variables. There are $\binom{n}{2}$ possible quadratic monomials, and n possible linear monomials, for a total of $(n^2 + n)/2$. Suppose further that one renames all the monomials with new names, so that each monomial is a unique new variable. This is best illustrated by example:

$$x_1 + x_2 x_3 = 1$$
$$x_1 x_2 + x_1 x_3 + x_1 = 0$$
$$x_2 x_3 + x_2 = 0$$
$$x_1 x_2 + x_1 + x_3 + x_2 = 0$$
$$x_1 + x_1 x_2 + x_3 = 0$$
$$x_2 x_3 + x_1 + x_2 = 1$$

Now apply the renaming: $x_1 = y_1$, $x_2 = y_2$, $x_3 = y_3$, $x_1 x_2 = y_4$, $x_1 x_3 = y_5$, $x_2 x_3 = y_6$. One obtains the following:

$$y_1 + y_6 = 1$$

$$y_4 + y_5 + y_1 = 0$$
$$y_6 + y_2 = 0$$
$$y_4 + y_1 + y_3 + y_2 = 0$$
$$y_1 + y_4 + y_3 = 0$$
$$y_6 + y_1 + y_2 = 1$$

Now suppose there were a solution (x_1, x_2, x_3) to the original polynomial system. It would be trivial to compute the remaining values of y for this solution, as the values of the linear monomials (x_1, x_2, x_3) contain all the information needed. It is also easy to see that the resulting, longer, vector of y's would solve the linear system.

Performing Gaussian Elimination on this system results in

$$y_1 = 1$$
$$y_2 = 0$$
$$y_3 = y_5$$
$$y_4 = y_5 + 1$$
$$y_5 = \text{free}$$
$$y_6 = 0$$

This yields two possible solutions, namely $(1,0,0,1,0,0)$ and $(1,0,1,0,1,0)$. The second one, by inspection, solves the original polynomial system. The first one, however, is absurd. The first three terms dictate that $x_1 = 1, x_2 = 0, x_3 = 0$. But the next term says that $x_1 x_2 = 1$. With $x_2 = 0$, that is not possible. The reason that this can occur is that the linearization process destroys information. In this case, the fact that y_4 cannot be one unless both y_1 and y_2 are one, is no longer represented in the linear system.

Therefore, it is clear that if a solution to the original polynomial system of equations exists, then that solution will be a solution to its linearization (after calculating the correct values for the "new" variables). However, the converse is false. Many solutions to the linear system of equations may be non-solutions to the original polynomial system. These are called spurious solutions to the linear system, to signify that they are not solutions to the non-linear system.

With this in mind, one might ask what is the benefit of linearization? Suppose one knows that a solution exists (this is often the case with cryptanalysis, since a message was indeed sent). We know that linear systems in general, over $\mathbb{Q}, \mathbb{R}, \mathbb{C}$, have either no solutions, one solution, or infinitely many solutions. Linear systems over $\mathbb{GF}(2)$, on the other hand, have either $2^{n'-r}$ solutions or 0 solutions (see Theorem 27 on Page 87), where n' is the number of variables after linearization, and r is the rank of the set of equations.

Since the polynomial system of equations has a solution by assumption, and since linearization destroys no solutions, we know there is at least one linear system solution, so the linear system then definitely has exactly $2^{n'-r}$ solutions. If the rank of the set of equations is high enough, i.e. $r \approx n'$, then there will be only one solution, or perhaps a very small number of them. One can then check to see which are solutions of the original system. Note, that $r > n'$ is impossible since the column rank is always less than the number of columns, by definition.

Thus if one is given roughly $m = (n^2 + n)/2$ equations, then one can perform this linearization process and get a few "candidate" solutions. Essentially, having $m \approx n' \approx n^2/2$ results in there being "too much" information in the linear system to permit spurious solutions to be numerous.

12.4 The XL Algorithm

The XL algorithm was first mentioned in [82, 64], and is due to Nicolas T. Courtois. Suppose one gives you the polynomial system below, and asks you to find all possible solutions:

$$1 + x + y + z + wz + yz = 0$$
$$x + z + wx + wy + wz + xy + xz + yz = 1$$
$$w + y + wx + xz + yz = 0$$
$$x + wx + wy + wz + yz = 1$$

At first it would seem that linearization would be of no help in this case. There are four variables, so there are 10 monomials, of which 6 are quadratic and 4 are linear. However, we only have 4 equations, far less than 10.

Suppose now that we made the system cubic. At first this seems like a disaster. Cubic polynomials have

$$\binom{n}{3} + \binom{n}{2} + \binom{n}{1} = \frac{n^3}{6} + \frac{5}{6}n$$

possible monomials. In our case, however, there are only $\binom{4}{3} = 4$ additional cubic monomials, and so 14 and not merely 10 equations would be required. The XL algorithm will make the additional needed equations (in this case, we will end with 20, and not 4, equations).

The XL algorithm is actually quite simple and is summarized in Algorithm 22. One elects to increase the degree of the system from degree d to some larger degree D, but usually $D = d + 1$ or sometimes $D = d + 2$. Then one multiplies every equation by every possible monomial of degree $D - d$ or lower. In this case, $d = 2$ and it turns out that $D = 3$ will be sufficient. So we must multiply every equation by all the degree 1 and 0 monomials. In this case, the set of monomials is $\{w, x, y, z, 1\}$. Thus where we had 4 equations, we now have 20. Then, you linearize and solve.

The XL algorithm would, in our example, yield the following. The breaks every 5 equations show the grouping where each of the 4 sets of 5 equations comes from one of the four original equations.

INPUT: A system of m polynomial equations in n unknowns, of degree d.
OUTPUT: A solution or solutions to the system of equations, if the equations have sufficient rank.
1: A human selects a degree $D > d$. Usually $D = d + 1$.
2: Make a list L of all monomials of degree $D - d$ or less, including the monomial 1, which has degree 0.
3: Multiply all equations by every member of L. (Since there were m equations before this step, there are $m|L|$ equations after it).
4: Linearize the system. See Section 12.3 on Page 211.
5: Solve via linear algebra.

Algorithm 22: The XL Algorithm [Nicolas Courtois]

$$1 + x + y + z + wz + yz = 0$$
$$w + wx + wy + \cancel{wz} + \cancel{wz} + wyz = 0$$
$$\cancel{x} + \cancel{x} + xy + xz + wxz + xyz = 0$$
$$\cancel{y} + xy + \cancel{y} + \cancel{yz} + wyz + \cancel{yz} = 0$$
$$\cancel{z} + xz + \cancel{yz} + \cancel{z} + wz + \cancel{yz} = 0$$

$$x + z + wx + wy + wz + xy + xz + yz = 1$$
$$\cancel{wx} + \cancel{wz} + \cancel{wx} + wy + \cancel{wz} + wxy + wxz + wyz = w$$
$$\cancel{x} + \cancel{yz} + wx + wxy + wxz + xy + \cancel{yz} + xyz = \cancel{x}$$
$$\cancel{xy} + \cancel{yz} + wxy + wy + wyz + \cancel{xy} + xyz + \cancel{yz} = y$$
$$\cancel{yz} + \cancel{z} + wxz + wyz + wz + xyz + \cancel{yz} + yz = \cancel{z}$$

$$w + y + wx + xz + yz = 0$$
$$w + wy + wx + wxz + wyz = 0$$
$$\cancel{wx} + xy + \cancel{wx} + xz + xyz = 0$$
$$wy + y + wxy + xyz + yz = 0$$
$$wz + \cancel{yz} + wxz + xz + \cancel{yz} = 0$$

$$x + wx + wy + wz + yz = 1$$
$$\cancel{wx} + \cancel{wx} + wy + wz + wyz = w$$
$$\cancel{x} + wx + wxy + wxz + xyz = \cancel{x}$$
$$xy + wxy + wy + wyz + yz = y$$
$$xz + wxz + wyz + wz + yz = z$$

Upon linearization we get a linear system equivalent to the matrix:

$$
\begin{bmatrix}
w & x & y & z & wx & wy & wz & xy & xz & yz & wxy & wyz & wxz & xyz \\
0 & 1 & 1 & 1 & 0 & 0 & 1 & 0 & 0 & 1 & 0 & 0 & 0 & 0 \\
1 & 0 & 0 & 0 & 1 & 1 & 0 & 0 & 0 & 0 & 0 & 1 & 0 & 0 \\
0 & 0 & 0 & 0 & 0 & 0 & 0 & 1 & 1 & 0 & 0 & 0 & 1 & 1 \\
0 & 0 & 0 & 0 & 0 & 0 & 0 & 1 & 0 & 0 & 0 & 1 & 0 & 0 \\
0 & 0 & 0 & 0 & 0 & 0 & 1 & 0 & 1 & 0 & 0 & 0 & 0 & 0 \\
\\
0 & 1 & 0 & 1 & 1 & 1 & 1 & 1 & 1 & 1 & 0 & 0 & 0 & 0 \\
1 & 0 & 0 & 0 & 0 & 1 & 0 & 0 & 0 & 0 & 1 & 1 & 1 & 0 \\
0 & 0 & 0 & 0 & 1 & 0 & 0 & 1 & 0 & 0 & 1 & 0 & 1 & 1 \\
0 & 0 & 1 & 0 & 0 & 1 & 0 & 0 & 0 & 0 & 1 & 1 & 0 & 1 \\
0 & 0 & 0 & 0 & 0 & 0 & 1 & 0 & 0 & 1 & 0 & 1 & 1 & 1 \\
\\
1 & 0 & 1 & 0 & 1 & 0 & 0 & 0 & 1 & 1 & 0 & 0 & 0 & 0 \\
1 & 0 & 0 & 0 & 1 & 1 & 0 & 0 & 0 & 0 & 0 & 1 & 1 & 0 \\
0 & 0 & 0 & 0 & 0 & 0 & 0 & 1 & 1 & 0 & 0 & 0 & 0 & 1 \\
0 & 0 & 1 & 0 & 0 & 1 & 0 & 0 & 0 & 1 & 1 & 0 & 0 & 1 \\
0 & 0 & 0 & 0 & 0 & 0 & 1 & 0 & 1 & 0 & 0 & 0 & 1 & 0 \\
\\
0 & 1 & 0 & 0 & 1 & 1 & 1 & 0 & 0 & 1 & 0 & 0 & 0 & 0 \\
1 & 0 & 0 & 0 & 0 & 1 & 1 & 0 & 0 & 0 & 0 & 1 & 0 & 0 \\
0 & 0 & 0 & 0 & 1 & 0 & 0 & 0 & 0 & 0 & 1 & 0 & 1 & 1 \\
0 & 0 & 1 & 0 & 0 & 1 & 0 & 1 & 0 & 1 & 1 & 1 & 0 & 0 \\
0 & 0 & 0 & 1 & 0 & 0 & 1 & 0 & 1 & 1 & 0 & 1 & 1 & 0
\end{bmatrix}
\mathbf{x} =
\begin{bmatrix}
+C \\
1 \\
0 \\
0 \\
0 \\
0 \\
\\
1 \\
0 \\
0 \\
0 \\
0 \\
\\
0 \\
0 \\
0 \\
0 \\
0 \\
\\
1 \\
0 \\
0 \\
0 \\
0
\end{bmatrix}
$$

This matrix, over $\mathbb{GF}(2)$, is of full column-rank (i.e. the set of vectors formed by its columns is linearly independent), and the linear system has exactly one solution, corresponding to $w = 0$, $x = 1$, $y = 0$, $z = 1$, $wx = 0$, $wy = 0$, $wz = 0$, $xy = 0$, $xz = 1$, $yz = 0$, $wxy = 0$, $wyz = 0$, $wxz = 0$ and $xyz = 0$. Since this is consistent (i.e. we do not have something akin to $x_1 x_2 = 1$ but $x_2 = 0$ in the previous example), we know this is a solution. Since all solutions of the polynomial system are solutions of the linear system, we know we have found all solutions.

For a Gröbner Bases algorithm, or some other method that works with the polynomials as true polynomials, this operation is foolish. Knowing that $xf(x) = xy$ when it is already known that $f(x) = y$ is of no help in those cases. But, in the case of linearization, the image of the equations as a linear system will have $xf(x) = xy$ linearly independent from $f(x) = y$, in all but exceptional cases, and so these equations do provide new information.

12.4.1 Complexity Analysis

Naturally, it would be easier to simply check all 2^4 possibilities for $w, x, y,$ and z but that takes 2^n steps in general. The XL algorithm will be faster than that in general. For $d = 2$ and $D = 3$, an m equation quadratic system in n unknowns will yield $n^3/6 + 5n/6$ monomials. There will be $m(n + 1)$ equations, resulting in a $(mn + m) \times (n^3/6 + 5n/6)$ matrix. Even with naïve Gaussian Elimination this requires

$$(mn + n)(n^3/6 + 5n/6)\min(mn + n, n^3/6 + 5n/6)$$

field operations.

Using $c = m/n$ notation, this becomes

$$\sim (cn^5/6)\min(cn^2, n^3/6) \sim (c/6)n^7\min(c, n/6)$$

or if c is chosen so that $m \approx n^2/6$, that is to say $c = n/6$, this is $\sim n^9/216$. The approximation $m \approx n^2/6$ requires some explanation.

12.4.2 Sufficiently Many Equations

An interesting thought is to measure the point at which the matrix is square. If the matrix is full column rank, then we will get a unique solution provided that a solution exists. If there are fewer rows than columns, then the matrix cannot be of full column-rank. However, this condition of having the right number of rows is of course insufficient for full column-rank problems, because (for example) several of the rows might be identical. Nonetheless, this is a "ball-park estimate" for a unique solution. Furthermore, we prove that random matrices generated by fair coins tend to have very low nullity (dimension minus rank), in Theorem 59 on Page 144. Our matrices in algebraic cryptanalysis are not random necessarily, but the property is suggestive.

The matrix is square for $D = d + 1$ when

$$(n + 1)m = n^3/6 + 5n/6 \text{ or } m \approx n^2/6$$

Previously, we required $\approx n^2/2$ equations to get a unique solution (from linearization) and now only $\approx n^2/6$ are required.

12.4.3 Jumping Two Degrees

It is interesting to see what happens when using XL on an MQ problem, and having $D = 4$. This means every equation will be multiplied by all possible monomials of degree 2, degree 1, and the degree 0 monomial (i.e. the constant 1).

The number of monomials of degree 2, degree 1, and degree 0 is

$$\binom{n}{2} + \binom{n}{1} + \binom{n}{0} = n(n + 1)/2 + 1$$

So the number of equations will be $m(n^2+n+2)/2$. The number of monomials will be

$$\binom{n}{4}+\binom{n}{3}+\binom{n}{2}+\binom{n}{1}+\binom{n}{0}=\frac{n^4-2n^3+11n^2+14n+24}{24}$$

Thus the system will be exaclty defined (a square matrix) when

$$m(n^2+n+2)/2=\frac{n^4-2n^3+11n^2+14n+24}{24}$$

$$m=\frac{2(n^4-2n^3+11n^2+14n+24)}{24(n^2+n+2)}$$

$$m\sim n^2/12$$

And thus we need only $1/6$ as many equations, $n^2/12$ versus $n^2/2$, using Linearization alone. Thus running time in this case can be found.

The final matrix has $m(n^2+n+2)/2$ rows and $(n^4-2n^3+11n^2+14n)/24$ columns. The complexity of a Gaussian Elimination is $\sim mn\min(m,n)$. Substituting $m=cn$, one obtains a running time of $\sim cn^{10}\min(n/12,c)/96$. If the matrix is approximately square, $m\approx n^2/12$ or $c\approx n/12$, and the running time becomes $\sim n^{11}/1152$.

12.4.4 Fix-XL

Here, we simply apply techniques of a "guess-and-determine" attack (see Section 11.7 on Page 206) to the general problem of solving a system of polynomial equations. Suppose there are m equations and n unknowns and degree 2. If we guess g variables, then only $n-g$ variables remain. For $D=5$, $D=4$ and $D=3$, we would have

Operating Degree	Monomials	Equations	Unknowns
$D=5$	$\binom{n-g}{3}+\binom{n-g}{2}+\binom{n-g}{1}$	$\left[\binom{n-g}{3}+\binom{n-g}{2}+\binom{n-g}{1}\right]m$	$\binom{n-g}{5}+\cdots+\binom{n-g}{1}+1$
$D=4$	$\binom{n-g}{2}+\binom{n-g}{1}$	$\left[\binom{n-g}{2}+\binom{n-g}{1}\right]m$	$\binom{n-g}{4}+\binom{n-g}{3}+\binom{n-g}{2}+\binom{n-g}{1}+1$
$D=3$	$\binom{n-g}{1}$	$\left[\binom{n-g}{1}\right]m$	$\binom{n-g}{3}+\binom{n-g}{2}+\binom{n-g}{1}+1$

Table 12.1 The typical parameters used in solving a 50-variable, 50-equation quadratic system of equations, with XL.

Guessed	Variables Remaining	New Cubics	New Quadratics	New Linear	Total New Monomials	Total Linear Equations	Possible Quintics	Possible Quartics	Possible Cubics	Possible Quadratics	Possible Monomials	Iterations	status
10	40	9,880	780	40	10,701	535,050	658,008	91,390	9,880	780	760,099	1,024	underdefined
15	35	6,545	595	35	7,176	358,800	324,632	52,360	6,545	595	384,168	32,768	underdefined
16	34	5,984	561	34	6,580	329,000	278,256	46,376	5,984	561	331,212	65,536	underdefined
17	33	5,456	528	33	6,018	300,900	237,336	40,920	5,456	528	284,274	131,072	overdefined
20	30	4,060	435	30	4,526	226,300	142,506	27,405	4,060	435	174,437	1,048,576	overdefined

Guessed	Variables Remaining	New Cubics	New Quadratics	New Linear	Total New Monomials	Total Linear Equations	Possible Quintics	Possible Quartics	Possible Cubics	Possible Quadratics	Possible Monomials	Iterations	status
20	30	0	435	30	466	23,300	0	27,405	4,060	435	31,931	1,048,576	underdefined
24	26	0	325	26	352	17,600	0	14,950	2,600	325	17,902	16,777,216	underdefined
25	25	0	300	25	326	16,300	0	12,650	2,300	300	15,276	33,554,432	overdefined
30	20	0	190	20	211	10,550	0	4,845	1,140	190	6,196	1,073,741,824	overdefined

Guessed	Variables Remaining	New Cubics	New Quadratics	New Linear	Total New Monomials	Total Linear Equations	Possible Quintics	Possible Quartics	Possible Cubics	Possible Quadratics	Possible Monomials	Iterations	status
20	30	0	0	30	31	1,550	0	0	4,060	435	4,526	1,048,576	underdefined
30	20	0	0	20	21	1,050	0	0	1,140	190	1,351	1,073,741,824	underdefined
32	18	0	0	18	19	950	0	0	816	153	988	4,294,967,296	underdefined
33	17	0	0	17	18	900	0	0	680	136	834	8,589,934,592	overdefined
35	15	0	0	15	16	800	0	0	455	105	576	34,359,738,368	overdefined

If the matrix at the end of the linearization process is not full rank, then a unique solution will not be found. Therefore, we desire that there be more equations than unknowns, but only slightly. Furthermore, the expected number of iterations is 2^g, and so we would like g to be as small as possible.

In the example shown in Table 12.1 on Page 218, we chose 50 variables and 50 equations, and operating degrees $D = 3$, $D = 4$ as well as $D = 5$. And so the choices are to solve a linear system of size 900×834 about 8.5 billion times, a $16{,}300 \times 15{,}276$ matrix about 34 million times, or a $300{,}900 \times 284{,}274$ matrix about $131{,}072$ times. And naturally, faced with matrices of this size, we cannot rule out the possibility of brute-force checking all 2^{50} choices.

Generally, since matrix operations are cubic time or better, we should choose the option with the fewest iterations, provided that we can store the matrix in RAM.

12.5 ElimLin

The ElimLin algorithm is also a creation of Nicolas T. Courtois, and appeared in [76]. Here, we will give more detail. In this section, we are concerned with quadratic systems of polynomials over any field. Usually it will be $\mathbb{GF}(2)$, but it can even be $\mathbb{Q}$, $\mathbb{R}$, or $\mathbb{C}$.

Suppose one has such a system of equations, and then one performs linearization as specified in Section 12.3 on Page 211. Then, suppose further that one executes Gaussian Elimination to RREF upon that matrix, of size $m \times n$ after the linearization, with the objective of putting the system into reduced row-echelon form. Let r denote the rank of the set of row vectors of the matrix. The output one would obtain would be the following:

$$
\begin{bmatrix}
1 & 0 & 0 & \cdots & 0 & a_{1,r+1} & a_{1,r+2} & \cdots & a_{1,n-1} & a_{1,n} \\
0 & 1 & 0 & \cdots & 0 & a_{2,r+1} & a_{2,r+2} & \cdots & a_{2,n-1} & a_{2,n} \\
0 & 0 & 1 & \cdots & 0 & a_{3,r+1} & a_{3,r+2} & \cdots & a_{3,n-1} & a_{3,n} \\
\vdots & \vdots & \vdots & \ddots & \vdots & \vdots & \vdots & \ddots & \vdots & \vdots \\
0 & 0 & 0 & \cdots & 1 & a_{r,r+1} & a_{r,r+2} & \cdots & a_{r,n-1} & a_{r,n} \\
0 & 0 & 0 & \cdots & 0 & 0 & 0 & \cdots & 0 & 0 \\
0 & 0 & 0 & \cdots & 0 & 0 & 0 & \cdots & 0 & 0 \\
\vdots & \vdots & \vdots & \ddots & \vdots & \vdots & \vdots & \ddots & \vdots & \vdots \\
0 & 0 & 0 & \cdots & 0 & 0 & 0 & \cdots & 0 & 0
\end{bmatrix}
$$

We will place one more requirement on the matrix. The columns associated with the quadratic monomials will come first, and then the columns associated with the linear monomials (i.e. the original variables) will come after. Let the number of original variables be v. Then either $r > n - 1 - v$ or $r \leq n - 1 - v$. We will assume the former, and call it the "sufficient rank condition."

When this "sufficient rank condition" occurs, we obtain the following matrix:

$$
\left[
\begin{array}{ccccc|ccccc|ccccc|c}
1 & 0 & 0 & \cdots & 0 & 0 & 0 & \cdots & 0 & 0 & 0 & a_{1,r+1} & a_{1,r+2} & \cdots & a_{1,n-1} & b_1 \\
0 & 1 & 0 & \cdots & 0 & 0 & 0 & \cdots & 0 & 0 & 0 & a_{2,r+1} & a_{2,r+2} & \cdots & a_{2,n-1} & b_2 \\
0 & 0 & 1 & \cdots & 0 & 0 & 0 & \cdots & 0 & 0 & 0 & a_{3,r+1} & a_{3,r+2} & \cdots & a_{3,n-1} & b_3 \\
\vdots & \vdots & \vdots & \ddots & \vdots & \vdots & \vdots & \ddots & \vdots & \vdots & \vdots & \vdots & \vdots & \ddots & \vdots & \vdots \\
0 & 0 & 0 & \cdots & 1 & 0 & 0 & \cdots & 0 & 0 & 0 & a_{n-v-2,r+1} & a_{n-v-2,r+2} & \cdots & a_{n-v-2,n-1} & b_{n-v-2} \\
0 & 0 & 0 & \cdots & 0 & 1 & 0 & \cdots & 0 & 0 & 0 & a_{n-v-1,r+1} & a_{n-v-1,r+2} & \cdots & a_{n-v-1,n-1} & b_{n-v-1} \\
\hline
0 & 0 & 0 & \cdots & 0 & 0 & 1 & \cdots & 0 & 0 & 0 & a_{n-v,r+1} & a_{n-v,r+2} & \cdots & a_{n-v,n-1} & b_{n-v} \\
\vdots & \vdots & \vdots & \ddots & \vdots & \vdots & \vdots & \ddots & \vdots & \vdots & \vdots & \vdots & \vdots & \ddots & \vdots & \vdots \\
0 & 0 & 0 & \cdots & 0 & 0 & 0 & \cdots & 1 & 0 & 0 & a_{r-2,r+1} & a_{r-2,r+2} & \cdots & a_{r-2,n-1} & b_{r-2} \\
0 & 0 & 0 & \cdots & 0 & 0 & 0 & \cdots & 0 & 1 & 0 & a_{r-1,r+1} & a_{r-1,r+2} & \cdots & a_{r-1,n-1} & b_{r-1} \\
0 & 0 & 0 & \cdots & 0 & 0 & 0 & \cdots & 0 & 0 & 1 & a_{r,r+1} & a_{r,r+2} & \cdots & a_{r,n-1} & b_r \\
\hline
0 & 0 & 0 & \cdots & 0 & 0 & 0 & \cdots & 0 & 0 & 0 & 0 & 0 & \cdots & 0 & 0 \\
0 & 0 & 0 & \cdots & 0 & 0 & 0 & \cdots & 0 & 0 & 0 & 0 & 0 & \cdots & 0 & 0 \\
\vdots & \vdots & \vdots & \ddots & \vdots & \vdots & \vdots & \ddots & \vdots & \vdots & \vdots & \vdots & \vdots & \ddots & \vdots & \vdots \\
0 & 0 & 0 & \cdots & 0 & 0 & 0 & \cdots & 0 & 0 & 0 & 0 & 0 & \cdots & 0 & 0
\end{array}
\right]
$$

The reader may wish to pause at this point and verify that the indices are correct. These indices are crucial in what follows. The lines can be explained as follows. The leftmost vertical line indicates the transition from quadratic to linear monomials. All columns to the left of it represent quadratic monomials, and all columns to the right of it represent linear monomials. The other vertical line separates the linear monomials from where the constants are to be found. The equations which are described in the rows found between the horizontal lines then are entirely composed of linear and constant monomials.

This is actually pessimistic, because it assumes no degenerate (pivot-less) columns. See Section 12.5.6 on Page 226.

The reader can also see that this region has $r - (n - v) + 1 = r - (n - v - 1)$ equations. Thus, this region of all-linear equations will be non-empty if $r > n - v - 1$—which is precisely the "sufficient rank condition."

By the method outlined here, one can therefore easily find linear equations which are in the span of a system of quadratic polynomials, considered as a vector-space over the base field. It should be noted that the "sufficient rank condition" is sufficient for a non-zero number of such equations to be found, but it is not necessary. More on this will follow, in Section 12.5.6 on Page 226.

12.5.1 Why is this useful?

A linear equation found by this approach looks like the following:

$$x_i + a_{i,j}x_j + a_{i,k}x_k + \cdots + a_{i,n-1}x_{n-1} = b_i$$

where b_i is a constant from the base field, and the $a_{i,z}$ are likewise for all z. This means we can re-arrange to form:

$$x_i = (-a_{i,j})x_j + (-a_{i,k})x_k + \cdots + (-a_{i,n-1})x_{n-1} + b_i$$

which can be considered a new definition of x_i.

Likewise, all $r + v + 1 - n$ linear equations that are found can become redefinitions of one distinct variable each. These can be substituted where ever they are found in the other equations which make up the polynomial system. This completely removes the redefined variables from the system. As stated before, the number of variables is a key factor in the difficulty of a polynomial system of equations, and reducing the number of them is highly useful.

12.5.2 How to use ElimLin

The algorithm itself is described in Algorithm Box 23 on Page 222. It can be summarized as follows. First, linearize the system of equations, with the added feature that all the quadratic terms are in the leftmost columns and the linear terms are in the rightmost columns. Of course, if the system is not quadratic, then it should be reduced to quadratic, using the techniques found in Section 11.4 on Page 192 for example.

Next one performs Gaussian Elimination to obtain Reduced-Row Echelon form. Denote the number of linear equations found as ℓ. These will be a set of consecutive equations above the all-zero rows, but below all equations containing any quadratic terms. If none are found, the algorithm must terminate.

If some are found, then optionally one may elect to reduce the weight of these linear equations using the techniques of Algorithm 24 on Page 225. Next, for each equation, a distinct variable found inside it should be selected. (It is important that each selected variable be unique, but otherwise it is *completely unclear* how to best select which variable, of the several available. In practice, the author simply takes the first valid choice.) If not over $\mathbb{GF}(2)$, the equation should be manipulated so that the variable to be redefined is on one side of the equal sign, and the other terms are on the opposite side. Over $\mathbb{GF}(2)$, this is trivial.

Once this is done, everywhere that this variable is found, including in degree 2 monomials, it should be substituted with its definition. This effectively and completely eliminates the variable from the system. Nonetheless, the definition should be stored, so that when the system is finally solved, the original "eliminated" variables can be recovered. For this reason, each of the linear equations is added to the set D once we are done with it.

Once all the substitutions are complete, the system of equations will look different. In particular, it is likely not to be in Reduced Row Echelon form any longer. Therefore, the process should repeat, by performing Gaussian Elimination again, and continuing as before. Obviously, the Gaussian Elimination will be much faster than normal, as the matrix differs only slightly from its RREF.

It should be noted that it is not too optimistic to assume that a second or third iteration will produce $\ell \neq 0$. This is because v is reduced by ℓ, and so n is reduced by $(v+1)\ell$. Since $\ell_t \approx r + v + 1 - n$, in the next iteration we expect

$$
\begin{aligned}
\ell_{t+1} &= r_{t+1} + v_{t+1} + 1 - n_{t+1} \\
&= r_{t+1} + (v_t - \ell_t) + 1 - (n_t - (v_t + 1)\ell_t) \\
&= r_{t+1} + v_t - \ell_t + 1 - n_t + v_t \ell_t + \ell_t \\
&= r_{t+1} - r_t + (r_t + v_t + 1 - n_t) + v_t \ell_t \\
&= (r_{t+1} - r_t) + \ell_t + v_t \ell_t \\
&= \ell_t(v_1 + 1) - (r_t - r_{t+1})
\end{aligned}
$$

If we could predict how the rank would change, this would be very exact and useful, but the author cannot predict that at the moment. On the other hand, $\ell_t(v_1 + 1)$ is large. So initially, if the rank of the system does not change much, we expect the second run to produce more of these, not fewer. At least it is clear that reducing the number of variables makes it more likely, not less likely, that linear equations will be found, because the value of $c = m/n$ is improved.

This process can continue until either no linear equations are found in the span of the quadratic equations, or until all the variables have been eliminated.

Finally, it should be noted, that monomials of type xy and yz but not xz might exist, where the letters x, y, and z refer to classes of variables. See Section 12.5.6 on Page 226 and also Section 12.8 on Page 228.

INPUT: A system of degree 2 polynomial equations.
OUTPUT: Either, a solution or solutions to the system, if the equations have sufficient rank, or if not, then a reduced system of equations in fewer variables than the original, to be solved by some other method.
1: $D \leftarrow \{\}$
2: Linearize the system of equations. (See Section 12.3 on Page 211).
3: Perform Gaussian Elimination to result in Reduced Row Echelon Form.
4: Let ℓ be the number of all-linear equations found.

   1: If $\ell = 0$ then STOP.
   2: Else, $\ell > 0$, therefore begin
      1: (Optional) Apply a rule to reduce the weight of the ℓ equations.
      2: For $i = 1 \ldots \ell$ do
         1: Move all the variables and constants, but one, to one side of the equal sign.
         2: Substitute this redefinition of a variable into the other equations, thus eliminating one variable.
         3: Substitute this redefinition of a variable into the other definitions in D.
         4: Add the definition to D.
      3: Goto Step 3, "Perform Gaussian Elimination."

Algorithm 23: The ElimLin Algorithm [Nicolas Courtois]

12.5.3 On the Sub-Space of Linear Equations in the Span of a Quadratic System of Equations

Consider the span of a set of equations S, namely the set of all equations which are linear combinations of equations from S. Since a scalar multiple of an equation in the span is also in the span, and since the sum of two equations in the span is also in the span, the span forms a vector space over the coefficient field.

Moreover, the sum of two linear equations in the span is a linear equation, and the scalar product of a linear equation with an element of the coefficient field is also linear equation. Therefore, the set of linear equations L in the span S forms a subspace.

In the ElimLin method, because of the submatrix $a_{n-v,n-v}, \ldots, a_{r,r}$, (which is equal to the $(r-n+v+1)$-dimensional identity matrix), we know that the rows $n-v$ to r are linearly independent. They are therefore a basis for their span. They are also all linear, and so they form a basis for a subspace of L. It turns out that this subspace is L itself, which simply means that the span of these $r-n+v+1$ equations forms all of L. This is a statement which is perhaps obvious, but interesting to prove.

Theorem 88. *All linear equations in the span of a set of polynomial equations S, can be found in the span of the set of linear equations found by executing ElimLin on S.*

Proof. Let A' be the matrix formed after linearizing S, so that all quadratic terms are on the left and all linear terms on the right. Let A be the Reduced Row Echelon form of A'. Denote the number of linear monomials in S as v.

Denote the rows of the matrix A as $f_1, \ldots, f_m$.

To show that the rows $n-v, n-v+1, \ldots, r$ span all linear equations in the span of S, let ϕ be a linear equation in the span of S. Since it is formed by being a linear combination of the equations, or rows, of A then it is of the form $c_1 f_1 + c_2 f_2 + c_3 f_3 + \cdots + c_m f_m$. Let $\mathbf{c} = (c_1, c_2, \ldots, c_m)$. One can see that $\mathbf{c}A = \phi$. (Here $\mathbf{c}$ is a row vector whereas most vectors in this book are column vectors. Please forgive the abuse of notation). We will show ϕ is actually a linear combination only of rows $n-v, \ldots, r$. That is to say, it is a linear combination of the ℓ rows found by ElimLin.

But note that if ϕ were linear, then it has all zeroes as the coefficients of the quadratic monomials. The linear coefficients are found in columns $n-v$ to column $n-1$. (Observe that there are $(n-1)-(n-v)+1 = v$ such columns). Thus the first coefficient until the $(n-v-1)$th coefficient of ϕ are all zeroes. This means that $\mathbf{c}A = \mathbf{s}$ has $s_1, \ldots, s_{n-v-1}$ all equal to zero.

Now consider the $m \times n$ matrices B and C where rows $1, 2, \ldots, n-v-1$ of A are copied into B, which is otherwise all zeroes, and rows $n-v, n-v+1, \ldots, r$ of A are copied into C, which is also otherwise all zeroes. Note, rows $r+1, r+2, \ldots, m$ are all zeroes in A in any case. Surely $A = B+C$ and so $\mathbf{s} = \mathbf{c}A = \mathbf{c}B + \mathbf{c}C$.

Observe that C will have all zeroes in the first $n-v-1$ columns. This means that $\mathbf{c}C = \mathbf{u}$ will have $u_1, u_2, \ldots, u_{n-v-1}$ all equal to zero. Thus $\mathbf{s} - \mathbf{u}$ will also have its first $n - v - 1$ entries equal to zero, since s has $s_1, \ldots, s_{n-v-1}$ all equal to zero. We know $cA = s$ and $cC = u$ so $cA - cC$ has its first $n - v - 1$ entries as zero.

But $cB = cA - cC$. Therefore cB has its first $n - v - 1$ entries as zero. If the first $n - v - 1$ entries of c are not all zero, then this represents a non-trivial linear combination of the rows of B, equaling zero. Since the rows $n - v, n - v + 1, \ldots, m$ of B are all zero, this non-trivial linear combination can be written so that all the non-zero coefficients are for the rows $1, 2, \ldots, n - v - 1$. This means the rows of B are linearly dependent. But the rows of B are in reduced row echelon form, and so are linearly independent. This is a contradiction.

Therefore, the first $n - v - 1$ entries of c are all zero, and ϕ is a linear combination of the rows of C alone. These are precisely the linear equations found by ElimLin.

$\square$

Corollary 89. *The linear equations L found by running one iteration of ElimLin on a quadratic system of equations S, forms a basis for the set of linear equations in the span of S.*

Proof. This follows directly from the previous theorem, after noting that the linear equations L are linear independent because all the rows of A are linearly independent (by virtue of being in reduced row echelon form). $\square$

12.5.4 The Weight of the Basis

However, a subspace has many bases, not just one basis. There is no reason to believe that the basis found by ElimLin is the lowest possible weight. In fact, that would be very lucky.

There are several hueristics which one can imagine that will result in a relatively low weight basis. However, since $\ell = r + v - n + 1$, the number of equations found is not likely to be very large at all, and therefore the following strategy is feasible, if expensive.

The method given in Algorithm 24 on Page 225 can be thought of as an expensive but completely effective Step 4(2)1 for ElimLin, as described in Algorithm 23 on Page 222.

If there is never a tie in Step 4, then this should produce a very low weight basis B. It is unclear if this "greedy algorithm" approach is optimal or not. But in the presence of ties (two vectors both equally of lowest weight), which is not to be completely unexpected, it is not optimal.

An alternative approach to this would be to swap rows at each iteration of the Gaussian Elimination. Prior to performing the Elimination on column i, the lowest weight row j of the matrix, among the rows $i, i + 1, i + 2, \ldots, m$ such that $a_{ji} \neq 0$, should be selected. This row j should be swapped with row i. This is called "naïve sparse Gaussian Elimination" or "lowest-weight row Gaussian Elimination."

Structured Gaussian Elimination is an old technique, effective but sub-optimal. For example, it is used currently in SAGE [7] for sparse $\mathbb{GF}(2)$-matrix elimination.

INPUT: A set of linear equations S, which are a basis for a subspace.

NOTE: When used with ElimLin, this set S is given by the rows $n-v, n-v+1, n-v+2, \ldots, r$ of the matrix. The sub-space is the set of linear equations in the span of a set of polynomial equations over any field.

OUTPUT: A low-weight basis for the span of S.

1: $T \leftarrow \{\mathbf{0}\}$
2: $B \leftarrow \{\}$
3: Enumerate the span of S by using the Gray Code and Rapid Subspace Enumeration (See Section 9.2 on Page 135). Call this set L.
4: Choose the lowest weight member of L, call it $\mathbf{u}$.
5: Insert $\mathbf{u}$ into B.
6: For each vector $\mathbf{t} \in T$ do

    1: Insert $\mathbf{u}+\mathbf{t}$ into T.
    2: Remove $\mathbf{u}+\mathbf{t}$ from L.

7: If L is non-empty, then go to Step 4, or else return B.

Algorithm 24: An Expensive but Effective Way to find a low Weight Basis [G. Bard]

The spike in matrix density normally associated with ordinary Gaussian Elimination is somewhat mitigated by this approach.

Thus we have listed two methods for reducing the weight of the basis. One is very expensive (see Algorithm 24 on Page 225), and the other (using Naïve-Sparse Gaussian Elimination instead of ordinary Gaussian Elimination) is almost free, being of linear expense per column or quadratic (and thus invisible) expense over all. But we have not stated why it is important that the basis be low weight.

When a linear equation (the definition) is substituted back into the original equations, several new terms are added to each old equation. Some will cancel out existing terms, but if the original was sparse, this cannot be expected often. Therefore, the weight of the row will increase. Also, the size of the matrix is decreasing. Since β is a ratio of these two things, the sparsity will increase significantly. Therefore it is crucial that the definitions be low weight. Naturally, being the bottom ℓ rows of the RREF, our definitions will be from the densest stage of the RREF.

12.5.5 One Last Trick for $\mathbb{GF}(2)$-only

In a $\mathbb{GF}(2)$-polynomial system, if it is ever found that some quadratic monomial $x_i x_j = 1$, then one can substitute this equation with the two equations $x_i = 1$ and $x_j = 1$. This in turn means that 1 can be substituted for x_i and x_j where ever they are found, and also $x_a x_i$ becomes x_a. Thus, even one such equation is bountiful in eliminating monomials and variables.

In pratice, the way to do this is after Step 4 in the ElimLin algorithm, detect the equation of the form $x_i x_j$, and then substitute this row with two rows, one indicating $x_i = 1$ and the other indicating $x_j = 1$. The rest of the ElimLin algorithm, in one or two iterations, will accomplish the remainder of the substitutions that come by consequence of this discovery.

12.5.6 Notes on the Sufficient Rank Condition

If the "sufficient rank condition" exists, then certainly a linear equation will be found. This can be seen by examining the matrix on Page 220. But this condition is not neccessary.

First, consider the possibility that, during the Gaussian Elimination, at column i, that there are no non-zero elements in $a_{i,i}, a_{i+1,i}, a_{i+2,i}, \ldots, a_{m,i}$. The elimination will resume with the next column. If the rank is r then the last element of the diagonal of ones will occur not at $a_{r,r}$ but at $a_{r,r+1}$ (assuming that this happens only once). Thus, "pivotless" columns, as these are called, will shift the diagonal strand of ones to the right.

Second, structural considerations matter. For example, it might be that certain variables never appear together in the same monomial. This means there is an entire column (associated with the forbidden monomial) which is all zeroes, for each possible pair of variables which never share a monomial.

Amplification

Suppose that there were two categories of variable in the polynomial system of equations, x's and y's. Suppose further that $x_i y_j$ terms never exist. Then when one goes to XL, no monomials of the form $x_i y_j y_k$ or $x_i x_j y_k$ exist. And so one has a monomial count of

$$2\binom{n/2}{3} + 2\binom{n/2}{2} + 2\binom{n/2}{1} \ll \binom{n}{3} + \binom{n}{2} + \binom{n}{1}$$

For example, a 256 variable system with 128 x's and y's would have 682,752 cubic, 16,256 quadratic and 256 linear terms, for a total of 699,264 terms. A general system of 256 variables would have 2,763,520 cubic terms, 32,640 quadratic and 256 linear terms. The β value at the quadratic level would be (for a random system)

$$\frac{16256 + 256}{32640 + 256} = 0.50194$$

but the cubic would be

$$\frac{682752 + 16256 + 256}{2763520 + 32640 + 256} = 0.25006$$

Another interesting point is that the overwhelming fraction of the monomial count comes from the highest degree alone. This is because

$$\binom{n}{3} \gg \binom{n}{2}$$

and so these improvements are expected.

A View from the Point-of-View of Randomness

If one considers the coefficients of an equation as random variables, linear equations are indeed expected. For example, if the coefficients come from the ciphertext of some encrypted message, they should be 1 and 0 with nearly equal probability. Thus, a few equations, roughly 2^{-p} out of the total, will have p particular coefficients as zero. If there are 2^m equations, 2^{m-p} will have those p particular coefficients as zero (in expectation). It is easy to see that a "lucky" equation might have so few quadratic coefficients that these can be eliminated by Gaussian Elimination, and so will result in a linear equation in the span.

Interestingly, it should be noted that in a cryptanalytic attack of the cipher Keeloq, there was one equation which had only variables from the secret key in it, even though only 64 of 384 variables in the system were elements of the secret key (see [84]).

12.6 Comparisons between XL and F4

There have been several discussions comparing the XL algorithm and the F4 algorithm. In some ways, this is wasteful, because preprocessors and heuristics for improving one should work for improving the other, as they are so similar. The scholarly time and energy spent on rivalry would be better applied to cooperation. In any case, the discussion has been decided by two pivotal papers, one in the theoretical direction, and the other in the practical direction. Note, the F4 algorithm is by Faugère [109], and is a Gröbner Bases method, and XL is described in Section 12.4 on Page 213. Meanwhile, the ElimLin algorithm is described in Section 12.5 on Page 219 and the XL-II algorithm is described in [80].

In theoretical terms, the XL algorithm is very simple, but wasteful. Somewhat improved are ElimLin, and XL-II. Then the F4 algorithm is remarkably similar to XL-II, provided that in XL-II one does not first multiply an equation by x_i as well as x_j, only to later multiply those two by x_j and x_i respectively—thus making two equations which are both $x_i x_j$ times the original and therefore identical and redundant. Some sort of tagging system can exclude this overhead. Of course, if $D - d = 20$, for example in [232], then this is a non-trivial problem.

The F4 algorithm is a bit more sophisticated and the F5 algorithm is more sophisticated still. The set of actions of each of these essentially form a subset-superset relationship, with the more complex algorithms doing everything that the simpler algorithms do. For an extremely detailed coverage of this topic, see [26]. It is an excellent paper to read because the argumentation cements the precise details of the various algorithms into one's head.

But in practice, it is precisely this sophistication that leads to speed differences and memory-usage differences. Yang, Chen, Bernstein and Chen settled this issue in [232], where those authors had a 20-variable system over $\mathbb{GF}(256)$ which was solved using the XL-II algorithm, and F4 implemented by Magma simply crashed due to lack of memory when examining even the 15-varible case—with 16 giga-bytes of RAM available. Note a 15 or 20 variable system over $\mathbb{GF}(256)$ is roughly equivalent to a 120 or 160 variable system over $\mathbb{GF}(2)$. It should be noted that Yang, et al, use the Block Wiedemann of Don Coppersmith for the linear algebra (see Section 5.2.3.1 on Page 72, and Appendix D on Page 323).

One interesting historical note is that both these algorithms were foreshadowed by Lazard [158]. In fact, Lazard noticed a connection between Gaussian Elimination after Linearization and Gröbner Bases via the Buchberger algorithm in 1983. For comparison, the first XL paper was [82] in 2000, and the first F4 paper in 1999 [109]. This connection was pointed out to the author by Dan Bernstein, and is also mentioned in [232].

12.7 SAT-Solvers

The use of SAT-solvers to solve polynomial equations over finite fields was first carried out by the author and Nicolas Courtois in [33]. However, it was proposed by Nicolas Courtois in his PhD dissertation [64], under the suggestion of Jacques Patarin. Since then, it has proven to be a useful technique. In particular, this author uses MiniSAT [6] [104].

The idea is that all NP-Complete problems are reducible to each other, and in particular, the NP-Complete problem called SAT is well-studied. There are annual competitions in which cash prizes are given for the fastest programs which can solve this problem for example cases. Strictly speaking, unless $P = NP$, these algorithms cannot run in polynomial time. However, for specific given problems, they might well run "fast enough."

The conversion between a $\mathbb{GF}(2)$ system of equations and the input to a SAT solver (a conjunctive normal form expression) will be given in Chapter 13. A rough and high-level introduction to the operation of SAT solvers will be given in Chapter 14.

12.8 System Fragmentation

Very rarely, a system of polynomial equations can be separated into two or more sets of equations, such that no equation in one set shares a variable with any other equation from any other set. We call such a system separable. We do not expect this to happen very often, but rather we will show how to force it to occur, when possible.

In this chapter, we present an algorithm for determining separability, as well as performing the actual separation. While cryptographers use systems of polynomials over finite fields, the methods in this note will work on any field or ring. Therefore, let $\mathfrak{R}$ be any ring.

12.8.1 Separability

Suppose one has two systems of multivariate polynomial equations over the same ring. In particular, $f_1(x_1, \ldots, x_n) = 0$, $f_2(x_1, \ldots, x_n) = 0$, $\ldots$, $f_m(x_1, \ldots, x_n) = 0$ and $g_1(y_1, \ldots, y_t) = 0$, $g_2(y_1, \ldots, y_t) = 0$, $\ldots$, $g_s(y_1, \ldots, y_t) = 0$.

Obviously this is one system of m equations in n unknowns labeled, $x_1, \ldots, x_n$, and another system of s equations in t unknowns labeled $y_1, \ldots, y_t$. What is also clear, is that it is a sytem of $m + s$ equations in $n + t$ unknowns, labeled $x_1, \ldots, x_n$, $y_1, \ldots, y_t$. Since the difficulty of solving polynomial systems of equations is worse than linear in the number of equations and number of variables, (if $P \neq NP$, it is worse than polynomial), solving the larger system will be much worse than solving the smaller systems individually. The property of separability is obvious in that it can be determined by inspection in this example. Given a random sparse system of equations, it can be hard to determine separability by mere examination.

What is less obvious is that if you "scramble" the equations of the large system, then the separability becomes obscure. Consider an example where the polynomials $f_1, \ldots, f_n$ and $g_1, \ldots, g_s$ are linear. Then we can write a matrix F and a matrix G such that $F\mathbf{x} = \mathbf{b_1}$ and $G\mathbf{y} = \mathbf{b_2}$. To further simplify matters, let $m = n$ and $s = t$, in other words, let F and G be square. The larger system becomes

$$\begin{bmatrix} F & 0 \\ 0 & G \end{bmatrix} \begin{bmatrix} \mathbf{x} \\ \mathbf{y} \end{bmatrix} = \begin{bmatrix} \mathbf{b_1} \\ \mathbf{b_2} \end{bmatrix}$$

The structure of the zeroes of this arrangement will make the separability of the system very clear, and it will be obvious how to write down the disjoint systems of equations from the larger matrix. But now consider a random $n + s$-dimensional permutation matrix P. Multiplying the above equation by P on both sides will "scramble" the larger matrix, and while there will still be many zeroes, it is not clear that the system is separable, or futhermore, how to separate it.

The only clue will be that out of $(n + s)^2$ entries, there will be at least $2ns$ zeroes. The remaining $n^2 + s^2$ entries may or may not be zero, and so the sparsity of the matrix is at least $2ns/(n + s)^2$. Many matrices have this sparsity but are not separable.

Definition 90. Let $\mathfrak{R}$ be any ring. Let $f_1(x_1, \ldots, x_n), \ldots f_n(x_1, \ldots, x_n)$ be a set of polynomial functions $\mathfrak{R}^n \to \mathfrak{R}$. If there exists a partition of $f_1, f_2, \ldots, f_n$ into two non-empty disjoint subsets A and B, such that if $f_i \in A$ and $f_j \in B$ implies that f_i

and f_j do not share any variables with non-zero coefficients, then the system of equations $f_1(\mathbf{x}), \ldots, f_n(\mathbf{x})$ is separable.

In cryptography, if the system has thousands, if not millions, of equations, then one cannot "look at" the equations at all, as the required piece of paper to write them would be huge. Therefore, automation is required.

12.8.2 Gaussian Elimination is Not Enough

In the linear example above, Gaussian Elimination would resovle the mystery. This is an expensive operation, taking $\Theta((n+s)^3)$ operations, but it also solves the system itself. For a polynomial system of equations, one can "linearize", as explained in Section 12.3 on Page 211. This means creating a new "dummy" variable for each monomial in the system. Thus $x_1 x_5 x_9$ might be mapped to y_a, while $x_1 x_9$ and $x_5 x_9$ might be mapped to totally unrelated y's. Yet, if f_i contained $x_1 x_5 x_9$, and f_j contained $x_5 x_9$, but were otherwise disjoint, they would be a non-separable pair of equations in truth, but appear to be separable in the linearization. In fact, f_i and f_j must appear in the same set when the f's are partitioned. This is invisible in Linearization followed by Gaussian Elimination.

12.8.3 Depth First Search

We will solve this problem by constructing a graph, and then performing an algorithm called a "Depth First Search." One vertex v_i should be made for each variable x_i. An edge should be drawn between v_i and v_j if and only if x_i and x_j ever appear in the same equation, anywhere in the system, regardless if in the same monomial or not. This graph tends to be rather dense.

The algorithm called "Depth First Search" (or DFS) is well known, and can be found, for example, in [63, Ch. 22]. Basically, one keeps an array of flags for each vertex, initially set to false. These are marked true when a vertex is visited. First, one visits some arbitrary initial vertex, and sets its flag to true. Then, one visits each of its unflagged neighbors, one at a time, setting those flags to true. Of course, upon the first visitation of the first neighbor, all of that neighbor's neighbors will have to be visited, and so the second neighbor of the original vertex is only visited after all the neighbors of the first neighbor are visited. Thus the name "depth first search." When this process is fully complete, if no unflagged vertices remain, then the graph is connected. Otherwise, it is disconnected.

At each visitation, one must check the flags of each of the neighbors. There are d neighbors (the average degree of the graph) on average. This is $O(|V|d)$ work, since each vertex is visited exactly once. Note that $|V|d = 2|E|$, but the notation $O(|E|)$ while absolutely correct, does not really express how the algorithm works.

If a Depth First Search concludes that the graph is not connected, then that means there is a set of vertices A and set of vertices B such that no edge ever goes from any vertex of A to any vertex of B. Therefore, all the variables associated with the vertices of A and their equations form a totally unrelated system of equations from the variables and equations associated with the vertices of B.

The graph, if not connected, has several connected components, called "islands" in the slang of graph theory. It is trivial to convert the Depth First Search into an island enumerator. First, run the DFS. Any nodes visited form the first island. If any nodes remain, pick one of those that remain and do a DFS from that point, resetting the visitation flag. The vertices visited the second time comprise the second island. Repeat until all vertices have been assigned to an island.

12.8.4 Nearly Separable Systems

In the rest of this section we restrict ourselves to finite rings of size $|\mathfrak{R}| = q$. Suppose $G = (V, E)$ is a graph that is connected. Suppose further that there is some vertex $v_i \in V$ such that the graph induced by removing v_i is disconnected. By "removing a vertex", we mean that a new graph, $G' = (V', E')$ is created. The vertices, $V' = V - \{v_i\}$, is just the old set of vertices with v_i removed. The edges which contain v_i at either end are removed, but all others are retained. If this graph is disconnected, and the original was connected, then we say that v_i is a bridge-vertex.

Since solving a polynomial system of equations in $n + s$ variables is much slower than solving one in n and another in s, it would be advantageous to "remove" the bridge vertex, and obtain two smaller systems.

Since the equations are over a finite ring (usually $GF(2)$ in fact), one can simply guess x_i. For example, over $GF(3)$, one would substitute $x_i = 0$, $x_i = 1$, and $x_i = 2$, to obtain 3 new systems of equations, each of which was separable into two smaller pieces. Solving these six new small systems yields the answers to the original system of equations.

This suggests the following algorithm:

INPUT: A system of polynomial equations.
OUTPUT: Either the indication that the system is separable, or if not, then a set of variables, such that if the value of one of them alone is known, the system becomes separable. This set may be empty.
1: Generate the graph, $G = (V, E)$, according to Section 12.8.3 on Page 230.
2: Run DFS to see if G is separable.
3: If not, for each vertex v_i:

- Remove v_i.
- Run DFS to see if G is separable.
- If yes, report that v_i is a bridge vertex, and restore v_i.
- If not, restore v_i.

Algorithm 25: Simple Vertex Removal Algorithm [Classic]

Assuming the original system is not separable, this would take $O(v^2d = ve)$ time, where $v = |V| = n$, $e = |E|$ and d is the average degree of the vertices of the graph. The total running time is $|V| = n$ times as much as before, and thus still polynomial.

12.8.5 Removing Multiple vertices

Alternatively, one could search all graphs formed by removing every possible pair, or triplet of vertices. If one searches for a set of r variables to remove to disconnect the graph, we say the algorithm is searching for r-tuples. The algorithm is listed as Algorithm 26 on Page 234.

The number of possible r-tuples is $\binom{|V|}{r}$. And the running time of the Depth First Search will be $|E|$ times that. While this is still polynomial for any fixed r, and solving the entire system of equations is not polynomial time if $P \neq NP$, eventually this would get expensive. Perhaps $r = 5$ is the limit of feasibility. Note that this algorithm is not polynomial time if r is not fixed. Usually, this would be stated as the algorithm is "pseudo-polynomial time in r" (see Section 11.4 on Page 192). The running time is

$$\Theta\left(\binom{|V|}{r}|V|d\right) = O(|V|^{r+1}d) = O(|V|^r|E|)$$

for finding a r-tuple. Now the complexity of solving the system after the separation must be considered.

If one has removed $r = 1, 2$, or 3, vertices, then one must solve $2q$, $2q^2$, or $2q^3$, systems of equations instead of just one, or $2q^r$ in general. If q is small, this is quite manageable. Since we usually are concerned with $\mathbb{GF}(2)$, where $q = 2$, solving one system of 400 variables would be much worse than solving 16 systems of roughly 200 variables, using $n = 400$, $r = 3$, $q = 2$ as an example.

In other contexts, when a trade-off like this is found, one might say that perhaps r-tuples of size $r = \log n$ might be the right size, instead of always using triples or pairs. Unfortunately, checking for disconnection by removing all sets of $\log_2 n$ verteces (instead of single points, pairs, or triplets), does not result in a polynomial time algorithm, because $\binom{n}{\log n}$ is not polynomial to n.

12.8.6 Relation to Menger's Theorem

The vertex connection number of a graph is the smallest number of vertices that can be removed from it to disconnect it. Menger's theorem states that if there are exactly k vertex-distinct paths from v_i to v_j, then the vertex connection number of the graph is at most k. A corollary is that the vertex connection number is the minimum of the number of vertex distinct paths between v_i and v_j, for all $v_i, v_j \in V$.

This could be found with $|V|^2$ runs of a max-flow-min-cut algorithm. However, that is not useful in this problem. The reason is that we are interested in balanced cuts, namely cuts that leave the graph in roughly two equally sized parts in terms of the number of vertices. This is explained below.

12.8.7 Balance in Vertex Cuts

The issue of balance among the halves of the graph is crucial. Suppose the system of 500 variables was broken by removal of 2 variables to make a system of 248 variables and a system of 250 variables. Then it is possible that solving 4 systems of 248 and 4 systems of 250 is feasible, while solving a system of 500 variables might be infeasible. On the other hand, suppose the removal of 2 variables resulted in a system of 3 variables, and a system of 495 variables. It might be that solving a 495 variable system 4 times is better than solving a 500 variable system, but the difference is probably not enough to make an infeasible system feasible. It might even make matters worse, since the solving must occur four times.

Therefore, when searching for removal sets, it is useful to keep track of all removal sets encountered. The "difficulty" of a removal set should be the size of the largest island that remains after the removal. This also gives a strong advantage to removal sets which cut the graph into three rather than two distinct subsystems, as is quite reasonable. The vertex subset with the smallest "difficulty" should be chosen.

12.8.7.1 Infinite Fields and Large Finite Fields

At the time of this writting, Kenneth Wong, Robert Lewis and the author have joint work on an algorithm, using simulated annealing and other graph-theoretic methods, that takes this into account, and produces a balanced cut more directly [231]. There was not time to incorporate that work here, but see the cited paper.

Furthermore, for infinite fields, or large finite fields, it is not possible to simply guess the values of c removed variables in a vertex cut. However, Robert Lewis has shown how to resolve the matter using resultants. The variables are removed without guessing their values. See [231].

12.8.8 Applicability

This algorithm will be of interest to cryptographers from several areas. First, one can accelerate solving a system of equations. Second, those building stream ciphers, block ciphers, hash functions, and public key cryptosystems from random sparse systems of polynomials equations will benefit from checking those systems for separability (e.g. QUAD) [42]. If the random system is separable, it would be

---

INPUT: A system of polynomial equations, and a parameter r.
OUTPUT: A set of at most r variables, such that if the value of each is known, the system becomes separable, or a failure if no such set of variables (of size r or smaller) exists.

1: Generate the graph, $G = (V, E)$, according to Section 12.8.3 on Page 230.
2: Make a list L of all subsets of V of size r or less, including the empty set.
3: $C \leftarrow \{\}$
4: For each subset $S \in L$ do

     a. Remove the vertices in S from the graph, and any edges incident on them.
     b. Run a Depth First Search to see if the graph is disconnected.
     c. If DISCONNECTED then
         i. Measure the size of each island in the disconnected graph.
         ii. Let d be the size of the largest island.
         iii. Insert S into C, tagging S with the difficulty d.
     d. Restore the vertices from S and any edges incident on them.

5: If C is empty, return failure.
6: If C is non-empty, return the subset S tagged with the smallest d.

Algorithm 26: Searching for an r-tuple of variables to guess. [G. Bard]

easy to break. Third, those developing general purpose algorithms for solving systems of equations can use this method to make sure that sample problems are not "too easy" when generating random, highly sparse, systems.

Analogs for this method exist in the sparse linear system of equations world. See [203, Ch. 13.6]

12.9 Resultants

Prior to the discovery of Gröbner Bases by Bruno Buchberger in the middle of the 20th century, the method of resultants was the primary method for solving polynomial systems of equations. There are many cases in which resultants out perform Gröbner Bases [162].

12.9.1 The Univariate Case

The resultant of $f(x)$ and $g(x)$ is a value which is zero if and only if f and g have a root in common, in the algebraic closure of the field being discussed. First, this is useful in its own right, but also it is useful because f has a repeated root if and only if $f(x)$ and $f'(x)$ share a root. The resultant of $f(x)$ and $f'(x)$ is called the discriminant of $f(x)$.

The discriminant is important in its own right in algebraic geometry, because when it equals zero, the polynomial has a repeated root and is in some way "degenerate". For example, the discriminant of $ax^2 + bx + c$ is $b^2 - 4ac$, which is familiar with to anyone who knows the quadratic formula.

There are many methods for calculating resultants. One choice is the determinant of the Sylvester Matrix [228], for two polynomials f of degree d_1 and g of degree d_2, which is given by

$$\text{res}(a_{d_1}x^{d_1} + a_{d_1-1}x^{d_1-1} + \cdots + a_1 x + a_0, b_{d_2}x^{d_2} + b_{d_2-1}x^{d_2-1} + \cdots + b_1 x + b_0) =$$

$$= \begin{bmatrix}
a_{d_1} & a_{d_1-1} & a_{d_1-2} & \cdots & a_1 & a_0 & 0 & 0 & 0 & \cdots & 0 & 0 \\
0 & a_{d_1} & a_{d_1-1} & \cdots & a_2 & a_1 & a_0 & 0 & 0 & \cdots & 0 & 0 \\
0 & 0 & a_{d_1} & \cdots & a_3 & a_2 & a_1 & a_0 & 0 & \cdots & 0 & 0 \\
\vdots & \vdots & \vdots & \ddots & \vdots & \vdots & \vdots & \vdots & \vdots & \ddots & \vdots & \vdots \\
0 & 0 & 0 & \cdots & 0 & 0 & 0 & a_{d_1} & a_{d_1-1} & \cdots & a_1 & a_0 \\
b_{d_2} & b_{d_2-1} & b_{d_2-2} & \cdots & b_1 & b_0 & 0 & 0 & 0 & \cdots & 0 & 0 \\
0 & b_{d_1} & b_{d_2-1} & \cdots & b_2 & b_1 & b_0 & 0 & 0 & \cdots & 0 & 0 \\
0 & 0 & b_{d_2} & \cdots & b_3 & b_2 & b_1 & b_0 & 0 & \cdots & 0 & 0 \\
\vdots & \vdots & \vdots & \ddots & \vdots & \vdots & \vdots & \vdots & \vdots & \ddots & \vdots & \vdots \\
0 & 0 & 0 & \cdots & 0 & 0 & 0 & b_{d_2} & b_{d_2-1} & \cdots & b_1 & b_0
\end{bmatrix}$$

where the polynomial f is repeated over d_2 rows and the polynomial g is repeated over d_1 rows, to produce a $(d_1 + d_2) \times (d_1 + d_2)$ matrix. Note that in every case, the constant term of each polynomial will appear in the final column, exactly once, for both polynomials. This is a good memory-hook when doing the operation by hand. Another memory hook is that the Sylvester matrix is always square.

12.9.2 The Bivariate Case

Suppose we have two polynomials, $f(x,y) = x^2 + y^2 - 1$ and $g(x,y) = x^3 - x - y$, and we are curious where these curves intersect. This is a very basic question in the study of polynomials from $\mathbb{Q}[x,y]$.

We can write these polynomials as polynomials in y alone, whose coefficients are from $\mathbb{Q}[x]$, the set of polynomial functions in the variable x, with rational number coefficients. Then we obtain $f(y) = (1)y^2 + (0)y + (x^2 - 1)$ and also $g(y) = (-1)y + (x^3 - x)$. The resultant, found via the Sylvester Matrix, is therefore given by

$$\det \begin{bmatrix} 1 & 0 & (x^2-1) \\ -1 & (x^3-x) & 0 \\ 0 & -1 & (x^3-x) \end{bmatrix} = (1)\det \begin{bmatrix} (x^3-x) & 0 \\ -1 & (x^3-x) \end{bmatrix} + 0 + (x^2-1)\det \begin{bmatrix} -1 & (x^3-x) \\ 0 & -1 \end{bmatrix}$$

$$= (x^3-x)^2 + (x^2-1)(1-x^3+x)$$
$$= x^6 - x^5 - 2x^4 + 2x^3 + 2x^2 - x - 1$$
$$= (x-1)(x+1)(x^4 - x^3 - x^2 + x + 1)$$

The quartic factor has no real roots, and so we can concentrate on $x = 1$ and $x = -1$. We can see that this leads to the points of intersection $(1,0)$ and $(-1,0)$ by substituting into either of the original polynomials.

One can see many inefficiencies here, because we are taking a determinant of a matrix which might become large, and whose entries are from a polynomial ring. Even multiplying two elements of a polynomial ring is expensive. Therefore, other methods have been invented, including taking the determinant of the Bezout-Caley Matrix [160], Dixon's Resultant Method [161], and others.

12.9.3 Multivariate Case

Now suppose we have m polynomial equations $f_1, f_2, \ldots, f_m$ in n variables, $x_1, x_2, \ldots, x_n$. If a particular $\mathbf{x} = (x_1, \ldots, x_n)$ is a solution to the polynomial system, then it is a solution to each polynomial individually. This means that it is a common root of f_i and f_j for any i and j in $\{1, 2, \ldots, m\}$.

Suppose we rewrite all the polynomials as if they were not over $\mathbb{Q}[x_1, \ldots, x_n]$, but rather as univariate polynomials in terms of x_1, whose coefficients were from $\mathbb{Q}[x_2, \ldots, x_n]$. This creates m univariate polynomials. Accordingly, there are $\binom{m}{2}$ possible resultants, of f_i and f_j. For any solution $\mathbf{x}$, each of these polynomials (which are in terms of $x_2, \ldots, x_n$) must be satisfied.

Thus we have replaced m equations in n unknowns with $\binom{m}{2}$ equations in $n - 1$ unknowns. Also, we need not keep all the equations necessarily. Vaguely, there is some redundancy. Of course, if there are many more equations than unknowns, this opens up opportunities for other algorithms in this chapter.

We could say we have "eliminated" x_1 from the system of equations, because we have a new system without it. But perhaps it would have been wiser to eliminate x_n or some other variable. The methods and techniques related to this are, accordingly, called "Elimination Theory" [86, Ch. 3].

12.9.3.1 Variables versus Parameters

Suppose we wish to frequently find the intersection of three spheres. The equation of a sphere with center (c_x, c_y, c_z) and radius r is given by

$$(x - c_x)^2 + (y - c_y)^2 + (z - c_z)^2 - r^2 = 0$$

and so we could simply substitute the correct values in each case, and solve. But it might be more useful to solve the general case, keeping (c_x, c_y, c_z) as unknown.

There is (x, y, z), which we want to find out, and it is common to each of the 3 spheres. But the three coordinates of the center, plus the radius, is another 4 unknowns per equation or 12 total. Surely a system of polynomials with 15 unknowns is fairly complex, but this problem seems simple.

There is a further distinction that we have not yet made. In any particular *instance* of the problem, the coordinates of the centers and of the radius are known, and only take a single value. Meanwhile, (x, y, z) might have no value (if there are no solutions) or multiple values, and they are not known at the start of the problem. Thus unknowns are divided into "variables" and "parameters" in the theory of resultants and in Elimination Theory, whereas normally "variables" and "unknowns" are treated as synonyms.

A fascinating generalization of this is the Apollonius problem [224], where one wants to find one of the 8 circles tangent to three particular given circles. In three dimensions, this becomes a search for a sphere tangent to four given spheres. Usually we want to solve this problem in relation to molecular chemistry, for example in lock-and-key drug design, to discover what atom or molecule can best fit in a receptor to block it. Hence the term "blockers" for this wide category of drugs. See [163] for example.

12.9.3.2 Solving the System

The following is oversimplified, and much faster methods exist. The objective, however, is to demonstrate what sorts of advantages that methods for solving polynomial systems based on resultants have over the other methods in this book.

One method to solve the problem then would be to write the 3 equations, which now are in 3 variables with 12 parameters. We could eliminate x, which means rewriting each equation as a univariate polynomial in x, and taking all $\binom{3}{2}$ resultants.

Now we have 3 equations, each in y and z plus 12 parameters. Call these "the Stage 1 equations." Any solution to the original equations is a solution to these. We could eliminate y, which means writing each of the Stage 1 equations as a univariate polynomial in y, and taking all $\binom{3}{2}$ resultants.

At this point, we have 3 equations, each in z plus 12 parameters. Call these "the Stage 2 equations." Any solution to the originals has a z which satisfies each of these. Since z is a solution to every equation, then z is a solution to the gcd (greatest common divisor) of every pair of these equations. By taking the gcd of all the equations in Stage 2, we will have a relatively small polynomial, in terms of z (and the 12 parameters).

By whatever means are appropriate to the problem (numerical methods or algebraic methods), we can recover all the possible values of z from these Stage 2 equations. Then we can plug those into the Stage 1 equations to get the values of y. Finally, we can plug those into the original equations to get the values of x.

Lastly, there may be times when we do not care about the value of x, for some reason inherent in the problem. Eliminating these "boring" variables first would produce a system without them. We would have no obligation to back-solve and recovery their values, and we could stop early.

12.9.4 Further Reading

The use of resultants in cryptography is extremely rare. One example is the paper [215]. Another, by Kenneth Wong, Robert Lewis and this author, is referred to in Section 12.8.7.1 on Page 233, which is useful for separable systems over large finite fields, or infinite fields. One "removes" the variables via resultants rather than guessing their values. Nonetheless, we have not done the subject sufficient justice here. To learn more, read [86, Ch. 3.5–3.6] or [59, Sec. 3.3.2].

12.10 The Raddum-Semaev Method

Recently, a new approach to solving sparse polynomial systems of equations over $\mathbb{GF}(2)$ was introduced by Hårvard Raddum and Igor Semaev. Here, we give a general description but the details can be found in [196], [132], and [209]. More recently some proofs and interesting properties can be found in [210] and [197].

One model of sparsity could be that any particular equation typically only involves a subset of the variables of the entire system. This is somewhat distinct from the normal β view of sparsity, for obvious reasons. Ordinarily, a low β would imply a small fraction of the possible *monomials* are present, but perhaps most or all variables are present. Here, we discuss a form of sparsity when only a limited number of variables actually appear in each equation.

In cryptanalysis, this could happen, as variables from round i are usually related to those from round $i - 1$ and to round $i + 1$. In fact, Raddum and Semaev used this method to attack the Data Encryption Standard (DES) to 4 and 5 rounds. The 4 round method was much faster than brute force. In Bivium and Trivium (see Section 5.1.2 on Page 61), the equations each have only 6 or fewer variables in them.

12.10.1 Building the Graph

First, an undirected graph is constructed. Every vertex in the graph will have a few variables associated with it. Thus for any $v \in V$, the set $\mathrm{Var}(v)$ is a subset of the set of variables in the equation system. Each vertex will also have a list of binary strings. The length of those strings will be equal to the size of $\mathrm{Var}(v)$. Each one is a possible assignment of values 1 or 0 to each variable. As the algorithm proceeds, impossible assignments will be identified and deleted.

Each equation in the system is represented by a vertex, and associated with that vertex is an exhaustive list of all the settings for the variables which it contains, that would result in that polynomial being satisfied. If a polynomial constains ℓ variables, then there would be between 0 and 2^ℓ entries in this list, and typically $2^{\ell-1}$. These vertices are called the "upper set". The easiest way to construct this list to simply check each of the 2^ℓ possibilities.

Next, we will add a "lower set" of vertices. For any two vertices v_i and v_j in the upper set, first calculate $\text{Var}(v_i) \cap \text{Var}(v_j)$. If this is empty, do nothing and continue to the next pair of "upper set" vertices. If this is non-empty, then create a new vertex w. We will set $\text{Var}(w) = \text{Var}(v_i) \cap \text{Var}(v_j)$. The list associated with vertex w shall be initially all $2^{|\text{Var}(w)|}$ possibilities.

An edge shall now be drawn from v_i to w and also from v_j to w. There are to be no other edges in the graph. Moreover, since edges only go from the "upper set" to the "lower set" then the graph is bipartite.

One detail has been omitted. It is possible that

$$\text{Var}(v_i) \cap \text{Var}(v_j) = \text{Var}(v_k) \cap \text{Var}(v_m) = \mathscr{S}$$

and in fact this might happen quite frequently. If this is the case, then the first time the intersection is performed (namely v_i and v_j), a vertex is created representing $\mathscr{S}$ (call it w). During some later intersection (perhaps v_k and v_m), this same set may appear again. Do not create a new vertex w but instead draw edges from w to both v_k and v_m. This is very important, otherwise the "lower set" would be huge.

12.10.2 Agreeing

The primary vertex operation is "agreeing" and proceeds as follows: For any two vertices that are connected by an edge, a comparison is made of the list of possibilites for those two vertices. Any variable can either be true or false, but it cannot be both at the same time. Thus it is likely that some of the entries of one list are mutually exclusive with some of the entries of the other list. Since every polynomial must be satisfied in a solution, any entry of one list that cannot agree with any of the entries of the other list must be spurious. And thus a few settings can be ruled out, and therefore deleted.

The easy way to do this is to realize that in the graph, if an edge goes from v_i to v_j, then either $\text{Var}(v_i) \subset \text{Var}(v_j)$ or $\text{Var}(v_j) \subset \text{Var}(v_i)$. In fact, the "upper set" nodes contain the "lower set". Take the list to be a matrix, with each row being a possible set of assignments, and each column associated to one of the variables in $\text{Var}(v)$. Without loss of generality, let us assume that $\text{Var}(v_i) \subset \text{Var}(v_j)$. Make a copy of the matrix for v_j (call it M), and then delete all columns from M associated with variables not found in v_i. If there is a row found in M which is not found in the matrix of v_i, then delete that row from the matrix of v_j. Likewise, if there is a row found in the matrix of v_i that is not found in M, then delete that row from v_j. Once this is done, vertices v_i and v_j are said "to agree". This is because their lists share the same settings for any variables that concern both vertices.

12.10.3 Propigation

Initially, one can try to run agreeing on each edge. After that, whenever a vertex has any changes in its list, it should try to agree with each of its neighbors. This way, changes and newly discovered facts propagate through the system. There are probably more sophisticated ways of detecting when an "agreeing" should occur.

12.10.4 Termination

Two possible outcomes can terminate the algorithm. First, if each vertex has only one string in its list (one row in its matrix), then the algorithm has found the unique solution to the problem. Second, one or more vertices might have no strings in its list (an empty matrix). This means that the equations have been shown to be unsatisfiable.

Of course, it may come to pass that all possible "agreeing" operations have been executed, and the system has not been terminated. For this situation, two more operations exist.

Basically, a configuration not on a list is not possible. However, there are still impossible configurations which are on the lists. All deleted configurations are deleted because they cannot be part of a satisfying solution. By contrapositive, a satisfying configuration will never be deleted. But this is very different from saying all non-satisfying configurations will be deleted. The agreeing step, along with the gluing step to be presented momentarily, will remove many, but not necessarily all, spurious configurations.

12.10.5 Gluing

In general, given two sets X and Y, one can partition them into A, B, C such that $A \cap C = \{\}$ and $B = X \cap Y$. This is essentially the operation of drawing the Venn Diagram of X and Y, with B being the football shaped region in the overlap of the two circles, and A and C being the two cookie shaped regions on either side. See Figure 12.1.

Once all possible agreements have been made, consider two vertices v_i and v_j. We are going to merge these two vertices, and make a new vertex, call it v_w. We will calculate (A, B, C), the Venn partition of $\text{Var}(v_i) \cup \text{Var}(c_j)$. That is to say, $B = \text{Var}(v_i) \cap \text{Var}(v_j)$ while $A \cap C = \{\}$. Then set $\text{Var}(w) = \text{Var}(v_i) \cup \text{Var}(v_j)$.

Each string in the list of v_i (or row in the matrix of v_i) consists of settings for variables in A and in B. Next, calculate a tag for this string/row representing the settings for variables in B alone (that is, deleting the settings for A). Do this also for every string/row in the list/matrix of v_j.

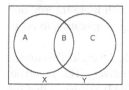

Fig. 12.1 Venn Diagram

Now, for every string/row in the list/matrix of v_i, make a new string/row in the list/matrix of w for each string/row in the list/matrix of v_j whose tag is the same as the tag of the row/string we are considering from v_i. (We will give an example momentarily). The entries for the A variables come from v_i, and the entries for the C variables come from v_j. By virtue of the fact that the tags are the same, we know that these two strings/rows agree on the B variables.

This requires an example. Suppose v_i has $\mathrm{Var}(v_i) = x_1, x_2, x_3$ and $\mathrm{Var}(v_j) = x_3, x_4, x_5$. Then $A = \{x_1, x_2\}$ while $B = \{x_3\}$ and of course $C = \{x_4, x_5\}$. Now suppose (to use the matrix analogy)

$$M_i = \begin{bmatrix} x_1 & x_2 & x_3 \\ 0 & 1 & 0 \\ 1 & 1 & 0 \\ 0 & 0 & 1 \\ 1 & 1 & 1 \\ 1 & 0 & 1 \end{bmatrix} \quad ; \quad M_j = \begin{bmatrix} x_3 & x_4 & x_5 \\ 0 & 1 & 1 \\ 0 & 0 & 0 \\ 0 & 0 & 1 \\ 1 & 0 & 0 \\ 1 & 0 & 1 \end{bmatrix}$$

then we should finish with

$$M_w = \begin{bmatrix} x_1 & x_2 & x_3 & x_4 & x_5 \\ 0 & 1 & 0 & 1 & 1 \\ 0 & 1 & 0 & 0 & 0 \\ 0 & 1 & 0 & 0 & 1 \\ 1 & 1 & 0 & 1 & 1 \\ 1 & 1 & 0 & 0 & 0 \\ 1 & 1 & 0 & 0 & 1 \\ 0 & 0 & 1 & 0 & 0 \\ 0 & 0 & 1 & 0 & 1 \\ 1 & 1 & 1 & 0 & 0 \\ 1 & 1 & 1 & 0 & 1 \\ 1 & 0 & 1 & 0 & 0 \\ 1 & 0 & 1 & 0 & 1 \end{bmatrix}$$

which while hard to describe in words is easy to calculate. Now w describes 12 out of 32 possible configurations for $\{x_1, x_2, x_3, x_4, x_5\}$.

An edge should be drawn from w to each neighbor of v_i and each neighbor of v_j. This new vertex w contains all the information from v_i and v_j, and so we can delete

v_i and v_j after w is finished being created. But w contains much more information that v_i and v_j did collectively, and one expects much agreeing propagation to occur afterward.

In our example, M_w now has 12 possibilities, but v_i and v_j had 5 each. This means that one would have anticipated 25 ways to combine the data naïvely. The gain from gluing is the removal of those 13 spurious configurations. The best case is that the number of rows in M_w is the maximum of the number of rows in M_i and M_j. The worst case is the product of M_i and M_j.

Thus, to avoid using too much memory, one must glue rarely, and particularly vertices with short lists.

12.10.6 Splitting

In the Guess-and-Determine paradigm (see Section 11.7 on Page 206) one simply picks some variables (perhaps g of them) and runs the system once for each of the possible 2^g guesses of the values of those variables. What is done here is more general.

One picks a vertex and splits its list or matrix in half, taking half of the strings/rows. Then let the algorithm continue. Perhaps it terminates with a solution. If it terminates with no solution, then restore the state at the moment of the guess, and repeat with the other half of the strings/rows. Clearly, guessing a variable is merely a special case of this.

If instead of terminating with a solution or no solutions, we might end up with getting stuck with no more argeeing/gluing operations possible. Then we must make another guess. If a contradiction occurs (i.e. no solution is found) then we should undo the most recent splitting first. This makes a search-tree.

This idea is essentially identical to the Davis-Putnam [92] method of backtracking, described in Section 14.3 on Page 269, but more general.

12.10.7 Summary

Therefore, one can see that this is a very novel and interesting way of solving systems of equations. It requires that each equation contain only a small fraction of the possible variables, but this could easily occur in practice. There are many more details in the two papers which we will not describe here. In any case, this is a very new and very different approach, and it may turn out to be, in future decades, the principle method for solving systems in practice. Alternatively, it might not be useful at all. No one can predict the future.

Also, the methods presented in Section 12.8 on Page 228 would seem to be an excellent pre-processor for the Raddum-Semaev method.

12.11 The Zhuang-Zi Algorithm

Another algorithm for solving polynomial systems of equations over finite fields is called the Zhuang-Zi algorithm, and was first proposed by Ding, et al, in 2006. See [95] and [96]. In this case, one lifts the problem of solving a set of multivariate polynomial equations over a small finite field to solving a set of single variable equations over an extension field. It is related to the Berlekamp-Massey algorithm [225].

This algorithm is also mentioned in [97, Ch. 7.6]. We are not aware of any uses of the algorithm other than by its authors, but the algorithm is extremely new, and this might change in future years.

12.12 Homotopy Approach

A notable class of algorithms are the homotopy methods. They have not, to the author's knowledge, been used in a cryptographic problem at the time of the writing of this text (late 2007). But, perhaps this is a new research opportunity.

The very classic example of a homotopy function is that function which slowly changes a coffee cup into a donut. (The reader, if unfamiliar with this, should ask any topologist). Here, we present how the algorithm works over $\mathbb{R}$ variables.

Consider a polynomial system over the reals, $g_1(x_1,\ldots,x_n)$, $g_2(x_1,\ldots,x_n)$, ..., $g_m(x_1,\ldots,x_n)$, that is difficult to solve. Imagine we have some $f_1(x_1,\ldots,x_n)$, $f_2(x_1,\ldots,x_n)$, ..., $f_m(x_1,\ldots,x_n)$, that we do know how to solve—usually because we know all the solutions to start. For example, each f is univariate, or something of this sort. Then define

$$h_i(x_1,\ldots,x_n) = \alpha g_i(x_1,\ldots,x_n) + (1-\alpha)f_i(x_1,\ldots,x_n)$$

Since the f's can be solved, then the h's can be solved when α is zero. Since each h_i is continuous, (in fact it is smooth), an infinitesimal variation in α only moves the roots infinitesimally. Therefore, a series of α's, going from 0 to 1, each very slightly different from the one before it, will produce a series of systems of equations, whose roots vary only slightly. Furthermore, each root of one system is a "good starting guess" for a multi-dimensional Newton's Method approach or Gradient Descent approach on the next.

The classical example of this would be the g's being a perturbed or slightly modified quadratic form, and the f's being a diagonal quadratic form, i.e. $x_1^2 = 1$, $x_2^2 = 1$, ... so that the 2^n roots initially are of the form $x_i = \pm 1$.

Applicability

It should be noted that while $\overline{\mathbb{GF}(2)}$, the algebraic closure of $\mathbb{GF}(2)$, is infinite, clearly $\mathbb{GF}(2)$ is not. Therefore, it is hard to imagine precisely what α would be in our finite field case. Also, note that finite fields cannot be ordered so it is unclear what a "sufficiently small" change would be for α.

Chapter 13
Converting MQ to CNF-SAT

13.1 Summary

The computational hardness of solving large systems of sparse and low-degree multivariate equations is a necessary condition for the security of most modern symmetric cryptographic schemes. Notably, most cryptosystems can be implemented with relatively inexpensive hardware, and thus have moderatley low gate counts, resulting in a sparse system of equations, which in turn renders such attacks feasible. Keeloq, described in the first part of this book, is an excellent example of this. On one hand, numerous recent papers on the XL algorithm and more sophisticated Gröbner-bases techniques [82, 81, 109, 110] demonstrate that systems of equations are efficiently solvable when they are sufficiently overdetermined or have a hidden internal algebraic structure that implies the existence of some useful algebraic relations.

On the other hand, most of this work, as well as most successful algebraic attacks, involves dense-oriented, not sparse-oriented algorithms, at least until linearization by XL or a similar algorithm. The natural sparsity, arising from the low gate-count, is thus wasted during the polynomial stage, even if it is taken advantage of in the linear algebra stage by the Block Wiedemann Algorithm or Lanczos's Algorithm . See also Appendix D on Page 323. No polynomial-system-solving algorithm we are aware of except the very recently published methods of Raddum and Semaev [132], demonstrates that a significant benefit is obtained from the extreme sparsity of some systems of equations.

In this chapter, we study methods for efficiently converting $\mathbb{GF}(2)$ systems of multivariate polynomial equations into a conjunctive normal form satisfiability (CNF-SAT) problem, for which excellent heuristic algorithms have been developed in recent years. Please note that much of this information can be found in the paper by the author, Chris Jefferson, and Nicolas Courtois [33], which also has some preliminary experiments.

The sparsity of a system of equations, denoted β, is the ratio of coefficients that are non-zero to the total number of possible coefficients. For example, in a quadratic system of m equations in n unknowns over $\mathbb{GF}(2)$, this would be

$$\beta = \frac{\kappa}{m\left(\binom{n}{2} + \binom{n}{1} + \binom{n}{0}\right)}$$

where κ is the number of non-zero coefficients in the system, sometimes called the "content" of the system of equations.

A direct application of this method gives very efficient results: we find that sparse multivariate quadratic systems (especially if over-defined) can be solved much faster than by exhaustive search if $\beta \leq 1/100$. In particular, our method requires no additional memory beyond that required to store the problem, and so often terminates with an answer for problems that cause MAGMA [2] and SINGULAR [9] to crash. On the other hand, if MAGMA or SINGULAR does not crash, then they tend to be faster than our method, but this case includes only the smallest sample problems.

Specific details of running times are subject to change from year to year. Therefore, it is not wise to include them in a book, and therefore, we recommend the reader who is interested in precise running times to look at papers on the topic in the published literature. For the running times that were used to write this chapter, see [33]. However, we have retained the experiments that are used to justify our model (the Gibrat distribution) for the running-time, and for modeling the effects of the cutting number (a parameter) and preprocessing.

13.2 Introduction

It is well known that the problem of solving a multivariate simultaneous system of quadratic equations over $\mathbb{GF}(2)$ (the MQ problem) is NP-hard (See Section 11.5). Another NP-hard problem is finding a satisfying assignment for a logical expression in several variables (the SAT problem) [148]. Inspired by the possibility that either could be an efficient tool for the solution of the other, since all NP-Complete problems are polynomially equivalent, we began this investigation.

There exist several off-the-shelf SAT-solvers, such as MINISAT [104] [6], which can solve even relatively large SAT problems on an ordinary PC. We investigate the use of SAT-solvers as a tool for solving a random MQ problem. In particular, we find that if the system of equations is sparse or over-defined, then the SAT-solver technique works faster than brute-force exhaustive search. If the system is both sparse and over-defined, then the system can be solved quite effectively (see Section 13.5 on Page 255).

In Section 13.2.1 we describe how this work applies to algebraic cryptanalysis. We define some notation and terms in Section 13.3, and describe the method of conversion of MQ problems into CNF-SAT problems in Section 13.4. We review previous work in Section 13.7.1. Finally, we note possible applications to cubic systems in Section 13.6. A brief overview of SAT solvers is given in Section 14.

While we take as given the NP-Completeness of these two problems in this chapter, note it is proven in Section 11.5.

13.2.1 Application to Cryptanalysis

As mentioned earlier, Algebraic Cryptanalysis can be summarized as a two-step process. First, given a cipher system, one converts it into a system of equations. Second, the system of equations is solved to retrieve either a key or a plaintext. Furthermore, note that all systems of equations over finite fields can be written as polynomial systems, as a special case of Theorem 72 on Page 190.

As pointed out by Courtois and Pieprzyk [81], this system of equations will be sparse, since efficient implementations of real-world systems require low gate counts. In practice, the systems are very sparse—the system used to break six rounds of DES in [76] has 2900 variables, 3056 equations and 4331 monomials appear somewhere in the system. There would be $\binom{2900}{2} + \binom{2900}{1} = 4,206,450$ possible monomials, and those authors report less than 15 monomials per equation, or $\beta = 3.57 \times 10^{-6}$.

It is also known that any system of any degree can be written as a degree 2 system. This is done by using the following step, repeatedly:

$$\{m = wxyz\} \Rightarrow \{a = wx; b = yz; m = ab\}$$

Note: this process is described in detail in Section 11.4 on Page 192.

Finally, it is usually the case that one can write additional equations by assuming that many plaintext-ciphertext pairs arc available. While the number of pairs is not literally unbounded, as many stream ciphers have a limit of 2^{40} bits before a new key is required, generally one has an over-abundance of equations. Therefore, we include in this study only systems where the number of equations is greater than or equal to the number of unknowns.

13.3 Notation and Definitions

An instance of the MQ problem is a set of equations

$$f_1(x_1, \ldots, x_n) = y_1, f_2(x_1, \ldots, x_n) = y_2, \ldots, f_m(x_1, \ldots, x_n) = y_m$$

where each f_i is a second degree polynomial over $\mathbb{GF}(2)$. By adjusting the constant term of each polynomial, it becomes sufficient to consider only those problems with $y_j = 0$ for all j. Note that n is the number of variables and m is the number equations.

If we define $\gamma = m/n$ or $\gamma n = m$, then $\gamma = 1$ will imply an exactly defined system, $\gamma > 1$ an over-defined system and $\gamma < 1$ an under-defined system. We will not consider under-defined systems here. The value of γ will be called "the over-definition" of a system, and it can also be denoted "c". Let M denote the number of possible monomials, including the constant monomial. Since we consider only quadratic polynomials (except for Section 13.6 on Page 258 on cubics),

$$M = \binom{n}{2} + \binom{n}{1} + 1$$

The system will be generated by flipping a weighted coin for each of the M coefficients for each equation. The value $\beta \in (0, 1]$ will be called the sparsity, and is the probability that a randomly selected coefficient is non-zero (equal to one). If $\beta \ll 1/2$ the system is considered sparse.

An instance of the Conjunctive Normal Form SAT or CNF-SAT problem is a set of clauses. Each clause is a large disjunction (OR-gate) of several variables, which can appear negated or not negated. If a set of values for all n variables makes every clause evaluate to true, then it is said to be a satisfying assignment. In this way, the set of clauses can be thought of as one long logical expression, namely a conjunction (AND-gate) of all the clauses.

13.4 Converting MQ to SAT

13.4.1 The Conversion

The conversion proceeds by three major steps. First, some preprocessing might be performed to make the system more amenable to this conversion (more detail will follow). Next, the system of polynomials will be converted to a (larger) linear system and a set of CNF clauses that render each monomial equivalent to a variable in that linear system. Lastly, the linear system will be converted to an equivalent set of clauses.

13.4.1.1 Minor Technicality

The CNF form does not have any constants. Adding the clause consisting of (T), or equivalently $(T \vee T \vee \cdots \vee T)$, would require the variable T to be true in any satisfying solution, since all clauses must be true in any satisfying solution. Once this is done, the variable T will serve the place of the constant 1, and if needed, the variable $\bar{T}$ will serve the place of the constant 0. Otherwise constants are unavailable in CNF.

Step One: From a Polynomial System to a Linear System

Based on the above technicality, we can consider the constant term 1 to be a variable. After that, every polynomial is now a sum of linear and higher degree terms. Those terms of quadratic and higher degree will be handled as follows.

The logical expression

$$(w \vee \bar{a})(x \vee \bar{a})(y \vee \bar{a})(z \vee \bar{a})(a \vee \bar{w} \vee \bar{x} \vee \bar{y} \vee \bar{z})$$

is tautologically equivalent to $a \iff (w \wedge x \wedge y \wedge z)$, or the $\mathbb{GF}(2)$ equation $a = wxyz$. Similar expressions exist for equations of the form $a = w_1 w_2 \cdots w_r$, for any $r > 1$.

Therefore, for each monomial of degree $d > 1$ that appears in the system of equations, we shall introduce one dummy variable. One can see that $d + 1$ clauses are required, and the total length of those clauses is $3d + 1$.

Obviously, if a monomial appears more than once, there is no need to encode it twice, but instead, it should be replaced by its previously defined dummy variable. On the other hand, in a large system, particularly an over-defined one, it is likely that every possible monomial appears at least once in some equation in the system. Therefore we will assume this is the case, but in extremely sparse systems that are not very over-defined, this is pessimistic, particularly for high degree systems.

At the risk of laboring over a minor point, note that in the production code we have a check-list, and never encode the same monomial twice, and only encode a monomial once it has appeared in the system. But, this algorithm can be encoded into LOGSPACE by simply enumerating all the possible monomials at the start, exactly once, and then continuing with the next step. We should note that, for a fixed degree, there are polynomially many monomials. If the degree is allowed to change, there are exponentially many.

Step Two: From a Linear System to a Conjunctive Normal Form Expression

Each polynomial is now a sum of variables, or equivalently a logical-XOR. Unfortunately, long XORs are known to be hard problems for SAT solvers [87]. In particular, the sum $(a + b + c + d) = 0$ is equivalent to

$$(a \vee b \vee c \vee d)(a \vee b \vee \bar{c} \vee \bar{d})(a \vee \bar{b} \vee c \vee \bar{d})(a \vee \bar{b} \vee \bar{c} \vee d) \qquad (13.1)$$
$$(\bar{a} \vee b \vee c \vee \bar{d})(\bar{a} \vee b \vee \bar{c} \vee d)(\bar{a} \vee \bar{b} \vee c \vee d)(\bar{a} \vee \bar{b} \vee \bar{c} \vee \bar{d})$$

which is to say, all arrangements of the four variables, with 0, 2, or 4 negations, or all even numbers less than or equal to four. For a sum of length ℓ, where $2 \lfloor \ell/2 \rfloor = j$, this requires

$$\binom{\ell}{0} + \binom{\ell}{2} + \binom{\ell}{4} + \cdots + \binom{\ell}{j} = 2^{\ell-1}$$

clauses, which is exponential compared to ℓ.

To remedy this, cut each sum into subsums of length c. (We will later call c the cutting number). For example, the equation $x_1 + x_2 + \cdots + x_\ell = 0$ is clearly equivalent to

$$x_1 + x_2 + x_3 + y_1 = 0$$
$$y_1 + x_6 + x_7 + y_2 = 0$$

$$\vdots \quad \vdots \quad \vdots$$

$$y_i + x_{4i+2} + x_{4i+3} + y_{i+1} = 0$$

$$\vdots \quad \vdots \quad \vdots$$

$$y_h + x_{\ell-2} + x_{\ell-1} + x_\ell = 0$$

if $\ell \equiv 2 (\bmod\ c)$. If $\ell \not\equiv 2 (\bmod\ c)$ then the final sum is shorter, and this is more efficient because a sum or XOR of shorter length requires fewer clauses. Therefore it is safe to be pessimistic and assume all equations are of length $\ell \equiv 2 (\bmod\ c)$. In either case, one can calculate $h = \lceil \ell/c \rceil - 2$. Thus there will be $h + 1$ subsums, and each will require 2^{c-1} clauses of length c each, via Equation 13.2 on Page 249.

13.4.2 Measures of Difficulty

Three common measures of the size of a CNF-SAT problem are the number of clauses, the total length of all the clauses, and the number of variables. It is not known which of these is a better model of the difficulty of a CNF expression. Initially we have n variables, and 0 clauses of total length 0.

Step 1 is preprocessing which will be described later. For a quadratic system of polynomials, the cost for each unique monomial in Step 2 of the conversion is 1 dummy variable, 3 clauses, of total length 7. This needs to be done for all possible $M - n - 1$ quadratic monomials. The constant monomial requires 1 dummy variable, and 1 clause of length 1.

The cost in Step 3 requires an estimate of the expected value of the length of each equation. Since there are M possible coefficients, then this is equal to $M\beta$. For the moment, assume the cutting number is $c = 4$. There will be (in expected value) $M\beta/2 - 1$ subsums per equation, requiring $M\beta/2 - 2$ dummy variables, $4M\beta - 8$ clauses and total length $16M\beta - 32$.

This is a total of

- Variables: $n + 1 + (M - n - 1)(1) + m(M\beta/2 - 1)$.
- Clauses: $0 + 1 + (M - n - 1)(3) + m(4M\beta - 8)$.
- Length: $0 + 1 + (M - n - 1)(7) + m(16M\beta - 32)$.

Substituting $m = \gamma n$ and $M = n^2/2 + n/2 + 1$, one obtains

- Variables: $\sim n^2/2 + \gamma n^3 \beta/4$.
- Clauses: $\sim (3/2)n^2 + 2\gamma n^3 \beta$.
- Length: $\sim (7/2)n^2 + 8\gamma n^3 \beta$.

Furthermore, so long as $\beta > 1/m$ then the first term of each of those expressions can be discarded. If $\beta < 1/m$ then $(n+1)/(2\gamma)$ monomials are found in each row. So long as γ is not too large, this is quite possible, but rather sparse indeed, and so we do not consider this case further. These expressions are summarized, for several values of cutting number, in Table 13.1 on Page 251

Table 13.1 CNF Expression Difficulty Measures for Quadratic Systems, by Cutting Number

Cutting Number	Variables	Clauses	Tot. Length	Avg. Length
Cut by 3	$\sim \gamma n^3 \beta/2$	$\sim 2\gamma n^3 \beta$	$\sim 6\gamma n^3 \beta$	3
Cut by 4	$\sim \gamma n^3 \beta/4$	$\sim 2\gamma n^3 \beta$	$\sim 8\gamma n^3 \beta$	4
Cut by 5	$\sim \gamma n^3 \beta/6$	$\sim (8/3)\gamma n^3 \beta$	$\sim (40/3)\gamma n^3 \beta$	5
Cut by 6	$\sim \gamma n^3 \beta/8$	$\sim 4\gamma n^3 \beta$	$\sim 24\gamma n^3 \beta$	6
Cut by 7	$\sim \gamma n^3 \beta/10$	$\sim (6.4)\gamma n^3 \beta$	$\sim 44.8\gamma n^3 \beta$	7
Cut by 8	$\sim \gamma n^3 \beta/12$	$\sim (32/3)\gamma n^3 \beta$	$\sim (256/3)\gamma n^3 \beta$	8

13.4.2.1 Bounds on SAT

The worse-case estimates for solving SAT on a CNF problem are commonly given in terms of the number of clauses K, the number of variables n, and the total length of all the clauses L. Another commonly used metric is the average length of a clause L/K. Currently, the following bounds are known. From the number of clauses, the bound is $O(1.27202^K)$ [112]. From the total length, the bound is $O(1.073997^L)$ [137].

The situation for n, the number of variables, is a bit messier. If a conjunctive normal form problem can be written so that the longest clause is length k, then we say the problem is a k-CNF-SAT problem. All problems can be written in 3-CNF-SAT form, but this involves the introduction of many additional variables. That, in turn, would affect the n.

If the problem can be written in 4-CNF-SAT or 5-CNF-SAT affects the worse-case running time. For 3-CNF-SAT it is $O(2^{0.410n})$ given by [138]. For 4-CNF-SAT it is $O(2^{0.562n})$ given by [188]. Also in [188] are worse-case times for 5-CNF-SAT at $O(2^{0.650n})$ and 6-CNF-SAT at $O(2^{0.711n})$.

A fascinating effect occurs if a solution is known to exist and is unique; this is often the case with cryptanalysis [188]. In fact, this effect occurs even if there are multiple solutions, but they are very far apart in Hamming Distance. In the case of

3-CNF-SAT, the worse-case moves to $O(2^{0.386n})$ and for 4-CNF-SAT it moves to $O(2^{0.554n})$, both given in [188]. This is not a huge change for 4-CNF-SAT, but for 3-CNF-SAT it is very significant.

Last but not least, recall that 2-CNF-SAT is polynomial time (see Section 14.2.2 on Page 265).

13.4.3 Preprocessing

It is clear from the above expressions that n is the crucial variable in determining the number of dummy variables, clauses, and total lengths of clauses. With this in mind, we devised the following preprocessing scheme, based on the idea of Gaussian Elimination. It is executed as Step 1 of the conversion. For any specific polynomial one can reorder the terms as follows

$$x_{a_0} = x_{a_1} + x_{a_2} + \cdots + x_{a_n} + (\text{quadratic terms}) + (+1)$$

where the $+1$ term is optional, and $a_i \in \{1, \ldots, n\}$. This is, in a sense, a re-definition of x_{a_0}, and so we add this equation to every polynomial in the system where x_{a_0} appears (except the first which is now serving as its definition). Afterword, x_{a_0} will appear nowhere in the system of equations, except in its definition, effectively eliminating it as a variable. Since SAT-solvers tend to choose the most-frequently-appearing variables when deciding which cases to branch on (except in a constant fraction of cases when they select randomly, e.g. 1% of the time), x_{a_0} will not be calculated until all other variables have been set. See also, Section 12.5 on Page 219.

If there are t equations of short length in the system, then, after preprocessing, these t variables only appear in their own definitions (not even the definitions of each other), and so far as the main system is concerned, there are now $n - t$ variables. In practice, the effect of this is slightly less than a doubling of performance (see [33]).

We only consider a polynomial for elimination if it is of length 4 or shorter (called "light massage") or length 10 or shorter (called "deep massage"). The reason for the length limit is to minimize the increase of β that occurs as follows.

When Gaussian Elimination is performed on an $m \times n$ sparse $\mathbb{GF}(2)$ matrix A, in the ith iteration, the β in the region $A_{i+1,i+1} \ldots A_{m,n}$ will tend to be larger (a higher fraction of ones) than that of $A_{i,i} \ldots A_{m,n}$ in the previous iteration (See [30] or [93, Ch. 7]). Even in "Naïve Sparse Gaussian Elimination", when the lowest weight row is selected for pivoting at each step, this tends to occur (see Appendix D on Page 323). By adding two rows, the new row will have as many ones as the sum of the weights of the two original rows, minus any accidental cancellations. Therefore, by only utilizing low weight rows, one can mostly mitigate the increase in β. See the experiments in Section 13.5.4 on Page 257, and Table 13.2 on Page 259, for the effect.

The Reverse Massage

Interestingly, when cryptanalyzing Bivium, the authors of [176] did the reverse of this. When massaging, we sacrifice sparsity in the sense of making the equations longer, but we eliminate variables. What they did (see Section 5.1.3 on Page 64, where we give details) accomplished the reverse. It added variables, but shortened equations, thus increasing sparsity.

13.4.4 Fixing Variables in Advance

Since cryptographic keys are generated uniformly at random, it makes sense to generate the x_i's as fair coins. But suppose g of these are directly revealed to the SAT solver by including the short equations $x_1 = 1, x_2 = 0, \ldots, x_g = 1$, and that a satisfying solution is found in time t_{SAT}. A real world adversary would not have these g values of course, and would have to guess them, requiring time at most $2^g t_{SAT}$, or half that value for expected time. As in algebraic cryptanalysis [82] it turns out that $g = 0$ is not the optimal solution. In our experiments on actual cryptographic systems, we manually tried all g within the neighborhood of values which produced t_{SAT} between 1 second and 1 hour, to locate the optimum (the value of g which yielded the lowest running time).

Since exhaustive search requires checking 2^{n-1} possible values of $x_1, \ldots, x_n$ on average, then this method is faster than brute force if and only if t_{ver}, the time required to check one potential key, satisfies

$$t_{ver} > t_{SAT} 2^{-(n-g)}$$

This method is useful for the cryptanalysis of a specific system, e.g. DES [76]. In addition to having fewer variables, note that $m/n > m/(n-g)$, and so the "over-definition" or γ will increase, yielding further benefit to fixing variables. This is sometimes called the "guess-and-determine" method.

However, for random systems of quadratic equations, fixing variables g and substituting their values results in another system, which is an example of a random system with m equations and $n - g$ unknowns, but with slightly different sparsity.

Therefore, for random systems of equations, there is no need to do this, but for any particular real problem, it should be very useful. in fact, the author wishes to strongly encourage those who attack actual cipher systems to attempt to fix variables. This was found to be effective in practice, and is described in Section 11.7 on Page 206. For more on this topic, see Section 12.4.4 on Page 217 and Section 11.7.2 on Page 207.

13.4.4.1 Parallelization of SAT

Suppose g bits are to be fixed, and 2^p processors (for some p) are available, with $p < g$. Then of the 2^g possible values of the g fixed bits, each processor could be assigned 2^{g-p} of them. After that, no communication between processors is required, nor can processors block each other. Therefore parallelization is very efficient. If interprocess communication is possible, then the "learned clauses" (explained in Section 14.5.1 on Page 275) can be propagated to all running SAT-solvers.

In the event that thousands of volunteers could be found, as in the DES challenge of 1997, or DESCHALL Project [88], then the low communications overhead would be very important.

13.4.5 SAT-Solver Used

The solver used in this chapter is MINISAT [6], a minimalist open-source SAT solver. In fact, MINISAT has won a series of awards including the three industrial categories in the SAT 2005 competition and first place in SAT-Race 2006. MINISAT is based on Chaff, but the algorithms involved have been optimized and carefully implemented. Also, Mini-SAT has carefully optimized variants of the variable order heuristics and learned clause removal heuristics.

13.4.5.1 Note About Randomness

The program MINISAT is a randomized algorithm in the sense of using probabilistic reasoning. At first the following phenomenon was observed. If one randomly shuffles the clauses of a CNF file, the performance of the SAT-solver changes dramatically. However, running the same input file several times yields the same running time, to within 1%, each time. Obviously, this "locked" randomness maybe a lucky choice, or an unlucky one. Since the actual performance of MINISAT on these problems is log-normal (see Section 13.5.3 on Page 256), the consequences of an unlucky choice are drastic. Therefore, one (in testing) should generate 20–50 CNF files of the same system, each perhaps different by fixing a different subset of g of the original n variables, or perhaps by reordering the clauses in a random shuffle.

The latter is very cheap computationally, but the former is better, as casual experimentation has shown there are definitely "lucky" and "unlucky" choices of variables to fix. More precisely, the running time is not dependent on g alone, but also on the specific g out of n monomials chosen to be fixed. The expected value of the running time in practice can then be calculated as the mean of the running times of the 20–50 samples, each with a distinct random choice of fixed variables.

The author has paper with Robert Lewis and Kenneth Wong [231] which identifies a method based on graph theory (similar to Section 12.8 on Page 228) that offers an excellent way to choose precisely which variables should be fixed.

Error in Dissertation

At first, this phenomenon was thought to be caused by the random number generator inside the SAT-solver MINISAT being seeded by a hash of the input file. Therefore, surely shuffling the clauses would produce a new hash. This is what the author stated in his dissertation [31, Ch. 3.4.5.1]. However, this turns out not to be the case. Instead, when making assumptions and back-tracking, see Section 14.3 on Page 269, the MINISAT SAT-solver will be influenced by the order of the clauses in the file in determining which branches of the tree to take [174]. Therefore, shuffling the clauses randomly over several runs continues to be necessary for drawing statistics, and sufficient for removing the luck factor, but the underlying reason for this action is slightly different.

13.5 Experimental Results

13.5.1 The Source of the Equations

In cryptanalysis, we always know that a message was indeed sent, and so we know at least one solution exists to our equations. But, in generating a random system of equations, if over-defined, we must take care, because many systems of equations will have no solution. Therefore we used the following technique.

We started with a random system of m polynomial equations in n variables. Each coefficient was set by a weighted coin, but independently and identically distributed. By moving all the terms to the same side of the equal sign, one can easily see this as m functions on n variables, or a map $F : \mathrm{GF}(2)^n \to \mathrm{GF}(2)^m$. Then we generated a random vector $\mathbf{x}$ in $\mathrm{GF}(2)^n$ by flipping fair coins. It is easy to calculate $F(\mathbf{x}) = \mathbf{y}$. Finally we gave our tools the job of finding $\mathbf{x}$ given only $\mathbf{y}$ and F.

13.5.2 Note About the Variance

In general, the running times are *highly* variable. We propose that the log-normal distribution, sometimes called Gibrat's distribution, is a reasonable model of the running time for a given system. This implies merely that the running time t is distributed as e^x, where x is some random variable with the normal (Gaussian) distribution. In practice, however, this presents an experimental design challenge.

The distributions of the running times vary so wildly that, at absolute minimum, 50 experiments must be performed to get an estimate of the expectation. Also, minor improvements, such as parameters of massaging, are only statistically significant after hundreds of repeated trials—which makes careful tuning of the massaging process impossible.

13.5.3 The Log-Normal Distribution of Running Times

Examine Figures 13.5.3 and 13.5.3 on Page 257, which plot the probability distribution of the running time, and its natural logarithm, respectively. One can observe that the second figure "looks normal", in the sense of being a bell curve that has had its right end truncated. For comparison, Figure 13.5.3 does not look like much of a distribution at all.

The kurtosis of a random variable is a measure of "how close to normal" it is, and takes values in $[-2, \infty)$. The normal distribution has a kurtosis of zero, and positive kurtosis implies a leptokurtic distribution (one with values near the mean being more common than in the Gaussian) and negative kurtosis implies a platykurtic distribution (one with values near the mean less common than the Gaussian). The plot of running times suggests an exponential of some kind, and so upon taking the natural logarithm of each point, a set of values with very low kurtosis (0.07) was found. The plot of the logarithm is close to a bell curve, and is from 442 data points, 15 of which were longer than the manually set 1800 sec time out, and 427 of which were plotted. Since $\log_e(1800) \approx 7.496$, this explains why the graph seems truncated at $\log_e t > 7.50$.

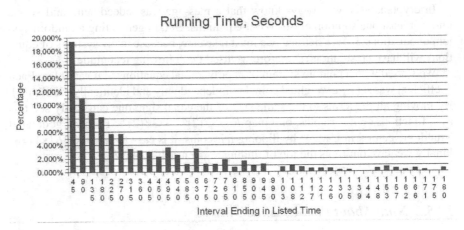

Fig. 13.1 The Distribution of Running Times, Experiment 1

A common (minimal) criteria for accepting statistical results as valid is that the standard deviation must never exceed the mean in a strictly-positive variable. Naturally, running times are strictly positive. When we look at the average and standard deviations of the running times themselves, we see that the standard deviation quite often exceeds the average, or is extremely close to it otherwise. (See Table 13.2 on Page 259, "Naïve Average" and "Naïve Standard Deviation") Therefore, these measurements should be rejected in favor of the mean and standard deviation of the logarithm of the running time.

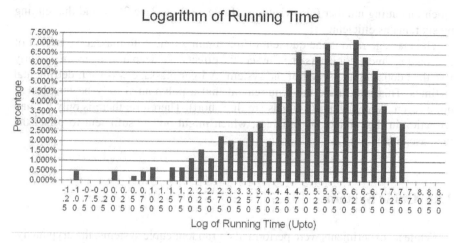

Fig. 13.2 The Distribution of the Logarithm of Running Times, Experiment 1

When looking at "Average(ln)" and "StDev(ln)", one can see that the standard deviation is always much less than the average. This gives us confidence that one should estimate the logarithm of the running time as a normal variable, which means that the running time is governed by the Gibrat distribution.

To further test the Gibrat-distribution hypothesis, let us look at the Kurtosis. Normally, a Kurtosis of ±1 is considered "somewhat" normal, with ±0.5 being "reasonably normal". We have 15 experiments. The kurtoses in 9 of the 15 experiments of the running time itself were outside of the ±1 window, and only three of the 15 experiments were within the ±0.5 window. (See "Naïve Kurtosis" in Table 13.2 on Page 259). On the other hand, taking the logarithm of the running times, only two experiments failed to be within ±1, and in fact 9 of them are within ±0.5.

Also, it should be noted that since the Kurtosis is the fourth moment of the probability distribution, it is very hard to measure from experiment.

13.5.4 The Optimal Cutting Number

See Table 13.2 on Page 259. The system solved here is identical to that in the previous experiment, except different cutting numbers and massaging numbers were used during the conversion. Also, only 50 experiments were run. The result shows that deep massaging is a worthwhile step, as it cuts the running time by half and takes only a few seconds. Furthermore, it shows cutting by six is optimal, at least for this system. Note, cutting by 8 would produce extremely large files (around 11 Mb)—those for cutting by 7 were already 5.66 Mb. Both in this case, and in casual experiments with other systems of equations, the running time does not depend too much on cutting number (also visible in Table 13.2 on Page 259), and that cutting by six remains efficient.

The massage ratio is the quotient of the running time with massaging to that of the running time without. As one can see, the effects of a deep massage were slightly less than doubling the speed of the system. A light massage was even detrimental at times. This is because the requirement that a polynomial only be length 4 is quite severe (very few polynomials are that short). Therefore, there is only a small reduction in the number of variables, which might not be sufficient to offset the increase in β.

13.6 Cubic Systems

While no experiments were performed on random cubic systems, the cryptanalysis of the first 6-rounds of the Data Encryption Standard by Courtois and Bard [76] was carried out using the method in this chapter. It was much faster than brute force; however, it was necessary to perform a great deal of human-powered preprocessing. See that paper for details.

In particular, the conversion for cubics proceeds identically to quadratics, with two exceptions. First, the number of possible monomials is much higher. Second, below, is the question if all possible monomials appear.

13.6.1 Do All Possible Monomials Appear?

Intuition implies that the assumption that every monomial is probably present might not be true. However, degree does not, in fact, affect the probability that a given monomial is present somewhere in the system, for a fixed β. The probability any particular monomial is present in any particular equation is β. Since there are m equations, the probability that a monomial is present anywhere is $1 - (1 - \beta)^m$. Degree has no role to play in that equation. Since this is obviously equal to the expected fraction of monomials missing, it is interesting to compute what β would need to be in order for a fraction r or less of the expected monomials to be present:

$$(1 - (1 - \beta)^m) \leq r$$

Since this would be a small β (for $r < 1/2$) we can approximate $(1 - \beta)^m \approx 1 - m\beta$, or $m\beta \leq r$.

It would not be worth the overhead to keep a checklist unless perhaps 3/4 or more of the monomials never actually appear. So it is interesting to discover what β, in a cubic and quadratic example, would result in that level of monomial absences (i.e. $r < 1/4$).

Table 13.2 Running Time Statistics in Seconds

	Cut by 3	Cut by 4	Cut by 5	Cut by 6	Cut by 7
			No Massaging		
Naïve Average	393.94	279.71	179.66	253.15	340.66
Naïve StDev	433.13	287.33	182.18	283.09	361.04
μ/σ	1.10	1.03	1.01	1.12	1.06
Naïve Kurtosis	0.93	5.12	0.79	1.16	2.47
Average(ln)	5.11	4.96	4.55	4.72	5.2
StDev(ln)	1.63	1.46	1.35	1.51	1.27
μ/σ	0.32	0.29	0.30	0.32	0.24
Kurtosis(ln)	0.51	0.8	0.43	-0.5	-0.32
			Light Massaging		
Naïve Average	413.74	181.86	269.59	217.54	259.73
Naïve StDev	439.71	160.23	301.48	295.88	237.52
μ/σ	1.06	0.88	1.12	1.36	0.91
Naïve Kurtosis	0.04	0.08	3.68	6.85	0.01
Massage Ratio	1.05	0.65	1.5	0.86	0.76
Average(ln)	5.3	4.64	4.84	4.52	4.87
StDev(ln)	1.39	1.29	1.5	1.47	1.5
μ/σ	0.26	0.28	0.31	0.33	0.31
Kurtosis(ln)	-0.38	0.07	0.09	-0.14	0.52
			Deep Massaging		
Naïve Average	280.22	198.15	204.48	144.94	185.44
Naïve StDev	363.64	292.21	210.53	150.88	49.53
μ/σ	1.30	1.47	1.03	1.04	0.27
Naïve Kurtosis	5.67	9.24	3.74	0.62	4.69
Massage Ratio	0.71	0.71	1.14	0.57	0.54
Average(ln)	4.82	4.34	4.54	4.07	4.33
μ/σ	0.31	0.39	0.36	0.43	0.36
StDev(ln)	1.48	1.68	1.63	1.73	1.54
Kurtosis(ln)	1.1	2.41	0.75	-0.06	-0.23

Cubic Example Consider $n = 128$, $m = 128\gamma$, a number of monomials $\approx 128^3/6 \approx 349525$. This would require $\beta \leq 1/512\gamma$. This means the average length of an equation would be $\leq 683/\gamma$. This could easily occur if the system is not highly overdefined, i.e. $\gamma \approx 1$. It is also easy to imagine systems where this would not occur.

Quadratic Example $n = 128$, $m = 128\gamma$, number of monomials $128^2/2 \approx 8192$. This would require $\beta \leq 1/512\gamma$. This means the average length of an equation would be $\leq 16/\gamma$. Therefore, the system would have to be rather sparse with very short equations in order for this to happen.

13.6.2 Measures of Efficiency

There are $\binom{n}{3} \sim n^3/6$ cubic monomials possible, each requiring 1 dummy variable, 4 clauses of total length 10. There are as before $\binom{n}{2} \sim n^2/2$ quadratic monomials possible, each requiring 1 dummy variable, 3 clauses of total length 7. The total number of monomials possible is thus

$$M = \binom{n}{3} + \binom{n}{2} + \binom{n}{1} + \binom{n}{0} \sim n^3/6$$

The expected length of any polynomial is $\beta M \sim \beta n^3/6$. Taking cutting by four as an example, this would require $\sim \beta n^3/12$ dummy variables, and $\sim (2/3)\beta n^3$ clauses of total length $\sim (8/3)\beta n^3$, for each of the m equations. Therefore, the cost of converting the monomials is negligible compared to that of representing the sums, as before.

Table 13.3 CNF Expression Difficulty Measures for Cubic Systems, by Cutting Number

Cutting Number	Variables	Clauses	Tot. Length	Avg. Length
Cut by 3	$\sim \gamma n^4 \beta/6$	$\sim (2/3)\gamma n^4 \beta$	$\sim 2\gamma n^4 \beta$	3
Cut by 4	$\sim \gamma n^4 \beta/12$	$\sim (2/3)\gamma n^4 \beta$	$\sim (8/3)\gamma n^4 \beta$	4
Cut by 5	$\sim \gamma n^4 \beta/18$	$\sim (8/9)\gamma n^4 \beta$	$\sim (40/9)\gamma n^4 \beta$	5
Cut by 6	$\sim \gamma n^4 \beta/24$	$\sim (4/3)\gamma n^4 \beta$	$\sim 8\gamma n^4 \beta$	6
Cut by 7	$\sim \gamma n^4 \beta/30$	$\sim (32/15)\gamma n^4 \beta$	$\sim (224/15)\gamma n^4 \beta$	7
Cut by 8	$\sim \gamma n^4 \beta/36$	$\sim (32/9)\gamma n^4 \beta$	$\sim (256/9)\gamma n^4 \beta$	8

13.7 Further Reading

13.7.1 Previous Work

The exploration of SAT-solver enabled cryptanalysis is often said to have begun with Massacci and Marraro [172, 171, 173, 135], who attempted cryptanalysis of DES with the SAT-solvers Tableau, Sato, and Rel-SAT. This was successful to three rounds. However, this was a head-on approach, encoding cryptographic properties directly as CNF formulas. A more algebraic approach has recently been published by Courtois and Bard [76], which breaks six rounds (of sixteen). Fiorini, Martinelli and Massacci have also explored forging an RSA signature by encoding modular root finding as a SAT problem in [114]. This was less successful.

The application of SAT-solvers to the cryptanalysis of hash functions, or more correctly, collision search, began with [143] which showed how to convert hash-theoretic security objectives into logical formulas. The paper [177], by Mironov and Zhang, continued the exploration of hash functions via SAT-solvers by finding collisions in MD4 and MD5.

We believe this is the first successful application of SAT-solvers to solving systems of equations over finite fields. However, the approach was mentioned in [64], upon the suggestion of Jacques Stern.

13.7.2 Further Work

Quite a few researchers have used the author's technique presented in this chapter. Here is a sampling, in no particular order.

- "Algebraic Cryptanalysis of the Data Encryption Standard" by Nicolas Courtois and Gregory Bard [76], published in the proceedings of the IMA conference on coding theory and cryptography.
- "Attacking Bivium using SAT Solvers" by Tobias Eibach, Enrico Pilz and Gunnar Völkel [105], published in the proceedings of SAT'08.
- "Periodic Ciphers with Small Blocks and Cryptanalysis of Keeloq" by Nicolas Courtois, Gregory Bard, and Andrey Bogdanov [83], published in the Tatra Mountains Mathematical Publication, the mathematics journal of the Slovak Academy of Sciences.
- "Secure PRNGs from Specialized Polynomial Maps over any F_q" by Feng-Hao Liu, Chi-Jen Lu and Bo-Yin Yang [167], published in the proceedings of PQC'08.
- "Extending SAT-Solvers to Low-Degree Extension Fields of GF(2)" by Gregory Bard [32], presented at CECC'08.
- "Guess-and-Determine Algebraic Attack on the Self-Shrinking Generator" by Blandine Debraize and Louis Goubin, published in the proceedings of FSE'08.
- "Analysis of Lightweight Stream Ciphers", the PhD Dissertation of Simon Fischer [115].
- "Algebraic and Slide Attacks on Keeloq" by Nicolas Courtois, Gregory Bard, and David Wagner [84], published at FSE'08.
- "Algebraic Attacks on the Crypto-1 Stream Cipher in MiFare Classic and Oyster Cards", by Nicolas T. Courtois, Karsten Nohl and Sean O'Neil [85], published as an IACR e-print.
- "An Algebraic Analysis of Trivium Ciphers based on the Boolean Satisfiability Problem", by Cameron McDonald, Chris Charnes and Josef Pieprzyk [176], published as an IACR e-print and in the proceedings of BFCA'2008.

- "Algebraic cryptanalysis of symmetric primitives" by Carlos Cid, Martin Albrecht, Daniel Augot, Anne Canteaut, and Ralf-Philipp Weinmann [58], an ECRYPT technical report.
- "Boolsche Gleichungssysteme, SAT Solver und Stromchi?ren", the Bachelorarbeit (Bachelor's Thesis) of Enrico Pilz [191], from Universität Ulm, Institut für Theoretische Informatik.
- "Algebraic Cryptanalysis of SMS4", REU report of Jermey Erickson of Taylor University [107].
- "Algebraic Techniques in Differential Cryptanalysis" by Martin Albrecht and Carlos Cid [16].

13.8 Conclusions

The problem of solving a multivariate polynomial system of equations over $\mathbb{GF}(2)$ is important to cryptography. We demonstrate that it is possible to efficiently covert such a problem into a CNF-SAT problem. Also, the massaging method of preprocessing has been shown to be useful. On most problems of even intermediate size, Gröbner-Bases-oriented methods, like MAGMA and SINGULAR, crash due to a lack of sufficient memory. Our method, on the other hand, in experiments given in [33] but not here, requires little more memory than that required to store the problem. In examples where MAGMA and SINGULAR do not crash, these tools are faster than our methods. However, our method is still much faster than brute force approximately when $\beta \leq 1/100$.

Chapter 14
How do SAT-Solvers Operate?

The purpose of this chapter is to explain how SAT-solvers operate (at least at the time of the writing of this book, late 2008). Two major families will be described in this section. The family of SAT-solvers used by the author are based on the Chaff Algorithm [182], culminating in MINISAT, a popular SAT-solver [104] [6]. This gives insight into Chapter 13 on Page 245 in particular, by highlighting why the number of variables per clause, number of clauses, and number of variables, are taken as the three general barometers of difficulty for a particular SAT problem. As a contrast, we will first describe Walk-SAT, a very different type of solver, to show how these methods and approaches differ and are similar. At this time, many SAT-solvers are in use, most of them of the Chaff/MINISATtype. However, that could someday change, perhaps even soon.

Besides the Chaff family and the Walk-SAT family, many other SAT algorithms have been proposed in previous years, and also many preprocessing techniques, none of which will be described below. While the author makes frequent and extensive use of SAT-solvers, he is not an expert on their internals. The following is meant to be informative and general, and so many details are omitted.

14.1 The Problem Itself

Given a logical sentence over certain variables, does there exist a set of assignments of true and false to each of those variables so that the entire sentence evaluates as true? This question is the "SAT" problem, and is the root of the theory of NP-Completeness.

The term "logical sentence" in this book refers to an expression composed of variables, as well as the operators from predicate calculus (AND, OR, NOT, IMPLIES, and IFF), arranged according to the grammar of predicate calculus. There are no universal quantifiers (i.e. $\forall$), existential quantifiers (i.e. $\exists$), or any functions. An example of such a sentence is

G.V. Bard, *Algebraic Cryptanalysis*, DOI: 10.1007/978-0-387-88757-9_14

$$(D \wedge \overline{B} \wedge A) \Rightarrow (B \vee C)$$

which is satisfied by (for example) setting all the variables to true.

It is a basic fact from digital circuit theory that any logical sentence can be written as a product of sums (Conjunctive Normal Form or CNF) or as a sum of products (Disjunctive Normal Form or DNF). These terms refer to the semiring first introduced in Section 6.2 on Page 81, where addition is logical-OR and multiplication is logical-AND.

Algebraic Normal Form (ANF) is a sum over $\mathbb{GF}(2)$, or logical-XORs, each input of which is a product (logical-AND). Polynomials over $\mathbb{GF}(2)$ are thus ANFs, and we desire to make them CNFs, for use with the SAT-solver.

14.1.1 Conjunctive Normal Form

A logical sentence in CNF is a set of clauses. Each clause is combined into a large conjunction or AND-gate. Thus the sentence is true if and only if each clause is true. The clauses are themselves OR-gates, or disjunctions. Each variable in the clause can appear negated, or not negated.

Product of Sums or Conjunctive Normal Form has been selected as the universal notation for SAT-solvers for many reasons. One reason is that all predicate calculus sentences can be written in CNF. Another interesting reason is that some sentences can be written with two or fewer variables per clause, and others require three variables at least for some clauses. There does not exist a logical sentence which cannot be written with the restriction of at most three variables per clause. Solving the SAT problem on CNF sentences with at most two variables per clause (2-CNF-SAT) is possible in polynomial time [63, Ch. 34.4]. For CNF sentences with up to three variables per clause (3-CNF-SAT), SAT is NP-Complete. In fact, SAT itself is the "mother problem" of NP-Completeness, in the sense that problems are proven to be NP-Complete by being reduced to SAT.

While one could write any logical sentence in 3-CNF-SAT notation, it is not required for SAT solvers that the author is aware of. The logical sentence need merely be in CNF form.

14.2 Solvers like Walk-SAT

In order to explain what the Chaff/Grasp family of SAT-solvers is about, it is worthwhile to investigate a different family, for contrast. The system described here, called Walk-SAT, is by Selman, Kautz and Cohen, and was the leading method for SAT problems for about a decade (very roughly speaking). For certain types of problems, it is still very good. In many ways, Walk-SAT and MINISAT are polar opposites.

Walk-SAT is simple, heuristic, randomized, without any form of learning, and never outputs UNSATISFIABLE. On the other hand, MINISAT is complex, based on logical principles, and learns far more clauses than it can use. MINISAT is equally able to output satisfiable solutions when they exist as it is to certify the non-existance of a solution with an UNSATISFIABLE output. There is no doubt that MINISAT is randomized but from the author's viewpoint, the randomization is an extremely secondary part of the system, whereas in Walk-SAT, it is very central.

The ability to handle unsatisfiable problems is very important, and so we will repeat it. The Grasp and Chaff family (which includes MINISAT) can output a satisfying solution, can output UNSATISFIABLE, or can run for a very long time. Instead, Walk-SAT can only output a satisfying solution or run forever. MINISAT will always terminate.

14.2.1 The Search Space

Consider a set of variables in some order. Any particular assignment to them would be a sequence or vector of 1s and 0s, each being a value for a particular variable. Obviously, if there are n variables, there are 2^n possible vectors. We can define a distance between two vectors as the number of positions in which they differ (this is the Hamming distance—named for Richard Wesley Hamming). Obviously the distance is always a non-negative integer.

A neighborhood of distance 1 about a vector is the set of all vectors which differ in exactly one position, as well as the original vector. Thus any one of the n variables can be flipped to produce such a vector, for a size of $n + 1$. The neighborhood of distance 2 would be all of these, as well as any formed by flipping any pair of variables, for a size of $n^2/2 + n/2 + 1$.

In Walk-SAT, we will be looking for vectors in the distance 1 neighborhood of the previous vector. We aim to reduce the number of unsatisfied clauses at each step. When we can improve in this way, we do. When we cannot, we voluntarily pick a less-good vector in the hopes that something in its distance 1 neighborhood will be better. Thus, in a way, we are searching the distance 2 neighborhood when the distance 1 has no vectors which improve the number of unsatisfied clauses.

14.2.2 Papadimitriou's Algorithm

The following algorithm, published by Papadimitriou in [186], can in $O(n^2)$ expected steps reach a satisfying solution with probability 1, but only for problems which can be written in 2-CNF-SAT form. This means that every clause can be written with at most two variables. Note that all CNF problems can be written in 3-CNF-SAT form [63, Ch. 34.4]. For a definition of CNF form see Section 13.3 on Page 248.

Start with a random initial guess, and pick a variable from among those that occur in unsatisfied clauses and flip its truth assignment. Repeat this until all clauses are satisfied. Amazingly, this simple approach simply works for 2-CNF-SAT problems. Whenever a variable is flipped, some satisfied clauses will become unsatisfied, and some unsatisfied clauses will become satisfied. This is the inspiration for "Greedy SAT" or GSAT.

INPUT: A set of clauses C and variables V, and two parameters n_{flip} and $n_{restart}$.
OUTPUT: Either a satisfying solution or an abort.
1: For each variable in V, choose either 0 or 1 with equal probability.
2: For $i = 1 \ldots n_{flip}$ do

- If all clauses are satisfied then output result and halt.
- For each variable $v \in V$ do
  - Compute the number of formerly satisfied clauses newly violated if v is flipped, denote this a_v.
  - Compute the number of formerly unsatisfied clauses newly satisfied if v is flipped, denote this b_v.
  - Let $k_v \leftarrow a_v - b_v$
- Choose to flip the v which minimizes the k_v. Break ties randomly.

3: If $n_{restart} = 0$ then abort else decrement $n_{restart}$.
4: Restart the algorithm.

Algorithm 27: The SAT-Solver GSAT or "Greedy SAT" [Selman, Levesque, and Mitchel]

14.2.3 Greedy SAT or G-SAT

Again, we start with a random assignment, and we see how many clauses are unsatisfied. Now we must, for every variable, determine how many clauses will flip their status between satisfied and unsatisfied depending on if that variable is flipped between true and false. We choose the variable which will reduce the total count of unsatisfied clauses the most. Then we repeat.

Therefore, this is the Greedy Algorithm. At each stage, we simply choose the best one variable to flip, and we keep doing that until we get a solution. This algorithm was published by Selman, Levesque, and Mitchel in [208], and there are several categories (listed in the paper) of problem for which it dramatically out performed the solvers of its time. These include the N-queens problem (finding locations for N chess queens on an $N \times N$ board so that none can attack each other) and graph coloring (see Appendix C on Page 315).

The algorithm has two parameters, $n_{restart}$ and n_{flip}. The $n_{restart}$ parameter is simply to make sure that the algorithm does not run forever (in fact, many textbooks require this before a procedure can be called an algorithm). However, in practice, if one really desires a solution, one would simply run the program a second time if $n_{restart}$ is exceeded. Thus, we can consider it to be ∞. On the other hand, n_{flip} is very important. The authors of [208] used $5n$, $10n$, and $15n$ where n is the number of variables. If it is set too low, then a very good guess must be made for the convergence to occur fast enough. This has a low probability. If it is set too high, then not enough of the possible initial conditions will be explored.

One important point that the paper [208] makes is that one must tolerate moves that do not make progress. In other words, sometimes a variable is flipped, and no increase in the number of satisfied clauses occurs. Recall, when considering flipping the variable v, that b_v is the number of newly satisfied clauses and a_v is the number of newly violated clauses. The algorithm seeks to minimize $a_v - b_v = k_v$. If $k_v \geq 0$ for all variables, then the flipping of v does not improve the number of satisfied clauses.

Call a move with minimal $k_v = 0$ to be "sideways" and $k_v > 0$ to be "uphill", and all other moves "downhill" (the analogy is to gradient descent). Normally, the vast majority of moves are downhill. The authors of [208] ran their algorithm in the form given here (permitting uphill, sideways, and downhill moves) and another version that would reset rather than make a sideways move. The performance was universally weaker, both in the solution time and in the fraction of problems solvable, when sideways moves are prohibited. Thus, the sideways and uphill moves are important.

In many algorithms of this form, a tie-breaker can be very important, as several variables might cause the same number of clause improvements if toggled. The heuristic chosen for a tie-breaker is to flip a random member of the tie. This way, very short loops of repeating the same variable settings over and over are avoided. This is particularly needed because "sideways" moves are valid. One should avoid cycles like $(x, y) \in \{(0, 0); (0, 1); (1, 1); (1, 0)\}$ which could otherwise repeat forever.

14.2.4 Walk-SAT

The Walk-SAT algorithm [207] has one more parameter than G-SAT, namely p_{noise}. Basically, the algorithm can be summarized as follows. First, make a random initial assignment. Then pick an unsatisfied clause. With probability p_{noise} we flip a random variable in that clause. Of course, the clause will now be satisfied but we may have done massive damage to the system, or perhaps not. With probability $1 - p_{noise}$, we will try flipping each of the variables, and we will go with the choice that reduces the number of unsatisfied clauses the most (the G-SAT step).

The case of flipping a randomly chosen variable seems very odd. However, it is based on the principle of simulated annealing. When making an alloy of two metals, via the process of annealing, a mixture is repeatedly heated and left to cool. During the heating phases the molecules are more free to move around, but in the cooling phases, they gravitate toward lower-energy states. Simulating annealing, on the other hand, randomly changes variables (the heating phase) and then has them move to lower energy situations (analogous to Gradient Descent or GSAT). An excellent introduction to simulated annealing can be found in [17].

14.2.5 Walk-SAT versus Papadimitriou

In the paper describing Walk-SAT [207], the authors describe it as a combination of Papadimitriou and G-SAT, using Papadimitriou's Algorithm with probability p_{noise} and using G-SAT otherwise.

However, they mention that even if $p_{noise} = 1$, then Walk-SAT does not quite become Papadimitriou's algorithm. The reason is that in Walk-SAT one first picks a random clause. Only then does one pick a random variable to flip. Thus a variable which appears in 100 unsatisfied clauses is much likelier to be chosen than a variable which appears in 1 unsatisfied clause. In contrast, Papadimitriou's algorithm does not have this voting, but instead one has a set of variables which appear in unsatisfied clauses, and a variable is chosen uniformly at random from there. Thus, even the $p_{noise} = 1$ move is "smart" rather than "random", in the sense of emphasizing frequently appearing variables.

14.2.6 Where Heuristic Methods Fail

Consider the following system, taken from [208], which has a fixed 5-clause part

$$(x_1 \vee \overline{x_2} \vee x_3)(x_1 \vee \overline{x_3} \vee x_4)(x_1 \vee \overline{x_4} \vee \overline{x_2})(x_1 \vee \overline{x_5} \vee x_2)(x_1 \vee x_5 \vee x_2)$$

as well as for some $n > 9$

$$(\overline{x_1} \vee \overline{x_6} \vee x_7)(\overline{x_1} \vee \overline{x_7} \vee x_8)(\overline{x_1} \vee \overline{x_8} \vee x_9) \cdots (\overline{x_1} \vee \overline{x_i} \vee x_{i+1}) \cdots (\overline{x_1} \vee \overline{x_{n-1}} \vee x_n)(\overline{x_1} \vee \overline{x_n} \vee x_6)$$

The first part is trivially satisfied if $x_1 = 1$. On the other hand, if $x_1 = 0$ then the last two clauses among the first five force that $x_2 = 1$, otherwise there is a contradiction arising from x_5. Given that $x_1 = 0$ and $x_2 = 1$, then the first clause causes $x_3 = 1$. Next, the second clause causes $x_4 = 1$, and finally the third clause is unsatisfiable. Thus, we know logically, that $x_1 = 1$ is required of any satisfying solution.

The second part, which can have variable length, creates a kind of loop. If $x_1 = 0$, then all $n - 5$ of these clauses in the second part are satisfiable, regardless of the value of $x_6, x_7, \ldots, x_n$. If $x_1 = 1$ on the other hand, then these $n - 5$ clauses are all satisfied if and only if the variables $x_6, x_7, \ldots, x_n$ are all equal.

Consider beginning the algorithm with a random assignment. If $x_1 = 0$ then all the clauses the second part are satisfied. Some subset of the first five are unsatisfied. Therefore, the clause selected by the algorithm has to be in the first part. Either x_1 is chosen or not. Most likely not, because flipping x_1 would move much of the second part to unsatisfied, and so would have a (very) positive k_v. As we stated earlier, negative k_v is the usual case. If x_1 is not chosen, we know that the problem will remain with the first part not all satisfied and the second part all satisfied, because $x_1 = 1$ in any satisfying solution and because $x_1 = 0$ trivially satisfies the second part. Now let us consider what happens if x_1 is in fact chosen, or if it begins as $x_1 = 1$.

If $x_1 = 1$ then the first five clauses are all satisfied, and so will not be selected. Then a clause of the form $(x_1 \vee \overline{x_i} \vee x_{i+1})$ will be chosen, where x_{n+1} can be thought of as a pseudonym for x_6. Flipping x_i or x_{i+1} will cause changes only to the neighboring clauses, because only the neighboring clauses contain them. On the other hand, flipping x_1 will cause $n - 5$ clauses to be satisfied. So surely x_1 will be chosen, and the algorithm returns immediately to the $x_1 = 0$ state.

Thus, for the whole time the algorithm runs, it will mostly consider the $x_1 = 0$ state, and it is "looking in the wrong places".

14.2.7 Closing Thoughts on Heuristic Methods

First, there are numerous enhancements to these algorithms which may have enormous impact in the final efficiency of them. Our purpose here is not to present the algorithms for use, but to explain them as a contrast to the Chaff/Grasp family of algorithm. Second, observe that these methods take no advantage whatsoever from the structure of the logical nature of CNF forms. Third, the fact that they never output UNSATISFIABLE, and may even never terminate (except for $n_{restart}$) means that we cannot calculate with confidence how long a calculation will take.

This has major consequences for the "guess-and-determine" approach (see Section 11.7 on Page 206), because there if we guess g variables then we must make 2^g subproblems. If only one solution actually exists, then $2^g - 1$ of the subproblems will be unsatisfiable.

The setting of the parameter p_{noise} is also problematic. Some suggest initially as high as 0.5 then lowering it slowly to 0.05 as the problem progresses. Others prefer a fixed value in the range $[0.05, 0.2]$.

14.3 Back-Tracking

We will now introduce the Davis-Putnam back-tracking algorithm, which lies at the heart of "Chaff"—the algorithm implemented in MINISAT. The basics of the back-tracking system is as follows. Suppose you have an algorithm that can solve n-variable SAT problems, and you wish to make one that can solve $n + 1$-variable SAT problems. By induction (since one-variable SAT problems are very easy to solve) you would then have an algorithm for all SAT problems. Let us call this n-variable SAT-solver f_n.

INPUT: A set of clauses C and variables V, and three parameters p_{noise}, n_{flip} and $n_{restart}$.
OUTPUT: Either a satisfying solution or an abort.
1: For each variable in V, choose either 0 or 1 with equal probability.
2: For $i = 1 \ldots n_{flip}$ do

- If all clauses are satisfied
  - then output result and halt,
  - else choose a violated clause $c \in C$, uniformly at random.
- For each variable $v \in c$ do
  - Compute the number of formerly satisfied clauses newly violated if v is flipped, denote this a_v.
  - Compute the number of formerly unsatisfied clauses newly satisfied if v is flipped, denote this b_v.
  - Let $k_v \leftarrow a_v - b_v$
- With probability p_{noise} choose to flip a random variable in c.
- With probability $1 - p_{noise}$ choose to flip the v which minimizes the k_v.

3: If $n_{restart} = 0$ then abort else decrement $n_{restart}$.
4: Restart the algorithm.

Algorithm 28: The SAT-Solver "Walk-SAT" [Selman, Kautz, and Cohen]

Given some $n + 1$-variable CNF problem, we could always just simply guess some variable. First, we make a backup-copy of the problem. Then, suppose we guess x_i is true. Then any clause containing x_i is now satisfied, and can be deleted. Now consider a clause containing $\overline{x_i}$. If that is the only term in the clause, then that clause cannot be satisfied, and we know our assumption that "x_i is true" is false; call this a "discovered contradiction". If the clause contains other terms, then the clause will be satisfied if and only if any of the remaining terms are. This is the same thing as writing the clause with the $\overline{x_i}$ term removed. And so, we delete the $\overline{x_i}$ term in that clause. This new system is satisfiable if and only if the original one was satisfiable after the restriction that $x_i = 1$. Baring a "discovered contradiction", since the new system has one fewer variable, we call f_n on the new system. If there is a "discovered contradiction", then we know that the guess is wrong, and we repeat the process as if we had guessed instead that $x_i = 0$. If this too is a "discovered contradiction", then we return that the problem is UNSATISFIABLE.

Either f_n will return UNSATISFIABLE, or a satisfying solution. If it returns a satisfying solution, we will simply append our guess (which we now know is good), and return that as the satisfying solution to the original system. If f_n returns UNSATISFIABLE for x_i is true, then we know that was a bad guess, and we will repeat the process guessing x_i is false. If that too results in a problem—either because we encounter a clause with only $\overline{x_i}$ in it (i.e. a discovered contradiction), or because a lower level of the recursion returns UNSATISFIABLE—then we will return UN-SATISFIABLE.

That is the process called back-tracking, but it can be made more efficient by using a stack instead of recursive calls. Whenever an assumption is made, it is pushed on to an assumption stack. The algorithm proceeds until it discovers a contradiction (i.e. a single-variable clause whose variable has been guessed the wrong way).

If this is the case, then we "pop" the most recent assumption off of the stack. If the popped assumption were that x_i is true, then we now assume x_i is false, and check to see if the contradiction remains. If the contradiction remains, we keep popping assumptions until the contradiction no longer remains. At least one x_i, if not several variables, have now changed their state, and so a great deal of rewinding must take place. Due to clever data structure design, akin to a journaling file-system, the rewinding can be made very efficient.

If the assumption stack becomes empty, and a contradiction remains, then the original problem was unsatisfiable. In this case, we output "UNSATISFIABLE" and we are done. Now assume that the contradiction can be repaired by popping off one, some, or all of the assumptions, and the algorithm then continues as before, with new guesses.

Some care is needed to make sure an infinite loop does not result but this is easily taken care of with flag variables. Once a variable setting has resulted in a contradiction (e.g. $v_5 = T$), and its negation is attempted (e.g. $v_5 = F$), if that also fails, the system should move further up the assumption stack, and not try the original (e.g. $v_5 = T$) a second time. Note, there is no reason whatsoever to guess true first. The best guess is usually to flip a fair coin.

Sometimes searches of a boolean space are described as a tree. Each assumption is analogous to taking one branch of the tree over another, slowly descending from the root to the leaves. Note that because more than one assumption can be popped off of the stack at once, it is possible to "lop off" large portions of the tree in a single move.

This "lopping off" of the tree also occurs when clauses are "learned" (see Section 14.5.1 on Page 275), a topic we have not yet introduced. Because it can "learn" clauses, Chaff is much faster than an ordinary tree search, no matter how cleverly implemented. For this reason, the true running time is conjectured to be $\sim c^n$, where n is the number of variables, and $1 < c < 2$ is a constant. So far, even $c < 2$ is not proven, and note that brute-force guessing all possibilities is possible in time equivalent to $c = 2$.

The reason that this "lopping off" occurs is that a CNF expression is like a product in a factored polynomial. When one factor of a polynomial is zero, the whole thing is zero. Thus if one clause is false, the conjunction of them all is false. For this reason, one need not investigate all the settings for the other clauses. Once a sub-tree has been identified as having the property of always forcing some particular clause to be false, that entire sub-tree can be ignored.

This idea is attributed to Davis and Putnam in the papers [92] and [91]. It is noteworthy that the DPLL algorithm (as well as many other SAT-solvers) can be used to create a proof-sketch of unsatisfiability via the backtracking process.

In [170], the idea of resolving the conflicts non-chronologically was introduced by João Maques-Silva and Karem Sakallah, with the algorithm "Grasp", which helped lead to "Chaff."

One can imagine that the back-tracking process can be improved. Between each guess, it might be quite logical to search for single-variable clauses. For each clause, if it is of the form (x_i), we can just set $x_i = 1$ and make the corresponding deletions.

If that results in a contradiction, then since setting $x_i = 0$ surely would have resulted in a discovered contradiction, we know some prior guess is wrong and we should immediately return UNSATISFIABLE (or pop the stack). If its of the form $\overline{x_i}$ we can likewise just set $x_i = 0$, because likewise $x_i = 1$ would cause a discovered contradiction. This can be dramatically generalized, and later became the "pure literal elimination" and "unit propagation" rules, which are the backbones of "Chaff."

In the absence of learned clauses, we can see that every case must be attempted before one can conclude that a problem is unsatisfiable. But given a satisfiable problem, especially one with many solutions, we can arrive at a solution quickly. Learned clauses, which will define in Section 14.5.1 on Page 275, will change this property dramatically.

14.4 Chaff and its Descendants

There is a large economic and financial incentive to make good SAT-solvers (see Section 14.6 on Page 276). For this reason, a series of competitions has been held each year since 2002 [46]. (See http://www.satcompetition.org/) The Chaff algorithm proposed by [182] is at the core of most currently competitive SAT-solvers.

Like most users of SAT-solvers, we treat the system as a black-box, worrying only on how to present it with our problem in a way that results in the most efficient search for a solution.

14.4.1 Variable Management

Every variable in the system will be given one of three values, namely true, false, and not-yet-known. Initially, all variables are set to not-yet-known. As mentioned earlier, variables in a clause can be negated or not-negated. The first step is to replace all the negated variables with new ones (thus doubling the number of variables). However, the original variables are identified by the positive integers. The negation of a variable has as its ID, the additive inverse of the original ID. Thus whenever the variable numbered x is set to true, then it is understood that $-x$ will be set to false, regardless of the original sign of x.

Because the terminology might get confusing, a useful distinction is made here. A "literal" is either $-x$ or x after the doubling has occurred. A "variable" refers to both x and $-x$, or in other words, a variable in the initial problem statement.

blarg

There are three consequences to this. First, none of the literals in the system are negated after this step; even though there are twice as many literals as variables, they are tied together as described; and when one literal is changed to true or false from "not-yet-known", its complement literal will be set accordingly.

There will be an array from $-n$ to n that contains all system literals. Each literal will have a list of clauses that contain it, and every clause will have a list of literals that it contains. This is vital in the efficient encoding of (and even the human understanding of) the algorithm.

14.4.2 Unit Propagation

Now each clause is a disjunction (OR-gate) of some particular literals. If any of those literals is at any time true, then the clause is satisfied. We will declare the clause "inactive" and it will be hidden from the algorithm. Thus the "active" clauses are those that are not yet satisfied. Likewise, if all of the literals are false, then satisfiability has become impossible, and back-tracking must take place. Back-tracking is discussed in Section 14.3 on Page 269.

Therefore, an active clause (except during backtracking) has no literals set to true—all of its literals are set to false or not-yet-known, with at least one of those being not-yet-known. But suppose, out of n literals, a clause were to have $n-1$ false literals and one not-yet-known. Clearly, in any satisfying assignment, that not-yet-known literal must be true, and so we can set it to true. This rule is called "unit propagation" or sometimes "derivation of implication."

14.4.3 The Method of Watched Literals

In practice, each clause has two pointers associated with it, which we will denote "fingers". Each finger must point to a literal, and since all variables/literals begin the system with the status of not-yet-known, in any clause they both point to a (distinct) not-yet-known literal. They may never point to the same literal. If the status of a fingered literal changes, then the finger will move. If the literal becomes true, then the clause is now inactive, and out of the algorithm. If the literal becomes false, then the finger will try to move to another literal in the same clause which is not-yet-known. If this is possible, it moves there. If not, then this means there is one not-yet-known literal (pointed to by the other finger) and all the other literals are false (because the clause is still active). As we stated before, this means that the remaining single not-yet-known literal must be set to true (precisely a unit propagation). And conveniently, we do not need to search for it, because the other finger is pointing to it. The clause is now satisfied and can be deleted. This is called the "Method of Watched Literals."

It is a good time to note there are two reasons that a separate list is kept for the set of clauses that use v and $-v$. Step 1 is that if unit propagation causes v to become true, the algorithm must do two things. First, it must simply deactivate any clauses containing v. Step 2, clauses containing $-v$ need to have their fingers moved. These are distinct operations. The other reason is to enable "pure literal" propagation, described in the next section.

14.4.4 Absent Literals

One additional rule is used. If a literal v is found somewhere in the entire system, and $-v$ is not, then it is safe to set v to true and $-v$ to false. This sounds like it might require a search. The beauty of the "Chaff" algorithm is that it uses pointers in a clever way to ensure that searches are not needed, except at setup. More precisely, each literal has a linked list to the clauses that use it. Likewise, if $-v$ is found in the system and not v, then it is safe to mark v false. This procedure is sometimes called "pure literal propagation".

When a literal is set to true, any clause that contains it is deactivated. Then for each of those newly deactivated clauses, the literals contained in them are notified to remove that clause from their list of clauses that contain them. If one of those lists becomes empty, the literal then is not found in the system. This means its complement can be marked true, and it can be marked false, with all the consequences that this paragraph requires from that marking. Once this is done, the complement of the original literal which was set to true can be set to false.

When a literal is set to false, all the clauses that contain it are notified. If that literal had one of the clause's two fingers pointing to it then that finger is moved to any literal in that clause that is currently marked not-yet-known. If no such literal is available, then the literal pointed to by the other finger is marked true, with all the consequences we described above. Of course, if an entire clause becomes false, the system has discovered a contradiction and must begin back-tracking (see Section 14.3 on Page 269). And if not already done so, the complement of the literal already set to false should be now set to true, with all the consequences that entails.

14.4.5 Summary

Thus, we start with a system with all the literals set to not-yet-known. We build the data-structures previously described. If any of the literals v fails to appear in the system, (i.e. the list of clauses containing v is empty), then we mark v false and mark $-v$ true, which hopefully sets off a flurry of activity. Then either the system will halt with all clauses inactive, which means we have found a satisfying assignment, and print it out; or halt with a contradiction which means the original problem was unsatisfiable; or with overwhelming probability, knock-out only a few things and leave a problem that looks relatively unchanged.

At this time, we choose a not-yet-known literal to be "assumed." For example, with 1% probability, it could be a randomly chosen variable. Otherwise, with probability 99%, it is that variable which appears in the largest number of clauses (has the longest list of clauses associated). The variable selection is a heuristic and varies from implementation to implementation. This variable will now be changed to true or false, decided by a fair coin. Then that assumption will be pushed on to an "assumption stack." Hopefully, this also sets off a flurry of activity and either results in a satisfying assignment or a smaller problem. If a satisfying assignment has resulted, we print the answer and declare victory. If a smaller problem results, we guess another variable.

14.5 Enhancements to Chaff

Once the previous description is understood, the following enhancements make the system very efficient.

14.5.1 Learned Clauses

There is one last element of this algorithm, namely learning new clauses. Suppose the assumption stack has five assumptions on it (without loss of generality: v_1, v_2, ..., v_5 are true), and a contradiction results. We then know that

$$\overline{(v_1 \wedge v_2 \wedge v_3 \wedge v_4 \wedge v_5)}$$

is true, which by DeMorgan's Law is equivalent to

$$\overline{v_1} \vee \overline{v_2} \vee \overline{v_3} \vee \overline{v_4} \vee \overline{v_5}$$

which is conveniently a CNF clause! Thus we have "learned" a new clause from this contradiction, which we can safely toss into the system. These learned clauses might be quite long, (if the stack was large when the contradiction occurred) and there might be many of them (if many contradictions were found). They are added to the system but flagged as "learned." If a learned clause has not served a purpose in a long time (e.g. it hasn't changed activation status within $t \approx 10^5$ steps) then it can be deleted, but clauses that were part of the original problem are never deleted. This keeps a bound on the number of clauses.

Sometimes a set of clauses will simultaneously become all false, each alone enough to produce a contradiction. In this case, many clauses can be added at once.

14.5.2 The Alarm Clock

Finally, the algorithm has an alarm clock. If the algorithm hasn't found an answer after a certain length of time has elapsed, then it will completely reboot the entire system, *except that it retains any clauses that it learned.* The idea is that by starting over with this new information, a more efficient path might be taken. The initial timer is set quite short, and then increases after each time-out. In at least one implementation, the new time-out will be $\sqrt{2}$ times as long as the old one. This is better than a fixed timer of t seconds, because a problem that required $t + \varepsilon$ seconds would be unsolvable. In any case, this is all heuristic, and it seems to work in practice.

14.5.3 The Third Finger

Another variant, not universally employed, is to add a third finger to each clause. Like the first two fingers, it can only be attached to a literal which is not-yet-known, and is not currently pointed to by one of the other two fingers. Once this is no longer possible in an active clause, the system is aware that two literals are not-yet-known, and all the others are false in that clause. (Note, if there were a literal equal to "true" there, then the clause would be inactive). Thus the failure of the third finger to attach gives the system warning that this particular clause is "ready" or "armed" and about to trigger a unit propagation if any of its literals become false.

14.6 Economic Motivations

A digital circuit can be laid out as a collection of transistors or simple gates based on its logical description. In fact, CNF, ANF and DNF, are often used for this purpose. However, to find the most efficient layout is almost impossible. Yet, sometimes a circuit's size can be cut by a huge factor by using optimization tools. The reason for this is that large sub-circuits might be redundant or equivalent. In industry, SAT-solvers are quite often used to detect this condition.

For example, suppose a certain sub-circuit of a larger circuit is not satisfiable. Then surely one can replace it with a 0, saving many gates. If not, then suppose its negation is not satisfiable. Then the designer can replace it with a 1. For this reason, the CNF-SAT problem is crucial to the efficient implementation of microelectronics. (Though the problems frequently solved by SAT-solvers are usually much more complex than those two simple examples).

A non-trivial example would be to detect the equivalence of sub-circuits A and B. If

$$\left(A \wedge \overline{B}\right) \vee \left(\overline{A} \wedge B\right)$$

is unsatisfiable, then clearly A and B always agree. Since their outputs are always the same, there is no need to have both of them present. This is an oversimplification, because speed-of-light considerations might make it useful to have one at each end of a huge circuit. But logically speaking, it is not useful to have both present.

Over the years, there have been so many new and efficient SAT-solvers that the common way to solve many NP-Complete problems is to convert the problem into a CNF file, and then call a SAT-solver to find a satisfying assignment. In some ways it is amazing that this works, because much information is lost when performing the conversion. Yet it is a common practice, because SAT-solvers have been so carefully tuned by many researchers over several decades. The common applications are planning, AI, circuit-layout, and automated theorem proving.

14.7 Further Reading

The following further reading will be useful for the Walk-SAT family.

- "On Selecting a Satisfying Truth Assignment" by C. H. Papadimitriou [186], published in the proceedings of FOCS'91.
- "A new method for solving hard satisfiability problems" by B. Selman, H. J. Levesque, and D. G. Mitchell [208], published in the proceedings of AAAI'92.
- "Local Search Strategies for Satisfiability Testing" by Bart Selman, Henry Kautz, and Bram Cohen [207], published in the proceedings of the Second DIMACS Implementation Challenge, 1993.

The following further reading will be useful for the Chaff/MINISAT family.

- "A computing procedure for quantification theory" by Martin Davis and Hilary Putnam [92], published in the Journal of the Association of Computing Machinery in 1960.
- "A Machine Program for Theorem Proving" by Martin Davis, George Logemann and Donald W. Loveland [91], published in Communications of the ACM in 1962.
- "GRASP: A Search Algorithm for Propositional Satisfiability" by João P. Marques-Silva and Karem A. Sakallah [170], published in the IEEE Transactions on Computers in 1999.
- "Chaff: Engineering an Efficient SAT Solver", by Matthew W. Moskewicz, Conor F. Madigan, Ying Zhao, Lintao Zhang, and Sharad Malik [182], published in the proceedings of the 28th Design Automation Conference (DAC'01).
- "An Extensible SAT-solver" by Niklas Een and Niklas Srensson [103], published at SAT'03.
- "Effective Preprocessing in SAT through Variable and Clause Elimination" by Niklas Een and Armin Biere [102], published at SAT'05.

Chapter 15
Applying SAT-Solvers to Extension Fields of Low Degree

15.1 Introduction

In this book we have been discussing polynomial equations over finite fields of characteristic two for quite some time. However, the finite field in question has almost always been $GF(2)$. In cryptanalysis, this is usually the field of interest, with notable exceptions, including several ciphers: Rijndael [89] which later become the Advanced Encryption Standard (AES), which uses $GF(256)$; the cryptosystem called TTM [127], over the same field; the Courtois Toy Cipher (CTC) [14] [73], can be modeled over $GF(8)$ for certain settings. The stream cipher family QUAD [42] can operate over any finite field, but finite fields of characteristic two would be the natural setting (see Section 5.2 on Page 66).

As a step toward being able to operate on systems of polynomial equations over fields like $GF(256)$ or $GF(2^{32})$, we present the following research on smaller extension fields. In particular, we will show how to use a classic matrix representation of these fields to efficiently produce algebraic normal forms (logical sentences) for the multiplication operation of the extension field as viewed from $GF(2)$. These formulas can then be used to rapidly convert the system from over $GF(2^n)$ to over $GF(2)$, with n times as many variables. Here and in Appendix B on Page 309, we provide formulas for $GF(4)$ upto $GF(64)$. If extension fields are not familiar to the reader, see Section 15.10 on Page 295 for a review.

Once these formulas are found, the system can be solved with SAT-solvers (explained below) rather than with traditional Gröbner Bases methods, such as SINGULAR [9] and MAGMA [2]. Even for small systems, MAGMA (using F4) would crash due to a lack of memory, for example allocating 29.9 gigabytes for a system of 8 cubic equations in 8 unknowns over $GF(32)$. While Gröbner Bases algorithms like Faugére's F4 [109] and F5 [110], as implemented in MAGMA, are often faster, they require tremendous quantities of memory, and so problems of interest to cryptanalysis are impossible. Meanwhile, traditional Gröbner Bases methods like the original Buchberger algorithm, implemented in SINGULAR, are far slower than MAGMA. Thus SAT-solvers provide an interesting medium, for problems that are far too hard

G.V. Bard, *Algebraic Cryptanalysis*, DOI: 10.1007/978-0-387-88757-9_15

for the available memory, but simple enough to not exhaust the user's patience in running time. To help understand the balance between these three options, we have performed experiments, described below. Furthermore, we have found that if the number of equations is roughly half the number of unknowns, that SAT-solvers do quite well, much better than MAGMA or SAGE.

15.2 Solving $\mathbb{GF}(2)$ Systems via SAT-Solvers

As discussed in Chapter 12, several options exist when presented with a polynomial system of equations over a finite field, in n variables. If there are roughly $n^2/2$ equations, linearization will work. If one is close to but below this threshold, then the XL algorithm [82], will work as explained in Section 12.4 on Page 213.

However, if the number of equations is roughly the number of unknowns, Gröbner Bases methods are the usual solution, or brute force guessing all the variables. There is some debate as to whether Gröbner Bases algorithms are faster than brute force in general, but in typical cryptanalytic problems, they can be faster than brute force [111] [58]. The traditional Buchberger algorithm (used by SINGULAR [9], for example) for Gröbner Bases is available, as well as the algorithms F4 and F5 by Faugère (used by MAGMA [2], for example), which is very fast but requires a great deal of memory in practice.

In Chapter 13, we proposed that SAT-solvers are useful for solving polynomial systems of equations over $\mathbb{GF}(2)$, especially if the system is sparse or over-defined. In particular, the SAT-solver approach was slower than MAGMA if MAGMA did not crash for lack of memory. But it almost always crashed for this lack, for reasonable sized problems. Conversely, SAT-solvers were slower in practice but use a constant amount of memory, and so actually returned answers.

15.2.1 Sparsity

Recall, the sparsity of a polynomial system of equations is represented by β, which is the ratio of the number of non-zero coefficients to the number of possible coefficients. So for a cubic system of m equations over n variables, with c non-zero coefficients this would be

$$\beta = \frac{c}{m\left(\binom{n}{3} + \binom{n}{2} + \binom{n}{1} + \binom{n}{0}\right)}$$

if the system is over $\mathbb{GF}(2)$. This is because there is never a need for an exponent in any monomial for a polynomial over $\mathbb{GF}(2)$, as $x^2 = x$. However, in extension fields, this becomes false, as there are x such that $x^2 \neq x$.

For example, in $\mathbb{GF}(q)$ for $q > 3$, we would add $2n$ to the denominator in the previous example to reflect monomials that are squares and cubes of a variable, as well as n^2 terms of the form $x_i^2 x_j$. Over all, however, the values of β are comparable, as these corrections are small compared to the value of the denominator for large n. See also Theorem 16 on Page 47.

Finally, note that β is sometimes called the "density", rather than the "sparsity", to reflect that $\beta = 0.0001$ is rather sparse, and has a low density of non-zero terms.

15.3 Overview

The entire process can be thought of as follows. First, for any particular extension field, we construct a matrix model of that finite field. Second, we perform some matrix multiplication to get a matrix that represents the product of two, three, and four unknown field entries (these are large symbolic expressions, and the last one is gigantic). Third, we convert this to a formula over $\mathbb{GF}(2)$, using the remarkable property that there are some elements of the matrix M that are always 0 for M^n, except for one particular value of n, where they are 1, with $0 \leq n < q$. We call such entries "dead give-aways".

Now we have a formula for a double, triple, and quadruple product over the finite field in question, and this process, which is very fast, need never be performed again for that extension field. For example, we have done this for $\mathbb{GF}(4), \ldots, \mathbb{GF}(64)$, and chose $\mathbb{GF}(16)$ as an extended example in this book, and $\mathbb{GF}(32)$ for the experiments.

Fourth, for any polynomial system of equations over the finite field, we will convert any monomial $c x_i x_j$ using our triple product formula. Likewise, $c x_i x_j x_k$ will be converted with the quadruple product formula, and $c x_i$ with the double product formula. There are also formulas for terms like $c x_i^2$ described in Section 15.6.1 on Page 287. The monomials are now each q monomials over $\mathbb{GF}(2)$, where the original formula was over $\mathbb{GF}(2^q)$. Fifth, we add together all these formulas for all the monomials in a given polynomial. Now the entire original polynomial is a set of q polynomials over $\mathbb{GF}(2)$. Sixth, we remove monomials that contain multiplication by zero (there will be many—roughly half the monomials written so far). Seventh, we convert the polynomial system of equations over $\mathbb{GF}(2)$ (with q times as many variables and equations as the original over $\mathbb{GF}(2^q)$) into a CNF-SAT problem using the techniques of [33] and Chapter 13. Eighth, we solve the CNF-SAT problem using the SAT-solver MiniSAT, and this yields one solution to the original problem. In algebraic cryptanalysis, there is only one solution, so we are done.

15.4 Polynomial Systems over Extension Fields of $\mathbb{GF}(2)$

The extension field will be modeled as a quotient ring of $\mathbb{GF}(2)[\alpha]$, the ring of polynomials in a single variable α with coefficients in $\mathbb{GF}(2)$. The quotient will be in terms of the principle ideal generated by a particular polynomial, of the degree of the required extension, and irreducible over $\mathbb{GF}(2)$. Specifically, the following polynomials were used

$$
\begin{array}{ll}
\mathrm{GF}(4) & \alpha^2 + \alpha + 1 \\
\mathrm{GF}(8) & \alpha^3 + \alpha + 1 \\
\mathrm{GF}(16) & \alpha^4 + \alpha + 1 \\
\mathrm{GF}(32) & \alpha^5 + \alpha^2 + 1 \\
\mathrm{GF}(64) & \alpha^6 + \alpha + 1
\end{array}
$$

Of course, all finite fields of the same size are isomorphic, so the particular polynomial chosen is of little importance. Each of the polynomials above has weight three (the number of monomials), and choosing polynomials of non-minimal weight would artificially increase the densities of the formulas we produce. But there is no reason not to choose a polynomial of minimal weight.

15.4.1 Extensions of the Coefficient Field

We should stress here that we are interested only in solutions over the coefficient field. If one is interested in solutions over an extension field of the coefficient field, then one can carry-out all operations here as if one were in the extension field from the beginning. In other words, if the coefficients are in $\mathrm{GF}(8)$ and one is interested in $\mathrm{GF}(512)$ solutions, then one can treat the entire system as over $\mathrm{GF}(512)$, since all elements of $\mathrm{GF}(8)$ are elements of $\mathrm{GF}(512)$ for the right choice of bases.

15.4.2 Difficulty in Bits

Some cryptographers measure the difficulty of a problem in bits, meaning the logarithm base two of the number of possible solutions. In this case, for $\mathrm{GF}(2^k)$ polynomial systems with n variables, the difficulty is kn-bits. Thus, for $\mathrm{GF}(32)$, used in our trials, one can see that 7 variables was roughly the threshold of feasibility for one hour. But that is 35 bits, an otherwise significant difficulty, as 2^{35} is roughly 32 billion possibilities.

15.5 Finding Efficient Arithmetic Representations via Matrices

The technique below, of representing field elements by matrices, is classic and has been known for a long time. However, we will use a property of those particular cells of the matrix which happen to be 1 for one basis element, but 0 for all other basis elements, to help us rapidly generate formulas for the finite field arithmetic.

We will use $\mathrm{GF}(16)$ as an extended example, but the same arguments work for any other choice. For our experiments, we switched to $\mathrm{GF}(32)$. The companion matrix of $\alpha^4 + \alpha + 1$ has several properties one of which is that when substituted for α in that equation, it results in the zero matrix. Call this matrix A. We can calculate the powers of A, which are listed below:

$$A^0 = \begin{bmatrix} 1 & 0 & 0 & 0 \\ 0 & 1 & 0 & 0 \\ 0 & 0 & 1 & 0 \\ 0 & 0 & 0 & 1 \end{bmatrix} \quad A^1 = \begin{bmatrix} 0 & 0 & 0 & 1 \\ 1 & 0 & 0 & 1 \\ 0 & 1 & 0 & 0 \\ 0 & 0 & 1 & 0 \end{bmatrix} \quad A^2 = \begin{bmatrix} 0 & 0 & 1 & 0 \\ 0 & 0 & 1 & 1 \\ 1 & 0 & 0 & 1 \\ 0 & 1 & 0 & 0 \end{bmatrix}$$

$$A^3 = \begin{bmatrix} 0 & 1 & 0 & 0 \\ 0 & 1 & 1 & 0 \\ 0 & 0 & 1 & 1 \\ 1 & 0 & 0 & 1 \end{bmatrix} \quad A^4 = \begin{bmatrix} 1 & 0 & 0 & 1 \\ 1 & 1 & 0 & 1 \\ 0 & 1 & 1 & 0 \\ 0 & 0 & 1 & 1 \end{bmatrix} \quad A^5 = \begin{bmatrix} 0 & 0 & 1 & 1 \\ 1 & 0 & 1 & 0 \\ 1 & 1 & 0 & 1 \\ 0 & 1 & 1 & 0 \end{bmatrix} \cdots$$

By virtue of the polynomial $\alpha^4 + \alpha + 1$, we know that $A^4 = -A^1 - A^0$, which since we are in characteristic two is $A + I$. Therefore α^4 and all higher powers can be discarded, and one can show that I, A, A^2, A^3 is a basis for the field, when the field is considered a 4-dimensional vector space over $\mathbb{GF}(2)$.

Now we can represent the element $f_0 + f_1\alpha + f_2\alpha^2 + f_3\alpha^3$ as $f_0I + f_1A + f_2A^2 + f_3A^3$, a single 4×4 matrix. This is a very old technique, and is found in the free Interent textbook *Applied Abstract Algebra* by Joyner, *et al* [144, "Matrix constructions of finite fields"], and so we will not include proofs at this point.

Consider $M = f_0I + f_1A + f_2A^2 + f_3A^3$. Observe that the M_{11} cell is 1 for A^0 but 0 for all other powers of A. Likewise, the M_{14} is 1 for A and zero for the other powers. The equivalent cell for A^2 is M_{13} and for A^3 is M_{12}. Thus after converting some fields elements into matrices, and performing a series of matrix operations on them to obtain M, we can recover the field element in $f_0 + f_1\alpha + f_2\alpha^2 + f_3\alpha^3$ form by noting that

$$f_0 = M_{11}, \quad f_1 = M_{14}, \quad f_2 = M_{13}, \quad f_3 = M_{12}$$

This is significant because it might otherwise be hard to convert back from a single matrix to the four-dimensional notation. These cells that are zero for all powers of A with one exception, are "dead give-aways" of the normal field representation. We prove they will always exist in Theorem 92 on Page 298.

Now, we will obtain a logical formula for the triple product $p = abc$, where a, b, and c, are all elements of $\mathbb{GF}(16)$. One can define $X = a_0I + a_1A + a_2A^2 + a_3A^3$, where $a = a_0 + a_1\alpha + a_2\alpha^2 + a_3\alpha^3$ and likewise Y and Z for b and c. Note that the a_i are elements of $\mathbb{GF}(2)$, and since $\mathbb{GF}(16)$ is a $\mathbb{GF}(2)$-vector space of dimension 4, the addition operation $T = X + Y + Z$ is just vectorial addition, with $s_i = a_i + b_i + c_i$, and $T = s_0 + s_1\alpha + s_2\alpha^2 + s_3\alpha^3$.

For example, we have $a = a_0 + a_1\alpha + a_2\alpha^2 + a_3\alpha^3$ encoded as

$$X = \begin{bmatrix} a_0 & a_3 & a_2 & a_1 \\ a_1 & a_3 + a_0 & a_2 + a_3 & a_1 + a_2 \\ a_2 & a_1 & a_3 + a_0 & a_2 + a_3 \\ a_3 & a_2 & a_1 & a_3 + a_0 \end{bmatrix}$$

and so the sum of two of these is just a matrix sum, with the simple addition of the entries.

The multiplication operation is much more complex. Using one's favorite algebra software (the author used MAPLE), one can calculate $M = XYZ$ symbolically, as a matrix product. Then using the "dead give-away" entries M_{11}, M_{14}, M_{13}, and M_{12}, one can recover the following formulas, for p_0, p_1, p_2 and p_3 respectively, each of which is the entry in the specified matrix position.

$$
\begin{aligned}
p_0 ={}& c_0a_0b_0 + c_0a_3b_1 + c_0a_2b_2 + c_0a_1b_3 + c_1a_0b_3 + c_1a_3b_0 + c_1a_3b_3 + c_1a_2b_1 + c_1a_1b_2 \\
&+ c_2a_0b_2 + c_2a_3b_2 + c_2a_3b_3 + c_2a_2b_0 + c_2a_2b_3 + c_2a_1b_1 + c_3a_0b_1 + c_3a_3b_1 + c_3a_3b_2 \\
&+ c_3a_2b_2 + c_3a_2b_3 + c_3a_1b_0 + c_3a_1b_3 \\
p_1 ={}& c_0a_3b_1 + c_1a_0b_3 + c_0a_1b_3 + c_1a_0b_0 + c_2a_3b_2 + c_2a_2b_0 + c_2a_2b_3 + c_2a_1b_1 + c_2a_0b_2 \\
&+ c_3a_0b_1 + c_3a_3b_1 + c_3a_2b_2 + c_3a_1b_0 + c_3a_1b_3 + a_3b_3c_3 + a_0b_3c_2 + c_1a_1b_2 \\
&+ a_3b_0c_2 + a_2b_1c_2 + a_1b_2c_2 + a_0b_2c_3 + a_3b_2c_0 + a_2b_0c_3 + a_2b_3c_0 + a_1b_1c_3 + c_1a_3b_1 \\
&+ c_1a_2b_2 + c_1a_1b_3 + a_0b_1c_0 + a_1b_0c_0 + c_0a_2b_2 + c_1a_3b_0 + c_1a_3b_3 + c_1a_2b_1 \\
p_2 ={}& c_2a_1b_3 + a_3b_0c_3 + a_2b_1c_3 + a_1b_2c_3 + c_2a_3b_3 + c_3a_3b_2 + c_3a_2b_3 + a_0b_3c_2 + a_0b_3c_3 \\
&+ a_3b_0c_2 + a_2b_1c_2 + a_1b_2c_2 + a_0b_2c_3 + a_3b_2c_0 + a_2b_0c_3 + a_2b_3c_0 + a_1b_1c_3 + c_1a_3b_1 \\
&+ c_1a_2b_2 + c_1a_1b_3 + c_2a_0b_0 + c_2a_3b_1 + c_2a_2b_2 + a_0b_2c_0 + a_3b_3c_0 + a_2b_0c_0 + a_1b_1c_0 \\
&+ c_1a_0b_1 + c_1a_3b_2 + c_1a_2b_3 + c_1a_1b_0 \\
p_3 ={}& c_3a_0b_0 + c_2a_1b_3 + a_3b_0c_3 + a_2b_1c_3 + a_1b_2c_3 + c_2a_3b_2 + c_2a_2b_3 + c_3a_3b_1 + c_3a_2b_2 \\
&+ c_3a_1b_3 + a_3b_3c_3 + a_0b_3c_0 + a_3b_0c_0 + a_2b_1c_0 + c_2a_3b_1 + c_2a_2b_2 + c_1a_3b_3 + a_1b_2c_0 \\
&+ c_1a_0b_2 + c_1a_2b_0 + c_1a_1b_1 + c_2a_0b_1 + c_2a_1b_0 + a_3b_3c_0 + c_1a_3b_2 + c_1a_2b_3 + a_0b_3c_3 \\
M ={}& p_0 + \alpha p_1 + \alpha^2 p_2 + \alpha^3 p_3
\end{aligned}
$$

Likewise, for the double product, $p = ab$, one obtains

$$
\begin{aligned}
p_0 &= a_0b_0 + a_3b_1 + a_2b_2 + a_1b_3 \\
p_1 &= a_0b_1 + a_3b_1 + a_3b_2 + a_2b_2 + a_2b_3 + a_1b_0 + a_1b_3 \\
p_2 &= a_0b_2 + a_3b_2 + a_3b_3 + a_2b_0 + a_2b_3 + a_1b_1 \\
p_3 &= a_0b_3 + a_3b_0 + a_3b_3 + a_2b_1 + a_1b_2
\end{aligned}
$$

and for the quadruple product $abcd$,

$$
\begin{aligned}
p_0 ={}& d_0c_0a_0b_0 + d_0c_0a_3b_1 + d_0c_0a_2b_2 + d_0c_0a_1b_3 + d_0c_1a_0b_3 + d_0c_1a_3b_0 + d_0c_1a_3b_3 + d_0c_1a_2b_1 \\
&+ d_0c_1a_1b_2 + d_0c_2a_0b_2 + d_0c_2a_3b_2 + d_0c_2a_3b_3 + d_0c_2a_2b_0 + d_0c_2a_2b_3 + d_0c_2a_1b_1 + d_0c_3a_0b_1 \\
&+ d_0c_3a_3b_1 + d_0c_3a_3b_2 + d_0c_3a_2b_2 + d_0c_3a_2b_3 + d_0c_3a_1b_0 + d_0c_3a_1b_3 + d_1c_3a_0b_0 + d_1c_2a_3b_2 \\
&+ d_1a_1b_2c_3 + d_1a_1b_2c_0 + d_1a_2b_1c_3 + d_1a_2b_1c_0 + d_1a_3b_3c_3 + d_1a_3b_3c_0 + d_1a_3b_0c_3 + d_1a_3b_0c_0 \\
&+ d_1a_0b_3c_3 + d_1c_3a_1b_3 + d_1c_3a_2b_2 + d_1c_2a_1b_3 + d_1c_2a_1b_0 + d_1c_2a_2b_2 + d_1c_2a_3b_1 + d_1c_2a_0b_1 \\
&+ d_1c_1a_1b_1 + d_1c_1a_2b_3 + d_1c_1a_2b_0 + d_1c_1a_3b_2 + d_1c_1a_0b_2 + d_1c_1a_3b_3 + d_1c_3a_3b_1 + d_1c_2a_2b_3 \\
&+ d_1a_0b_3c_0 + d_2c_2a_3b_3 + d_2c_2a_0b_0 + d_2a_1b_2c_3 + d_2a_2b_1c_3 + d_2a_3b_3c_0 + d_2a_3b_0c_3 + d_3a_1b_1c_3 \\
&+ d_2a_0b_3c_3 + d_2c_3a_2b_3 + d_2c_3a_3b_2 + d_2a_2b_3c_0 + d_2a_2b_0c_3 + d_2a_2b_0c_0 + d_2a_3b_2c_0 + d_2a_0b_2c_3 \\
&+ d_2a_0b_2c_0 + d_2a_1b_2c_2 + d_2a_2b_1c_2 + d_2a_3b_0c_2 + d_2a_0b_3c_2 + d_2c_2a_1b_3 + d_2c_2a_2b_2 + d_2c_2a_3b_1
\end{aligned}
$$

$+d_2c_1a_2b_3 + d_2c_1a_3b_2 + d_2c_1a_1b_0 + d_2c_1a_2b_2 + d_2c_1a_3b_1 + d_2c_1a_0b_1 + d_2a_1b_1c_3 + d_2a_1b_1c_0$

$+d_2c_1a_1b_3 + d_3c_1a_1b_2 + d_3c_2a_1b_1 + d_3c_2a_2b_3 + d_3c_1a_1b_3 + d_3c_2a_2b_0 + d_3c_0a_2b_2 + d_3c_1a_3b_1$

$+d_3c_1a_0b_0 + d_3c_2a_3b_2 + d_3c_1a_0b_3 + d_3c_0a_1b_3 + d_3c_0a_3b_1 + d_3a_3b_3c_3 + d_3c_3a_1b_3 + d_3c_3a_1b_0$

$+d_3c_3a_2b_2 + d_3c_2a_0b_2 + d_3c_1a_3b_0 + d_3a_2b_3c_0 + d_3a_2b_0c_3 + d_3c_1a_2b_1 + d_3a_3b_2c_0 + d_3a_0b_2c_3$

$+d_3a_1b_2c_2 + d_3a_2b_1c_2 + d_3a_3b_0c_2 + d_3a_0b_3c_2 + d_3c_1a_3b_3 + d_3a_1b_0c_0 + d_3a_0b_1c_0 + d_3c_3a_0b_1$

$+d_3c_3a_3b_1 + d_3c_1a_2b_2$

$p_1 = c_1a_0b_0d_0 + a_3b_3c_3d_0 + a_2b_3c_0d_0 + a_2b_0c_3d_0 + a_3b_2c_0d_0 + a_0b_2c_3d_0 + a_1b_2c_2d_0 + a_2b_1c_2d_0$

$+a_3b_0c_2d_0 + a_0b_3c_2d_0 + a_1b_0c_0d_0 + a_0b_1c_0d_0 + c_1a_2b_2d_0 + c_1a_3b_1d_0 + a_1b_1c_3d_0 + c_1a_1b_3d_0$

$+c_1a_0b_2d_2 + c_1a_3b_3d_2 + c_3a_3b_1d_2 + c_2a_2b_3d_2 + a_0b_3c_0d_2 + c_2a_0b_0d_3 + a_1b_2c_3d_3 + a_2b_1c_3d_3$

$+a_3b_3c_0d_3 + a_3b_0c_3d_3 + a_0b_3c_3d_3 + a_2b_0c_0d_3 + a_0b_2c_0d_3 + c_2a_1b_3d_3 + c_2a_2b_3d_3 + c_2a_3b_1d_3$

$+c_1a_2b_3d_3 + c_1a_3b_2d_3 + c_1a_1b_0d_3 + c_1a_0b_1d_3 + a_1b_1c_0d_3 + d_0c_0a_3b_1 + d_0c_0a_2b_2 + d_0c_0a_1b_3$

$+d_0c_1a_0b_3 + d_0c_1a_3b_0 + d_0c_1a_3b_3 + d_0c_1a_2b_1 + d_0c_1a_1b_2 + d_0c_2a_0b_2 + d_0c_2a_3b_2 + d_2c_1a_1b_0$

$+d_0c_2a_2b_0 + d_0c_2a_2b_3 + d_0c_2a_1b_1 + d_0c_3a_0b_1 + d_0c_3a_3b_1 + d_0c_3a_2b_2 + d_0c_3a_1b_0 + d_2a_0b_3c_2$

$+d_0c_3a_1b_3 + d_1c_3a_0b_0 + d_1a_1b_2c_3 + d_1a_1b_2c_0 + d_1a_2b_1c_3 + d_1a_2b_1c_0 + d_1a_3b_3c_3 + c_1a_2b_0d_2$

$+d_1a_3b_3c_0 + d_1a_3b_0c_3 + d_1a_3b_0c_0 + d_1a_0b_3c_3 + d_1c_2a_1b_3 + d_1c_2a_1b_0 + c_2a_1b_0d_2 + c_2a_0b_1d_2$

$+d_1c_2a_2b_2 + d_1c_2a_3b_1 + d_1c_2a_0b_1 + d_1c_1a_1b_1 + d_1c_1a_2b_3 + d_1c_1a_2b_0 + d_1c_1a_3b_2 + d_1c_1a_0b_2$

$+d_1a_0b_3c_0 + d_2c_2a_3b_3 + d_2c_2a_0b_0 + d_2a_3b_3c_3 + d_2c_3a_2b_3 + d_2c_3a_3b_2 + d_2a_2b_3c_0 + c_1a_1b_1d_2$

$+d_2a_2b_0c_3 + d_2a_2b_0c_0 + d_2a_3b_2c_0 + d_2a_0b_2c_3 + d_2a_0b_2c_0 + d_2a_1b_2c_2 + d_2a_2b_1c_2 + d_2a_3b_0c_2$

$+d_2c_1a_2b_2 + d_2c_1a_3b_1 + d_2c_1a_0b_1 + d_2a_1b_1c_3 + d_2a_1b_1c_0 + d_2c_1a_1b_3 + d_3c_1a_1b_2 + d_3c_2a_3b_3$

$+d_3c_2a_1b_1 + d_3c_2a_2b_3 + d_3c_2a_2b_0 + d_3c_0a_2b_2 + d_3c_1a_0b_0 + d_3c_2a_2b_2 + d_3c_1a_0b_3 + d_3c_0a_1b_3$

$+d_3c_0a_3b_1 + d_3a_3b_3c_3 + d_3c_3a_1b_3 + d_3c_3a_1b_0 + d_3c_3a_2b_3 + d_3c_3a_2b_2 + d_3c_3a_3b_2 + d_3c_2a_0b_2$

$+d_3c_1a_3b_0 + d_3c_1a_2b_1 + d_3c_1a_3b_3 + d_3a_1b_0c_0 + d_3a_0b_1c_0 + d_3c_3a_0b_1 + d_3c_3a_3b_1 + d_1c_0a_0b_0$

$+d_1c_0a_3b_1 + d_1c_0a_2b_2 + d_1c_0a_1b_3 + d_1c_1a_0b_3 + d_1c_1a_3b_0 + d_1c_1a_2b_1 + d_1c_1a_1b_2 + d_1c_2a_0b_2$

$+d_1c_2a_3b_3 + d_1c_2a_2b_0 + d_1c_2a_1b_1 + d_1c_3a_0b_1 + d_1c_3a_3b_2 + d_1c_3a_2b_3 + d_1c_3a_1b_0 + c_3a_0b_0d_2$

$+c_2a_3b_2d_2 + a_1b_2c_0d_2 + a_2b_1c_0d_2 + a_3b_0c_0d_2 + c_3a_1b_3d_2 + c_3a_2b_2d_2$

$p_2 = a_2b_3c_0d_0 + a_2b_0c_3d_0 + a_3b_2c_0d_0 + a_0b_2c_3d_0 + a_1b_2c_2d_0 + a_2b_1c_2d_0 + a_3b_0c_2d_0 + a_0b_3c_2d_0$

$+c_1a_2b_2d_0 + c_1a_3b_1d_0 + a_1b_1c_3d_0 + c_1a_1b_3d_0 + c_1a_0b_2d_2 + a_0b_3c_0d_2 + c_2a_0b_0d_3 + a_2b_0c_0d_3$

$+a_0b_2c_0d_3 + c_1a_1b_0d_3 + c_1a_0b_1d_3 + a_1b_1c_0d_3 + d_0c_2a_3b_3 + d_0c_3a_3b_2 + d_0c_3a_2b_3 + d_1c_2a_3b_2$

$+d_1a_3b_3c_3 + d_1c_3a_1b_3 + d_1c_3a_2b_2 + d_1c_1a_3b_3 + d_1c_3a_3b_1 + d_1c_2a_3b_3 + d_2c_2a_3b_3 + d_2a_1b_2c_3$

$+d_2a_2b_1c_3 + d_2a_3b_3c_3 + d_2a_3b_3c_0 + d_2a_3b_0c_3 + d_2a_0b_3c_3 + d_2c_3a_2b_3 + d_2c_3a_3b_2 + d_2c_2a_1b_3$

$+d_2c_2a_2b_2 + d_2c_2a_3b_1 + d_2c_1a_2b_3 + d_2c_1a_3b_2 + d_3c_2a_3b_3 + d_3c_2a_2b_3 + d_3c_1a_1b_3 + d_3c_2a_3b_2$

$+d_3a_3b_3c_3 + d_3c_3a_1b_3 + d_3c_3a_2b_3 + d_3c_3a_2b_2 + d_3c_3a_3b_2 + d_3a_2b_3c_0 + d_3a_2b_0c_3 + d_3a_3b_2c_0$

$+d_3a_0b_2c_3 + d_3a_1b_2c_2 + d_3a_2b_1c_2 + d_3a_3b_0c_2 + d_3a_0b_3c_2 + d_3c_1a_3b_3 + d_3c_3a_3b_1 + d_3c_1a_2b_2$

$+d_3c_1a_3b_1 + d_3a_1b_1c_3 + d_1c_0a_3b_1 + d_1c_0a_2b_2 + d_1c_0a_1b_3 + d_1c_1a_0b_3 + d_1c_1a_3b_0 + d_1c_1a_2b_1$

$+d_1c_1a_1b_2 + d_1c_2a_0b_2 + d_1c_2a_2b_0 + d_1c_2a_1b_1 + d_1c_3a_0b_1 + d_1c_3a_1b_0 + c_3a_0b_0d_2 + a_1b_2c_0d_2$

$+a_2b_1c_0d_2 + a_3b_0c_0d_2 + c_2a_1b_0d_2 + c_2a_0b_1d_2 + c_1a_1b_1d_2 + c_1a_2b_0d_2 + d_2c_0a_0b_0 + d_2c_0a_3b_1$

$+d_2c_0a_2b_2 + d_2c_0a_1b_3 + d_2c_1a_0b_3 + d_2c_1a_3b_0 + d_2c_1a_2b_1 + d_2c_1a_1b_2 + d_2c_2a_0b_2 + d_2c_2a_2b_0$

$+d_2c_2a_1b_1 + d_2c_3a_0b_1 + d_2c_3a_1b_0 + c_3a_0b_0d_3 + a_1b_2c_0d_3 + a_2b_1c_0d_3 + a_3b_0c_0d_3 + c_2a_1b_0d_3$

$+c_2a_0b_1d_3 + c_1a_1b_1d_3 + c_1a_2b_0d_3 + c_1a_0b_2d_3 + a_0b_3c_0d_3 + c_2a_0b_0d_0 + a_1b_2c_3d_0 + a_2b_1c_3d_0$

$+a_3b_3c_0d_0 + a_3b_0c_3d_0 + a_0b_3c_3d_0 + a_2b_0c_0d_0 + a_0b_2c_0d_0 + c_2a_1b_3d_0 + c_2a_2b_2d_0 + c_2a_3b_1d_0$

$+c_1a_2b_3d_0 + c_1a_3b_2d_0 + c_1a_1b_0d_0 + c_1a_0b_1d_0 + a_1b_1c_0d_0 + d_1c_1a_0b_0 + d_1a_2b_3c_0 + d_1a_2b_0c_3$

$+d_1a_3b_2c_0 + d_1a_0b_2c_3 + d_1a_1b_2c_2 + d_1a_2b_1c_2 + d_1a_3b_0c_2 + d_1a_0b_3c_2 + d_1a_1b_0c_0 + d_1a_0b_1c_0$

$$+d_1c_1a_2b_2 + d_1c_1a_3b_1 + d_1a_1b_1c_3 + d_1c_1a_1b_3$$

$$p_3 = a_3b_3c_3d_0 + c_1a_3b_3d_2 + c_3a_3b_1d_2 + c_2a_2b_3d_2 + a_1b_2c_3d_3 + a_2b_1c_3d_3 + a_3b_3c_0d_3 + a_3b_0c_3d_3$$

$$+a_0b_3c_3d_3 + c_2a_1b_3d_3 + c_2a_2b_2d_3 + c_2a_3b_1d_3 + c_1a_2b_3d_3 + c_1a_3b_2d_3 + d_0c_1a_3b_3 + d_0c_2a_3b_2$$

$$+d_0c_2a_2b_3 + d_0c_3a_3b_1 + c_3a_0b_0d_0 + a_0b_3c_0d_0 + a_3b_0c_0d_0 + a_2b_1c_0d_0 + a_1b_2c_0d_0 + c_1a_2b_0d_0$$

$$+c_1a_1b_1d_0 + c_2a_1b_0d_0 + d_1c_2a_0b_0 + d_1a_0b_2c_0 + d_1a_2b_0c_0 + d_1a_1b_1c_0 + d_1c_1a_0b_1 + d_1c_1a_1b_0$$

$$+d_2c_1a_0b_0 + d_2a_0b_1c_0 + d_2a_1b_0c_0 + d_3c_0a_0b_0 + c_2a_0b_1d_0 + c_1a_0b_2d_0 + d_0c_3a_2b_2 + d_0c_3a_1b_3$$

$$+d_1a_1b_2c_3 + d_1a_2b_1c_3 + d_1a_3b_3c_0 + d_1a_3b_0c_3 + d_1a_0b_3c_3 + d_1c_2a_1b_3 + d_1c_2a_2b_2 + d_2a_3b_3c_3$$

$$+d_2a_2b_3c_0 + d_2a_2b_0c_3 + d_2a_3b_2c_0 + d_2a_0b_2c_3 + d_2a_1b_2c_2 + d_2a_2b_1c_2 + d_2a_3b_0c_2 + d_2a_0b_3c_2$$

$$+d_2c_1a_2b_2 + d_2c_1a_3b_1 + d_2a_1b_1c_3 + d_2c_1a_1b_3 + d_3c_1a_1b_2 + d_3c_2a_3b_3 + d_3c_2a_1b_1 + d_1c_1a_3b_2$$

$$+d_3c_2a_2b_0 + d_3c_0a_2b_2 + d_3c_1a_0b_3 + d_3c_0a_1b_3 + d_3c_0a_3b_1 + d_3a_3b_3c_3 + d_1c_2a_3b_1 + d_1c_1a_2b_3$$

$$+d_3c_3a_1b_0 + d_3c_3a_2b_3 + d_3c_3a_3b_2 + d_3c_2a_0b_2 + d_3c_1a_3b_0 + d_3c_1a_2b_1 + d_1c_1a_1b_3 + d_1a_1b_1c_3$$

$$+d_3c_3a_0b_1 + d_1c_2a_3b_3 + d_1c_3a_3b_2 + d_1c_3a_2b_3 + c_2a_3b_2d_2 + c_3a_1b_3d_2 + c_3a_2b_2d_2 + d_1c_1a_3b_1$$

$$+d_2c_0a_3b_1 + d_2c_0a_2b_2 + d_2c_0a_1b_3 + d_2c_1a_0b_3 + d_2c_1a_3b_0 + d_2c_1a_2b_1 + d_2c_1a_1b_2 + d_2c_2a_0b_2$$

$$+d_2c_2a_2b_0 + d_2c_2a_1b_1 + d_2c_3a_0b_1 + d_2c_3a_1b_0 + c_3a_0b_0d_3 + a_1b_2c_0d_3 + a_2b_1c_0d_3 + a_3b_0c_0d_3$$

$$+c_2a_1b_0d_3 + c_2a_0b_1d_3 + c_1a_1b_1d_3 + c_1a_2b_0d_3 + c_1a_0b_2d_3 + a_0b_3c_0d_3 + a_1b_2c_3d_0 + a_2b_1c_3d_0$$

$$+a_3b_3c_0d_0 + a_3b_0c_3d_0 + a_0b_3c_3d_0 + c_2a_1b_3d_0 + c_2a_2b_2d_0 + c_2a_3b_1d_0 + c_1a_2b_3d_0 + c_1a_3b_2d_0$$

$$+d_1a_2b_3c_0 + d_1a_2b_0c_3 + d_1a_3b_2c_0 + d_1a_0b_2c_3 + d_1a_1b_2c_2 + d_1a_2b_1c_2 + d_1a_3b_0c_2 + d_1a_0b_3c_2$$

$$+d_1c_1a_2b_2$$

which is shockingly worse. Therefore, one can see that it is essential that the process of obtaining these formulas be automated, at least partially, if one is to tackle 10th degree polynomials over $\mathbb{GF}(256)$.

15.6 Using the Algebraic Normal Forms

Each equation is to be independently converted into a set of $\mathbb{GF}(2)$ equations. Once this is done for all equations, the union of these sets of equations is converted into a SAT problem as described in Chapter 13. Then a SAT-solver is called to produce a solution or declare unsatisfiability.

We used $\mathbb{GF}(32)$ in our experiments. Each equation over $\mathbb{GF}(32)$, consists of a sum of terms (monomials). Each of these monomials is a product of some number of variables and some number of constants. Using the ANFs of the product operation, as found in the previous section, we have a logical formula for each of the "bits" or $\mathbb{GF}(2)$ terms of any $\mathbb{GF}(32)$ product. Thus, we simply apply these formulas to the constant(s) and variable(s) present.

Once this is done for each monomial in a polynomial, these formulas can be simply added, adding them component-wise as 5-dimensional vectors. This is because the addition operation over $\mathbb{GF}(32)$ is just 5-dimensional vector addition over $\mathbb{GF}(2)$. These sums of $\mathbb{GF}(2)$-polynomials are $\mathbb{GF}(2)$-polynomials, and there should be 5 sums per original polynomial.

These sums now form a $\mathbb{GF}(2)$ system of equations, with five times as many unknowns and equations. That, in turn, can be converted into a SAT-problem via the techniques found in Chapter 13. In fact, the author simply used his old code, without modification, to do so.

15.6.1 Remarks on the Special Forms

The product $cxyz$ is sufficient to encode the "special cases" of cx^3 and cx^2y. For higher degree formulas, naturally there are even more special cases. As it turns out, there is much to be gained by having a separate formula for special forms like cx^2. Consider that the formula for cxy in $\mathbb{GF}(16)$ given above has 22, 34, 31 and 27 terms in the expressions for the constant, linear, quadratic and cubic coefficients, or a total of 114 terms. Instead, if we generate formulas for cx^2 we obtain

$$p_0 = c_0y_0 + c_0y_2 + c_3y_2 + c_2y_3 + c_2y_1 + c_1y_3$$
$$p_1 = c_3y_3 + c_1y_0 + c_1y_2 + c_3y_1 + c_3y_2 + c_0y_2 + c_2y_1 + c_1y_3$$
$$p_2 = c_2y_0 + c_2y_2 + c_1y_2 + c_3y_1 + c_0y_3 + c_0y_1 + c_2y_3$$
$$p_3 = c_3y_0 + c_3y_2 + c_2y_2 + c_1y_3 + c_1y_1 + c_3y_3 + c_0y_3$$

which has 6, 8, 7, and 7 terms, or a total of 28 terms. When comparing the 114 to the 28, we must also recall that the terms in the general product were degree 3 terms, whereas here we have degree 2 terms. This is because $x^2 = x$ in $\mathbb{GF}(2)$, or more simply because $1 \times 1 = 1$ and $0 \times 0 = 0$. Thus, the fair comparison is 342 total symbols versus 56 total symbols.

15.6.2 Remarks on Degree

We chose quadratic equations, principally because the logical formula for a $\mathbb{GF}(32)$ quadruple product (multiplication of 4 variables) is huge. In fact, it was too large to copy down from MAPLE. However, for smaller fields, the 4 product formula was not as bad, and so one could write the converter for cubics as well. Recall, a quadratic term requires a triple product: a coefficient multiplied by two variables.

It is noteworthy to mention that, by the addition of new variables, any system of equations can be written as degree 2, regardless of the original degree of the polynomials. This can be done while introducing zero spurious solutions and destroying zero original solutions. See Section 11.4 on Page 192 for details.

For example, if one had originally $a + bcd = 1$, then one can say let $x = bc$, and one has $a + xd = 1$ and $x = bc$ as two equations, now of quadratic degree. One can show that if the degree of the original system of equations is fixed, that only polynomially many new variables are introduced. One can see that the order in which these new variables are introduced can have a tremendous impact on the number required.

15.6.3 Remarks on Coefficients

Unlike $GF(2)$ polynomials, the polynomials over $GF(2^k)$ (with $k > 1$) have co-efficients. Since a formula for a triple product is available, it makes sense for the first of the three multiplicands to be the coefficient. A shortcut makes the resulting system of equations smaller. One should simply delete any terms with coefficient zero. This may sound obvious but in a sparse system, it can be a large savings.

Second, one could imagine 31 extra variables, one for each of the 31 extra con-stants. These additional variables are fixed to the constants by the insertion of 31 additional equations of the form $x_i = c_i$ for the constant c_i. Then the entire sys-tem of polynomials can have its coefficients removed and be a series of products of variables. This is very inefficient as it turns out. Instead, when c_{i1} or c_{i2} is needed, signifying the first or second bit of the constant c_i, one should substitute "0" or "1" as required. Variables permanently set to 0 or 1 are easily handled by a SAT-Solver, as they are removed early on. See Chapter 14.

For example, if we had the monomial $(\alpha^2 + \alpha)xy$ then $c_0 = 0$, $c_1 = 1$, $c_2 = 1$, $c_3 = 0$, and $c_4 = 0$. Any monomial containing c_0, c_3 or $c_4 = 0$ in the product formulas will represent multiplying by 0. This is obviously equal to the 0 monomial, and so we can remove any term from the system that contains c_0, c_3 or c_4 in this monomial. In practice, we delete approximately half the monomials from the $GF(2)$ system of equations this way.

15.6.4 Solving with Gröbner Bases

When solving with a Gröbner Bases algorithm, one need merely state the poly-nomials to be solved. They form an ideal, and all polynomials in that ideal are zero on a set of points called the variety, or set of solutions. If there is only one point in this set, then the Gröbner Bases will look like $x_1 = 1$, $x_2 = 0$, etc... (or equivalently $x_1 + 1, x_2, \ldots$)

One problem is that we are only interested in solutions that have values in the coefficient field. Taking $GF(32)$ as an example, there are 31 elements in the mul-tiplicative group, which has 1 as its identity. Therefore, by group theory and, in particular, Lagrange's Theorem $x^{31} = 1$, or alternatively $x^{32} - x = 0$. The second equation has the added property that it is true for 0, the only field element excluded from the multiplicative group. (This is also sometimes called Fermat's Little Theo-rem).

Thus, an element $x \in \overline{GF(2)}$ is in the field $GF(32)$ if and only if $x^{32} - x = 0$. We can add an equation of this form for each variable in the polynomial system. Then, finally, we will be restricted to solutions in the coefficient field, as desired. However, 32 is a relatively high degree, and this is why the Gröbner Bases may have performed badly in these experiments.

In Gröbner Bases approaches, it is important to note the choice of variable order-ing. We used `degrevlex`, on the advice of M. Albrecht.

15.7 Experimental Results

The experimental results can be found in the table on Page 289.

Special Symbols in Results Table:

The following special indicators are used in the results table. First, "crashed" signifies that the software aborted due to a lack of memory, usually by attempting to allocate 30 gigabytes. Second, "> 70 mins" signifies that the software exceeded the time limit that etiquette requires on a shared machine. Third, "no trial" signifies that since a smaller version of the same problem either crashed or timed out, this size was not tried.

Num Vars	Num Eqns	β or Sparsity	MAGMA	SINGULAR	SAT
2	2	1.0	0.02 sec	0.01 sec	0.01 sec
3	3	1.0	0.04 sec	0.04 sec	0.07 sec
4	4	1.0	0.43 sec	423.48 sec	213.56 sec
5	5	1.0	4.32 sec	>75 mins	19278.9 sec
6	6	1.0	42.78 sec	no trial	>75 mins
7	7	1.0	1139.8 sec	no trial	no trial
8	8	1.0	crashed[a]	no trial	no trial
9	9	1.0	no trial	no trial	no trial
4	4	0.2	0.03 sec	0.08 sec	0.02 sec
5	5	0.2	0.55 sec	14.89 sec	61.89 sec
6	6	0.2	10.04 sec	6.74 sec	0.03 sec[b]
7	7	0.2	52.89 sec	>70 mins	4111.71 sec
8	8	0.2	crashed[c]	no trial	>75 mins
9	9	0.2	no trial	no trial	no trial
4	2	1.0	20.39 sec	>70 mins	0.14 sec
5	3	1.0	192.69 sec	no trial	20.51 sec
6	3	1.0	>70 mins	no trial	17.44 sec
7	4	1.0	no trial	no trial	5388.9 sec
8	4	1.0	no trial	no trial	no trial

Experimental Results: MAGMA, SINGULAR, and Mini-SAT.

[a] At one point the process had allocated 29.9 Gigabytes of RAM.
[b] This phenomenon remains unexplained, but is reproducible on repeated trials.
[c] At one point the process had allocated 24.9 Gigabytes of RAM.

Underdefined Systems of Equations

When there are fewer equations than unknowns, i.e. $m < n$, then one can expect many solutions over a finite field, and infinitely many over the rational numbers. This concept cannot be made precise, because imagine 10 equations over 100 unknowns, with an 11th equation being $x^2 = 2$. Then this 11th equation is not satisfiable by any rational number, and so there are no rational solutions to that equation, and as a result to the entire system of equations. But vaguely, the intuition remains that $n - m$ variables will remain "free," or unconstrained, and so we anticipate q^{n-m} solutions.

The $m = 7$ case took the SAT-solver just under 90 minutes, but the $m = 6$ and $m = 5$ cases were extremely trivial for the SAT-solver, both taking less than half a minute. For comparison, SINGULAR could not handle even the $m = 4$ case in less than 70 minutes. On the other hand, MAGMA took over 3 minutes for $m = 5$, and could not solve $m = 6$ in less than 70 minutes.

There are several reasons to expect SAT-solvers to do well in this circumstance. First, SAT-solvers are required to find only one solution, where as a Gröbner Basis encodes enough information to reconstruct all solutions. In $\mathbb{GF}(32)$, there might be very many solutions indeed. Second, the number of variables, number of clauses and total length of all the clauses would be reduced because m is reduced (see Section 13.1 on Page 251).

Third, for a Gröbner Bases method, after the "field equations" $x_i^q - x_i$ are introduced for all x_i, we are restricted to entries in the field $\mathbb{GF}(q)$, as desired. Thus, there are only q^n candidate solutions (possible assignments of values from $\mathbb{GF}(q)$ to n variables) and so we know we will finish with finitely many solutions. It is part of the folk-lore of Gröbner Bases methods that the most difficult problems are those that have finitely many but very many solutions. The Gröbner Basis used to describe such a solution set would necessarily be very complicated.

On the Efficacy of the Translation

Earlier, we claimed that this method is "efficient", and it is important to address what is meant by that. We know that very large polynomial systems are very difficult to solve in general, because of the NP-Completeness of the problem (see Section 11.5 on Page 199). Systems like MAGMA, and SINGULAR exist, and can solve problems of certain sizes. Here, we have shown that via this method of conversion, a SAT-solver can solve sizes that MAGMA and SINGULAR cannot. Furthermore, the conversion takes almost no time (a few seconds, far too short to measure accurately), versus the long solution times given in Table 15.7. There may be better methods.

Larger Fields

We can see that with $GF(32)$ there appears to be a threshold of feasibility at 5 variables and 7 variables, for the $\beta = 1$ and $\beta = 1/5$ cases, which implies 25 to 35 bits of unknowns. At 10 bits per variable, in $GF(1024)$, we can therefore expect the threshold of feasibility to be very roughly 3 variables—in other words the technique would be totally useless. It is unclear where the region of feasibility is, especially as both the SAT-solver communities and the Gröbner Bases communities continue to produce new algorithms, as well as refinements to old ones.

15.7.1 Computers Used

For MAGMA and SINGULAR, we used a a large-scale computer provided by the National Science Foundation for numerically-intensive research on SAGE. It is a special-purpose 64-bit computer built by Western Scientific that has 64GB of RAM and 16 AMD Opteron cores. For MiniSAT, we used 5 ordinary PCs with 1 gigabyte of RAM and one 2 GHz processor, running Linux, at the University of Maryland Mathematics Department.

15.7.2 Polynomial Systems Used

The polynomials were generated randomly, and were degree 2. Every coefficient was present with probability β, and if present, had a coefficient chosen uniformly at random from the 31 available non-zero coefficients in $GF(32)$. The numbers of equations, unknowns, and β are listed in the table.

First, we analyzed the $m = n$ case. This was done with $\beta = 1.0$ for dense systems, and $\beta = 0.2$ for sparse systems. While this may sound like a very pessimistic β for a sparse system, as linear systems often have $\beta = 1/1000$, one should note that in an n variable quadratic system, there are only $(n^2 + 3n + 2)/2$ possible coefficients. Thus with $n = 5$, we would have 21 possible coefficients, and a β much below 0.1 would allow for an all-zero equation somewhere in the system. As was the case with $GF(2)$, the SAT-solver method does better in the sparse case than the dense (see Table 15.7 on Page 289). This is also the case with Gröbner Bases approaches, in fact the difference is rather dramatic. For example, compare the running times for $m = 7 = n$ on MAGMA, namely 83 seconds versus 1140 seconds.

Note that SAT-solvers need only find one solution, while Gröbner Bases solutions find a basis for the set of solutions, which surely would be complicated if there are several solutions. For this reason, we decided to try $n = 2m$, a system with half as many equations as variables, rounded up. One would expect many solutions in this case, but still finitely many because we are in a finite field. As we mentioned earlier, SAT-solvers did much better than expected, but of course they only find 1 solution.

Also, SINGULAR did spectacularly badly with this problem, for example requiring more than 70 minutes for 4 variables and 2 equations!

Lastly, we should mention that we forced the existence of at least one solution via the technique described in Section 13.5.1 on Page 255.

15.8 Inverses and Determinants

We return to $\mathbb{GF}(16)$ and $M_4(\mathbb{GF}(2))$, i.e. the ring of 4×4 matrices with entries in $\mathbb{GF}(2)$, as our example. Though it is not necessarily relevant to polynomial systems of equations, the determinant or inverse of the matrix $M = a_0 I + a_1 A + a_2 A^2 + a_3 A^3$ can be calculated.

15.8.1 Determinants

The formula for the determinant is

$$\det M = a_0 a_1 a_2 a_3 + a_0 a_1 a_2 + a_0 a_1 a_3 + a_0 a_2 a_3 + a_1 a_2 a_3 + a_0 a_3 + a_0 a_1$$
$$+ a_0 a_2 + a_1 a_2 + a_1 a_3 + a_2 a_3 + a_0 + a_1 + a_2 + a_3$$

Further examination yields that this is always 1, unless $0 = a_0 = a_1 = a_2 = a_3$. However, this should make sense as we are in a field, and so each non-zero element is required to have a multiplicative inverse. Thus every non-zero element's matrix must have a determinant that is non-zero. Since the determinant is from the base field, it must be 1, the only non-zero element in $\mathbb{GF}(2)$.

15.8.2 Inverses

Now we can inquire as to the inverse of M, and we can do this in the form $(\det M)M^{-1}$, as this means multiplying by one, (otherwise M^{-1} does not exist). This trick of multiplying by $\det M$ simplifies the equations. The matrix formed by $(\det M)M^{-1}$ is sometimes called the "adjugate" or "classical adjoint" matrix, because the conjugate transpose of M is the usual definition of adjoint for matrices in $M_n(\mathbb{C})$. See Section 8.3.1 on Page 120.

We obtain the 4×4 matrix shown in Figure 15.1 on Page 293. Reading off the four cells that give us the basis coefficients from the usual spots, we learn that

$$\left(a_0 + a_1 \alpha + a_2 \alpha^2 + a_3 \alpha^3 \right)^{-1} = (a_0 + a_1 + a_2 + a_3 + a_1 a_2 + a_0 a_2 + a_0 a_1 a_2 + a_1 a_2 a_3)$$
$$+ (a_3 + a_0 a_1 + a_0 a_2 + a_1 a_2 + a_1 a_3 + a_0 a_1 a_3)\, \alpha$$

$$\begin{bmatrix}
a_0 + a_1 + a_2 + a_3 + a_1a_2 + a_0a_2 + a_0a_1a_2 + a_1a_2a_3 & a_1 + a_2 + a_3 + a_0a_3 + a_1a_3 + a_2a_3 + a_1a_2a_3 \\
a_3 + a_0a_1 + a_0a_2 + a_1a_2 + a_1a_3 + a_0a_1a_3 & a_0 + a_0a_2 + a_0a_3 + a_1a_2 + a_1a_3 + a_2a_3 + a_0a_1a_2 \\
a_2 + a_3 + a_0a_1 + a_0a_2 + a_0a_3 + a_0a_2a_3 & a_3 + a_0a_1 + a_0a_2 + a_1a_2 + a_1a_3 + a_0a_1a_3 \\
a_1 + a_2 + a_3 + a_0a_3 + a_1a_3 + a_2a_3 + a_1a_2a_3 & a_2 + a_3 + a_0a_1 + a_0a_2 + a_0a_3 + a_0a_2a_3 \\
\text{Column 1} & \text{Column 2}
\end{bmatrix}$$

$$\begin{bmatrix}
a_2 + a_3 + a_0a_1 + a_0a_2 + a_0a_3 + a_0a_2a_3 & a_3 + a_0a_1 + a_0a_2 + a_1a_2 + a_1a_3 + a_0a_1a_3 \\
a_1 + a_0a_1 + a_0a_2 + a_1a_3 + a_2a_3 + a_0a_2a_3 + a_1a_2a_3 & a_2 + a_1a_2 + a_0a_3 + a_1a_3 + a_0a_1a_3 + a_0a_2a_3 \\
a_0 + a_0a_2 + a_0a_3 + a_1a_2 + a_1a_3 + a_2a_3 + a_0a_1a_2 & a_1 + a_0a_1 + a_0a_2 + a_1a_3 + a_2a_3 + a_0a_2a_3 + a_1a_2a_3 \\
a_3 + a_0a_1 + a_0a_2 + a_1a_2 + a_1a_3 + a_0a_1a_3 & a_0 + a_0a_2 + a_0a_3 + a_1a_2 + a_1a_3 + a_2a_3 + a_0a_1a_2 \\
\text{Column 3} & \text{Column 4}
\end{bmatrix}$$

Fig. 15.1 The Inverse of the Matrix $M = a_0I + a_1A + a_2A^2 + a_3A^3$

$$+ (a_2 + a_3 + a_0a_1 + a_0a_2 + a_0a_3 + a_0a_2a_3)\,\alpha^2$$
$$+ (a_1 + a_2 + a_3 + a_0a_3 + a_1a_3 + a_2a_3 + a_1a_2a_3)\,\alpha^3$$

and therefore we have an efficient way of calculating the inverse of field elements.

15.8.3 Rijndael and the Para-Inverse Operation

In order to create an operation that is relatively complex, the Rijndael cipher [89], which later became AES [10], uses an operation like an inverse for elements of $\mathbb{GF}(256)$. For non-zero inputs, it is the inverse, and for the zero input, the output is zero. This is called inv0 by some authors in cryptanalysis, but we denote it here as the para-inverse. (Note that the term pseudo-inverse is already taken, and means $(A^TA)^{-1}A^T$ for rectangular matrices—see Section 7.7.1 on Page 102).

This operation can be represented by the inversion formulas in previous subsection, because if $a_0 = a_1 = a_2 = a_3 = 0$ is the input, the answer comes out 0, even though this violates the assumptions under which the formula was derived. Thus the previous formula is not only a $\mathbb{GF}(16)$ inverse, it also can serve as the para-inverse.

Niels Ferguson, Richard Schroeppel and Doug Whiting have made a continued fraction representation of the AES [113], which could be represented with this operation. As for Rijndael itself, we show here in our experiments a threshold of feasibility of roughly 25 to 35 bits depending on sparsity. But, over $\mathbb{GF}(256)$ this means 3–4.5 variables. Thus, there is at present no danger of the Rijndael/AES cipher being broken in this way.

On the other hand, Vincent Rijmen and Elisabeth Oswald [201] have a paper which shows representations of the AES over $\mathbb{GF}(16)$ and $\mathbb{GF}(4)$. The author has not explored the implications of this, but it seems to be the case that even in $\mathbb{GF}(4)$, only 12 to 18 variables are available and it would be extremely hard to imagine how to represent something as complex as a block cipher in that way.

15.9 Conclusions

The distinguishing features of this class of problems are that first, the polynomials are over fields of characteristic two, and second, that we are only interested in the base field solutions (sometimes called rational points). This is natural in cryptanalysis, as bits are $GF(2)$ elements and a fractional bit or irrational bit makes little sense. Without these properties, SAT-solvers would be of little use.

Another property is that we anticipate one solution (key) in cryptanalysis, and so finding the first available solution means finding all solutions. Rarely are there two keys under which the same plaintext will become the same ciphertext, so the risk of finding an undesired solution among many possible solutions is inapplicable. There is an interesting connection to SAT lower bounds, see Section 13.4.2.1 on Page 251.

In the case of finitely many but several solutions, a Gröbner Basis must describe all of the finitely many solutions, and so would be rather complex, whereas SAT-solvers stop at the first solution. Note that infinitely many solutions in a finite field is impossible, unless there are infinitely many variables, which we do not consider here. Of course, if $m \approx 2n$, then there are many solutions, and so we saw a greater performance gap in this case. Furthermore, if one merely wishes to classify a system of equations as "consistent" or "inconsistent", then one must either verify unsatisfiability or find one single solution—there is no need to find all solutions.

And so given these constraints, especially the requirement of adding the field equations

$$x_i^{2^n} - x_i = 0$$

for $GF(2^n)$, renders Gröbner Bases approaches disadvantaged compared to SAT-solvers.

While there is a wealth of theory about Gröbner Bases algorithms, far less is known about the Grasp algorithm and modern SAT-Solvers. Clearly, more work in this area is waiting to be done.

Despite the advantages of SAT-solvers, MAGMA was always faster when it did not crash for $m = n$. However, the memory required was substantial, and so for large problems where the user is patient, SAT-solvers will be slow but will work. On the other hand, MAGMA might require too much memory to operate. It is also noteworthy to realize that MAGMA is actually quite expensive, whereas SINGULAR and MiniSAT are free. The performance differences between SINGULAR and MAGMA in the table are interesting, and may justify the use of expensive software in research on this topic.

15.10 Review of Extension Fields

15.10.1 Constructing the Field

Recall that the field $GF(p^r = q)$ is built around a finite field $GF(p)$ where p is prime, and an "imaginary" element α which is the root of a monic irreducible polynomial of degree r. Let the polynomial be $\pi(x)$, and note that the coefficients of $\pi(x)$ come from $GF(p)$. This polynomial allows for an "arithmetic rule" for the field, or alternatively, one can think of working in the polynomial ring $GF(p)$ "mod" the principle ideal generated by $\pi(x)$.

Note that monic just means the leading coefficient is 1. If it were not 1, we could simply divide by it and get a monic polynomial. For a polynomial

$$\pi(x) = a_0 + a_1 x + a_2 x^2 + a_3 x^3 + \cdots + a_{r-1} x^{r-1} + x^r$$

the arithmetic rule is merely

$$\alpha^n = -a_0 - a_1 \alpha - a_2 \alpha^2 - a_3 \alpha^3 - \cdots - a_{r-1} \alpha^{r-1}$$

and multiplying both sides by α gives "rules" for higher powers of α. Thus any (division-free) expression in the field would look like a polynomial of degree at most $r-1$, because any α^r or higher degree terms can be rewritten with the above rule.

Example

To be very plain, we work with expressions that are degree $r-1$ polynomials, and with coefficients in the field $GF(p)$. A typical element might be $\alpha + \alpha^2$ and another one might be $\alpha + 1$. Multiplying them yields $\alpha^3 + 2\alpha^2 + \alpha$. If $r \geq 3$, then the "rules" defined above would allow us to rewrite the expression in a lower degree.

Thus we have defined a ring, since we can add, subtract and multiply. Can we define a division? Observe

$$0 = a_0 + a_1 x + a_2 x^2 + \cdots + a_{r-1} x^{r-1} + x^r$$
$$\Leftrightarrow a_0 = -a_1 x - a_2 x^2 - \cdots - a_{r-1} x^{r-1} - x^r$$
$$\Leftrightarrow 1 = \frac{-a_1}{a_0} x + \frac{-a_2}{a_0} x^2 + \cdots + \frac{-a_{r-1}}{a_0} x^{r-1} + \frac{-1}{a_0} x^r$$
$$\Leftrightarrow 1 = x \left(\frac{-a_1}{a_0} + \frac{-a_2}{a_0} x + \cdots + \frac{-a_{r-1}}{a_0} x^{r-2} + \frac{-1}{a_0} x^{r-1} \right)$$

we can simply substitute $x = \alpha$ and get an expression (in the large parenthesis) for the multiplicative inverse of α. In any commutative ring, the multiplicative inverse is[1] unique. Furthermore, this is an algorithm for inverting any particular element of the field. On the other hand, there may be simpler ways to invert elements, and one is given in Section 15.8.2 on Page 292. Of course, this was only possible because $a_0 \neq 0$. But if $a_0 = 0$ then $\pi(x)$ would not be irreducible (it would be divisible by x). To invert some other element $\beta = b_0 + b_1\alpha + \cdots + b_{r-1}\beta^{r-1}$, simply substitute β for x in the formula above. Whatever is found inside of the big parentheses is the multiplicative inverse of β.

Now we can expand our statement to say that not only is every division-free expression in the field representable as a polynomial of degree at most $r - 1$, but using the above method of carrying out division[2], we know that all expressions can be written as polynomials of this bounded degree.

Since we can add, subtract, multiply and divide, we now have a field. We have shown that a degree $r - 1$ representation for the field is sufficient, but we must also show it is necessary!

We now know that having $1, \alpha, \alpha^2, \ldots, \alpha^{r-1}$ is sufficient. How do we know that we cannot dispense with any of those? Suppose there was a formula to find α^{r-1} in terms of the others. For example,

$$k_0 + \alpha k_1 + \alpha^2 k_2 + \cdots + \alpha^{r-2}k_{r-2} = \alpha^{r-1}$$

then the function given by

$$f(x) = k_0 + xk_1 + x^2k_2 + \cdots + x^{r-2}k_{r-2} - x^{r-1}$$

has a root at $x = \alpha$.

Since both $\pi(\alpha) = 0 = f(\alpha)$ then 0 is a root of the gcd of $\pi(x)$ and $f(x)$, which we can denote $g(x)$. Since $g(x)$ has a root, it is not degree 0, and so it is some degree 1 or higher polynomial. Also, because it is the greatest common *divisor* this means that it divides $\pi(x)$. But since we required that $\pi(x)$ be irreducible, the only divisors are trivial—constants and scalar multiples of itself. We know $g(x)$ is at least degree 1 so it must be a scalar multiple of $\pi(x)$. But then $g(x)$ is degree r and $f(x)$ is degree $r - 1$, and yet because $g(x)$ is the gcd of $f(x)$ and $\pi(x)$ it must divide $f(x)$. How can a degree r polynomial divide a degree r polynomial? Thus, no such $f(x)$ may exist, and there is no formula for α^{r-1}. We cannot dispense with the α^{r-1} term. For any other term, e.g. α^d for $0 < d < r - 1$, we could simply replace $f(x)$ above with the arbitrary polynomial of degree d and the same proof would work. We cannot dispense with any of these terms.

Nonetheless, what if there were a different basis that had fewer terms? We've only shown there is no subset of this basis which works. It is well known that the basis of a vector space is the same size as any other basis of the same vector space. Since the extension field is a vector space over the finite field (simply by "forgetting" the multiplication between extension field elements and working with addition and

---

[1] Suppose $ab = 1$ and $ac = 1$. Then multiplying the second by b gives $bac = b$ or $1c = b$ thus $b = c$.

[2] More precisely, finding the multiplicative inverse of the denominator and just multiplying.

scalar multiplication by base field elements) then we know no basis is smaller then our basis.

How many objects are in our extension field? For all possible terms of our field we have p choices for the 0th degree coefficient, p choices for the first degree, $\ldots$, up to p choices for the $r-1$th degree coefficient, therefore we have p^r objects in our field. We invoke the theorem [101, Ch. 13.2] that all finite fields of the same size are isomorphic, and so we know we have constructed the finite field of size p^r, or $\mathrm{GF}(p^r)$.

Theorem 91. *Let $\mathrm{GF}(p)$ be the finite field of characteristic p and $\mathrm{GF}(p^r)$ an extension field of size p^r. Let $f(x)$ be an irreducible polynomial of degree r from $\mathrm{GF}(p)[x]$ and let α represent one of its roots. The field $\mathrm{GF}(p^r)$ is isomorphic to*

$$\mathrm{GF}(p)[x]/(f(x))$$

15.10.2 Regular Representation

The above is a construction of a finite field, and works for any finite field that is not of prime size. Of course, for prime sizes the field is simply $\mathbb{Z}_p$, and what could be simpler. And while the polynomial representation above is good for many purposes, there are other representations of finite extension fields. The one used in this chapter is that of $r \times r$ matrices over the base field $\mathrm{GF}(p)$, which we write $\mathrm{M}_r(\mathrm{GF}(p))$.

For a polynomial

$$\pi(x) = a_0 + a_1 x + a_2 x^2 + a_3 x^3 + \cdots + a_{r-1} x^{r-1} + x^r$$

the companion matrix is

$$M = \begin{bmatrix} 0 & 0 & 0 & \cdots & 0 & a_0 \\ 1 & 0 & 0 & \cdots & 0 & a_1 \\ 0 & 1 & 0 & \cdots & 0 & a_2 \\ 0 & 0 & 1 & \cdots & 0 & a_3 \\ \vdots & \vdots & \vdots & \ddots & \vdots & \vdots \\ 0 & 0 & 0 & \cdots & 1 & a_{n-1} \end{bmatrix}$$

and this matrix has several properties. First, its entries are considered over the field that the polynomial's coefficients come from, or in this case $\mathrm{GF}(p)$. Second, the minimal and characteristic polynomial of the matrix are both $\pi(x)$ provided that $\pi(x)$ is irreducible. This means that

$$a_0 I_{r \times r} + a_1 M + a_2 M^2 + a_3 M^3 + \cdots + a_{n-1} M^{r-1} + M^r = 0_{r \times r}$$

where $I_{r \times r}$ and $0_{r \times r}$ are the identity and zero matrices from $\mathrm{M}_r(\mathrm{GF}(p))$.

And therefore, mapping 1 in the field $\mathbb{GF}(p^r)$ to the identity matrix, and the primitive element α to the matrix M makes a map from $f : \mathbb{GF}(p^r) \rightarrow \mathbb{M}_r(\mathbb{GF}(p))$. This map is an isomorphism, which can be more plainly seen by realizing that

$$b_0 + b_1\alpha + b_2\alpha^2 + \cdots + b_{n-1}\alpha^{n-1} \mapsto b_0 I_{r\times r} + b_1 M + b_2 M^2 + b_3 M^3 + \cdots + b_{n-1} M^{r-1}$$

and that the addition and multiplication rules of matrices do what they should to the matrices.

What is amazing is that given any $r \times r$ matrix we can easily recover the field element that it represents.

15.11 Reversing the Isomorphism: The Existence of Dead Give-Aways

We now will prove the following theorem

Theorem 92. *Let $\mathbb{GF}(p)$ and $\mathbb{GF}(p^r)$ be finite fields of size p and size p^r, and let $\pi(x)$ be an irreducible polynomial of degree r over $\mathbb{GF}(p)$ given by*

$$\pi(x) = a_0 + a_1 x + a_2 x^2 + a_3 x^3 + \cdots + a_{r-1} x^{r-1} + x^r$$

and α a root of $\pi(x)$. Finally, let

$$M = \begin{bmatrix} 0 & 0 & 0 & \cdots & 0 & a_0 \\ 1 & 0 & 0 & \cdots & 0 & a_1 \\ 0 & 1 & 0 & \cdots & 0 & a_2 \\ 0 & 0 & 1 & \cdots & 0 & a_3 \\ \vdots & \vdots & \vdots & \ddots & \vdots & \vdots \\ 0 & 0 & 0 & \cdots & 1 & a_{n-1} \end{bmatrix}$$

Then the map given by

$$1 \mapsto I_{r\times r}, \quad \alpha \mapsto M, \quad \alpha^2 \mapsto M^2, \quad \ldots, \quad \alpha^{r-1} \mapsto M^{r-1}$$

will have, for each $d_1 \in 0, 1, 2, \ldots, r-1$ an entry ij such that for any $d_2 \in 0, 1, \ldots, r-1$,

$$(M^{d_1})_{ij} = 1 \text{ and } (M^{d_2})_{ij} = 0 \text{ provided that } d_2 \neq d_1$$

where we construct here that $ij = (d+1, 1)$. We call such an entry a "dead give-away".

The only reason this proof is challenging is that we are taking about a matrix M and its powers, but we don't know the values in the right-hand column. It turns out because d never gets very large (never exceeds $r - 1$), that the "mystery column" never affects the leftmost column, and this is the heart of the proof. Each of the steps may take a moment of thought, and so we have numbered them.

Proof. 1. First, let us define $\mathscr{I}_i$ as those $r \times r$ matrices over the base field which have the leftmost i columns as all zero. Thus $\mathscr{I}_r$ consists only of the zero matrix, and $\mathscr{I}_1$ those with the leftmost column as all zero. We define $\mathscr{I}_0$ to be all $r \times r$ matrices (over the base field), i.e. in $\mathbb{M}_r(\mathrm{GF}(p))$.

2. It is easy to see for any matrix $M_1 \in \mathbb{M}_r(\mathrm{GF}(p))$, and any matrix $M_2 \in \mathscr{I}_i$ that $M_1 M_2 \in \mathscr{I}_i$. Furthermore, $\mathscr{I}_i$ is closed on addition. Thus $\mathscr{I}_i$ is a right ideal.

3. The ideals $\mathscr{I}_x \subset \mathscr{I}_y$ if and only if $x \geq y$. More plainly, if $M \in \mathscr{I}_x$ then $M \in \mathscr{I}_{x-1}, \mathscr{I}_{x-2}, \ldots, \mathscr{I}_0$.

4. Decompose M as follows

$$
M = \begin{bmatrix} 0 & 0 & 0 & \cdots & 0 & a_0 \\ 1 & 0 & 0 & \cdots & 0 & a_1 \\ 0 & 1 & 0 & \cdots & 0 & a_2 \\ 0 & 0 & 1 & \cdots & 0 & a_3 \\ \vdots & \vdots & \vdots & \ddots & \vdots & \vdots \\ 0 & 0 & 0 & \cdots & 1 & a_{n-1} \end{bmatrix} = \underbrace{\begin{bmatrix} 0 & 0 & 0 & \cdots & 0 & 0 \\ 1 & 0 & 0 & \cdots & 0 & 0 \\ 0 & 1 & 0 & \cdots & 0 & 0 \\ 0 & 0 & 1 & \cdots & 0 & 0 \\ \vdots & \vdots & \vdots & \ddots & \vdots & \vdots \\ 0 & 0 & 0 & \cdots & 1 & 0 \end{bmatrix}}_{A} + \underbrace{\begin{bmatrix} 0 & 0 & 0 & \cdots & 0 & 0 \\ 1 & 0 & 0 & \cdots & 0 & 0 \\ 0 & 1 & 0 & \cdots & 0 & 0 \\ 0 & 0 & 1 & \cdots & 0 & 0 \\ \vdots & \vdots & \vdots & \ddots & \vdots & \vdots \\ 0 & 0 & 0 & \cdots & 1 & 0 \end{bmatrix}}_{B}
$$

5. $B \in \mathscr{I}_{n-1}$, and thus $B^j \in \mathscr{I}_{n-1}$ for all $j > 0$.

6. For any matrix M, the matrix AM is just the matrix M but shifted down one row, with a row of zeros on the top, and the last row deleted.

7. In particular, if $M \in \mathscr{I}_k$ then $A^i M$ is in $\mathscr{I}_k$ also, for $i \geq 0$.

8. For any matrix M, the matrix MA is just the matrix M but shifted left one column, with a row of zeros on the right, and the left-most column deleted.

9. In particular, if $M \in \mathscr{I}_k$ then MA^i is in $\mathscr{I}_{k-i}$ also, for $k \geq i \geq 0$.

10. Furthermore, $BA^k \in \mathscr{I}_{r-1-k}$, for $r - 1 \geq k \geq 0$

11. The product $(A + B)^d$ is equal to the sum of all possible length-d codewords from the alphabet $\{A, B\}$, each taken exactly once. Thus

$$
\begin{aligned}
(A + B)^4 &= AAAA + AAAB + AABA + ABAA + BAAA + AABB \\
&\quad + ABAB + BAAB + ABBA + BABA + BBAA + ABBB \\
&\quad + BABB + BBAB + BBBA + BBBB
\end{aligned}
$$

★ This is the metaphor for Pascal's Triangle over a non-commutative ring.

12. For any codeword of length d, if it contains a B, then let the right-most appearance of B be the position j, where 1 is the left-most and d is the right-most. Thus, the $d - j$ remaining entries are all A. Therefore, since BA^{d-j} is in $\mathscr{I}_{r-1-(d-j)} = \mathscr{I}_{r-d-1+j}$, then the codeword is in $\mathscr{I}_{r-d-1+j}$ also, because $\mathscr{I}_i$ is a right-ideal for all i.

13. For any codeword of length $0 < d < r$, if it contains a B, then the position of B is $0 < j < r$. And thus $r - d > 0$ or $n - r - 1 \geq 0$ and finally $r - d - 1 + j \geq 1$ since $1 \geq j \geq d$ so the codeword is in $\mathscr{I}_1$.

14. In the summation $(A + B)^d - A^d$, with $0 < d < r$ the "all A codeword" has been removed, and so every codeword contains a B. Since they all contain a B they are all in $\mathscr{I}_1$ and therefore their sum is also in $\mathscr{I}_1$.

15. Because for $0 < d < r$, we know $(A + B)^d - A^d$ has an all-zero left column (because it is in $\mathscr{I}_1$), then surely $(A + B)^d$ and A^d are equal in the left column.

16. Trivially, $(A + B)^0$ and A^0 are equal in the left column, because they are both the identity matrix.

17. Finally, we can conclude that $(A + B)^d$ and A^d are equal in the left column for $0 \leq d < r$.

18. The left column of A^d has a 1 in position $d + 1$ and a zero everywhere else, for $0 \leq d \leq r - 1$.

19. Thus, finally, we conclude that M^d has a 1 in position $(d + 1, 1)$ and a zero in all other entries of column 1.

$\square$

Thus, we see there will always be "a dead give-away" in each matrix for $1, \alpha, \alpha^2, \ldots, \alpha^{d-1}$ in the field $\mathbb{GF}(2^d)$.

Appendix A
On the Philosophy of Block Ciphers With Small Blocks

The purpose of this appendix is to discuss the philosophical point of whether or not attacks on block ciphers should be considered faster than brute force if they are faster than exhaustive search of the key space (which we believe) or alternatively, faster than exhaustive search of the key space or plaintext space, whichever is smaller. Related questions include

- If the "code-book" of the cipher is known (defined below), would anyone still be interested in finding the key?
- Are there block ciphers with very small plaintext spaces, but very large keys? And what must be true for them to be secure?
- And, can the application in which a block cipher is used change the standard for what is an interesting attack?

We hope that these few pages will stimulate some thought on the matter. They originally appeared in [83]. In the end, however, the conclusions are all just a matter of opinion.

A.1 Definitions

Abstractly, a block cipher is a function $E : K \times P \to C$ where K is the keyspace, P is the plaintext-space and C is the ciphertext-space. In practice, these are bit strings, and one can rewrite this as

$$E : \{0,1\}^{\ell_k} \times \{0,1\}^{\ell_P} \to \{0,1\}^{\ell_C}$$

The stereotype is that $\ell_k = \ell_P = \ell_C$, but this is almost never the case in practice, as shown by the examples in Table A.1.

The ciphers with $\ell_P < \ell_k$ have several interesting properties not shared by those with $\ell_P \geq \ell_k$. This question has not received much attantion in the cryptographic community so far, and the particularities of the case $\ell_P < \ell_k$ become important when

Table A.1 Block Ciphers, with their Block-Lengths and Key-Lengths

Cipher	$\ell_P = \ell_C$	ℓ_k
IDEA	64	128
DES	64	56
Two-key Triple-DES	64	112
AES	128	128 or 192 or 256
Two-Fish	128	128 or 192 or 256
Keeloq	32	64
Blowfish	64	32, 40, 48, ..., 432, 440, or 448

ℓ_P is small, for example in Keeloq. We believe that it is important to understand this somewhat curious situation better.

Let the *code-book* of a cipher E under a key k be the set of all 2^{ℓ_P} pairs (P,C) such that $E(k,P) = C$. If $2^{\ell_P} < 2^{\ell_k}$, it takes less time to compute the entire code-book than to do the exhaustive key search. Therefore, a natural question would be why, precisely, would one want to recover the key, if it is possible to have the entire code-book? From the point of view of theory and security models, this question was recently studied by Pornin and Granboulan in Section 5 of [128].

In this discussion we look at it in a similar way but from the point of view of practical real-life applications and their security. We will give several examples of such applications.

A.2 Brute-Force Generic Attacks on Ciphers with Small Blocks

There are two major points of view on block ciphers with very small blocks.

Point of View 1: Theoretical

In a theoretical perspective, we can assume that the adversary is very powerful and has oracle chosen-plaintext access to the cipher and very large (usually unrealistic) quantity of memory. Then if the block size is small, one can judge that the security of the block cipher is $2^{\min(\ell_k, \ell_P)}$, and once the adversary recovers and is able to store the whole code-book, one can consider that the adversary has no interest in actually recovering the original key, though from a scientific point of view, of course, the key-recovery process is interesting in its own right. In practice, even in this extreme scenario, the actual key recovery can be very valuable because it can lead to a master key (discussed below, but using one is a very common practice in industry) and having this key would compromise the security on a much wider scale.

Point of View 2: Practical

Consider a known-plaintext attack, and even if the block size is vey small the known-plaintext attack is not equivalent to a chosen-plaintext attack, not only because the privilege of choice might not actually exist in practice, but more importantly because not all plaintexts actually arise in real life (there is often some padding and a specific probability distribution of possible data). Here the adversary can recover a number of plaintext-ciphertext pairs, for example up to 50% of all possible pairs, but he cannot hope to recover all pairs. More importantly, the utility value of pairs he does not have may be very large, while the utility value of pairs he already has might be very small. Here the key recovery allows the adversary to have all possible pairs, some of which are potentially very valuable, or to recover a master key, which might be even more valuable.

Summary

In the first (theoretical) scenario the security of the block cipher is $2^{\min(\ell_k, \ell_P)}$, while in the second scenario, the security is 2^{ℓ_k} regardless of the block size. The next section tries to find real-world analogs for these ideas.

A.3 Key Recovery vs. Applications of Ciphers with Small Blocks

In this section, we present several practical application scenarios which illustrate the importance of key recovery for ciphers with small blocks and a larger key size. This is meant to motivate further detiled study of key recovery techniques in ciphers such as Keeloq.

For the reader in a hurry, one might want to skip to the last paragraph of Scenario Seven, which makes the argument most strongly.

Scenario One: LORI-KPA/LORI-CPA

Consider the notion of Left-or-Right-Indistinguishability in either the Known-Plaintext Attack, or Chosen-Plaintext Attack models [40]. There are two plaintexts, either known to the attacker, or chosen by the attacker, which we will denote as "active plaintexts". The attacker can then make "polynomially many" queries[1], submitting any plaintexts of his choice for encryption, but not the active plaintexts. We

---

[1] There must exist a polynomial $p(x)$ such that the number of queries is $\leq p(k)$, where k is a security parameter, almost always the length of the secret key.

can translate this definition to a "concrete security" treatment when the security parameter (key length) is fixed, and allow the attacker to request the encryption of any plaintext, except the two which are active. Therefore one can consider that the code-book is actually known to the adversary, for all but two values. Such a scenario is also explicitly considered in Section 5 of [128].

This is equivalent to a real-life cryptographic attacker knowing the entire code-book, except for two entries (the active plaintexts). One might ask, if in this situation, he or she would have any reason to engage in key-recovery.

We note that if a message has been observed in transmitted traffic, and it is not found in the code-book, then it is clearly one of the remaining two. This message can be of vital importance, yet it might not be possible to determine which of the remaining two it is. Key recovery would accomplish this.

If the reader doubts the practicality of this scenario, where most of a code-book is known and only a few values remain, consider the following. According to David Kahn [145], in 1942, the United States decrypted many messages encrypted with the famous "Purple" cipher, forecasting an attack at "AF". There were only a few possible targets, and so a very short list of candidates was made and Midway Island seemed the most reasonable choice. The Americans, however, needed confirmation to be 100% confident, because they planned to strike with every available aircraft carrier, and a mistake would be a tremendous waste of scarce resources. The US Navy decided to send a message about the water supply on Midway, using their own code that they knew to be broken by the Japanese. Very soon another message about "AF" was sent over Japanese channels, describing the problem with the water. Consequently, overwhelmeng force was sent to Midway and Japan's offensive power at sea was crippled, which had a pivotal impact on winning World War II.

Scenario Two: Manufacturer Sub-Keys

One usage of Keeloq in automobiles could be to take a 32-bit string called a "manufacturer key", and a 32-bit string called a "per-automobile" key, and concatenate them to form a 64-bit key for each automobile (see [234]). This means that the automobile manufacturer can produce a machine to recover the key for any particular vehicle in 2^{32} operations, but all other attackers cannot, if the total key remains unknown for every automobile. If the code-book is known for one automobile, and not its key, then that specific automobile can be stolen. But if a key recovery is then performed on a single automobile, both keys are recovered and thus every automobile of that manufacturer could then be much more easily stolen, using 2^{32} rather than 2^{64} test encryptions per automobile.

Incidentally, a more secure way of accomplishing the above is to generate a "manufacturer key" k_M randomly, and let the per automobile key be $k_s = E(k_M, s)$, where s is the serial number of the car. Here, there would exist no obvious attack, and key recovery against one automobile does not help on any other. Also, the manufacturer can easily recover the key of any particular automobile later.

Scenario Three: Short but Private Data

Suppose short strings must be encrypted, but with high security. In the USA, social security numbers (SSN's) are 9 digits, and this can be encoded in Binary Coded Decimal (BCD) with 36 bits (or 30 bits with pure binary). Of course, one can use AES (E with $\ell_k = \ell_P = \ell_C = 128$) and encrypt the 36 bits padded with 92 bits of zeroes or a fixed padding, or even with a padding that is a function of the SSN. Most padding functions are not keyed, but we could even imagine one that depended on the block cipher key. If $\pi_k(x)$ is the padding function, then we have

$$C = E(k, \pi_k(x))$$

but then this defines a function $E'(k,x)$, given by the above functional composition, with $\ell_P = 36$, $\ell_C = 128$, $\ell_k = 128$. We call this the induced block cipher. This is related to the idea of "nuggets" as presented in [35].

Scenario Four: Assigning Account Numbers

A bank or a stock-broker can assign random-looking account numbers to unique identifiers such as customer name plus date of birth or social security number, encrypted with a block cipher. As we saw in the previous example, short plaintexts, especially without keyed padding, induces a new block cipher with a tiny code-book but large key. In this application it is not clear that every single new plaintext-ciphertext pair is valuable to the attacker, and one single pair can be worth much more than any other pair (if, for example, a particularly wealthy customer can be targeted). One can imagine that knowing all of the code-book, except one particular entry, could be useless while knowing that one entry alone could be very valuable.

Scenario Five: Scratch Cards and Software Serial Numbers

Block ciphers with small blocks are used by industry to generate so-called scratch cards, that are used for example to obtain calling credit on a mobile phone. The permutation is used to associate random-looking and unique (hard to forge) numbers on scratch cards, to unique account identifiers that are typically an encryption of the numbers 0, 1, 2, 3, 4,... The same method is sometimes used to obtain unique serial numbers for software. This avoids keeping a database of all existing serial numbers which can be replaced by a short piece of code or a secure cryptographic hardware token with embedded key.

Scenario Six: Random Number Generation

A more complex version of the above follows. Suppose one wants to generate a series of random numbers, for a cryptographic or other use. One common technique is to generate a secret key k and initial value i_0. Then the series $a_i = E(k, i_0 + i)$ will be random in the sense that any algorithm that can distinguish it from random with certain success probability will distinguish E_k from a random permutation in similar time and with similar chance of success. These statements can be made more exact but we do not need that here.

It is better if i_0 is random, but often it is fixed in advance or known (e.g., the date). Often, it is simply all zeroes. In this case, if the highest n such that a_n is needed has $n \leq 2^m$, all the plaintexts lie in the range $\{0, 1, \ldots, 2^m - 1\}$. Thus only the m least significant bits of the plaintext matter (in this application) and the bits $m+1, m+2, \ldots, \ell_P$ are always zero. This induces a block cipher $E'(k, i)$ such that E' has ℓ_C and ℓ_k the same as E, but $\ell_P = m$. For any fixed i_0, the same is also true though the induced cipher is slightly different.

Scenario Seven: Fast Shuffling and Anonymity

Given a random permutation σ on the set of n elements, one can trivially shuffle a list of n objects. This is needed in many areas, most notably in scrambling data to preserve the privacy of patients in medical research. Note that sending each item i to the spot $\sigma(i)$ is sufficient for a random shuffle and takes $\Theta(n)$ time total; for a large n this is much better than assigning a random number to each item and then sorting, which would take $\Theta(n \log n)$ time. One can do this by using

$$\sigma(i) = E(k, i) \qquad \text{mod } n$$

But, especially if n is a power of 2, this induces a block cipher with high ℓ_k (to protect anonymity) but with small block size $\ell_P = \ell_C = \lceil \log_2 n \rceil$.

Surely even for 128 clients, a cipher with a 128-bit key used in this way has input and output of only 7 bits in length. Yet an attacker should be lower bounded by 2^{128} operations, not 2^7 operations.

A.4 The Keeloq Code-book—Practical Considerations

The original version of our Keeloq attack required on average about 60% of the code-book [31, Ch. 2], but as described in Section 3.4.1 on Page 21, we allow the fraction of the code-book available to be a variable η.

We have not touched upon the issue of how the code-book can be obtained in the case of Keeloq and automobile applications. Either it can be obtained from a remote encryption oracle, or simply harnessing the circuitry without being able to read the key in order to clone the device. While this may sound like a practical attack scenario, in practice the devices are simply too slow to obtain this. It is also noteworthy that since each plaintext is 2^5 bits long, and there are 2^{32} of them, the entire code-book is 2^{37} bits or 16 Gigabytes. This amount of RAM is already available on high-end PC's at the time of the writing of this book (or even a few years prior).

Oddly, the 64-bit key size implies that the exshaustive search is actually feasible in practice, and hackers and car thieves implement it with[2] FPGA's [184]. Such an attack requires only 2 known plaintexts (one known plaintext does not alone allow one to uniquely determine the key, which is another consequence of the unusually small block size). We note that while 2^{32} encryptions is difficult to obtain with the original chips that are quite inexpensive and slow, with FPGA's as much as 2^{64} encryptions is feasible. This is because the FPGA's are faster and are compatible with parallel processing.

Therefore, we do not know if it is possible to obtain the originally estimated $\eta = 0.6 \times 2^{32}$ plaintexts to mount the Keeloq attack. A smaller η might be slightly more feasible, but it might be cheaper to buy the car that you were hoping to steal. But, if the block cipher Keeloq is to be compared to any other block cipher, then using an encryption oracle, finding and storing the codebook is not difficult.

A.5 Conclusions

After looking over this appendix, we hope that the reader believes that attacks which recover a key, even after the entire code-book is known, are worthy of attention and are related to real-world situations. But, in the end, we must confess that this is a matter of opinion and not scientific argumentation.

---

[2] An FPGA is a Field Programmable Gate Array, basically a device which can be used to rapidly prototype integrated circuits, much faster than individual transistors but much less efficient than the very expensive process of making one's own chip.

Appendix B
Formulas for the Field Multiplication law for Low-Degree Extensions of $\mathbb{GF}(2)$

B.1 For $\mathbb{GF}(4)$

For the product $p = ab$

$$p_0 = a_0b_0 + a_1b_1$$
$$p_1 = a_1b_0 + a_0b_1 + a_1b_1$$

For the product $p = abc$

$$p_0 = a_0b_0c_0 + a_1b_1c_0 + c_1a_1b_0 + c_1a_0b_1 + c_1a_1b_1$$
$$p_1 = a_1b_0c_0 + a_0b_1c_0 + a_1b_1c_0 + a_0b_0c_1 + c_1a_0b_1 + c_1a_1b_0$$

For the product $p = abcd$

$$p_0 = d_0a_0b_0c_0 + d_0a_1b_1c_0 + d_0c_1a_1b_0 + d_0c_1a_0b_1 + d_0c_1a_1b_1 + d_1a_1b_0c_0 + d_1a_0b_1c_0$$
$$+ d_1a_0b_0c_1 + d_1c_1a_0b_1 + d_1c_1a_1b_0 + d_1a_1b_1c_0$$
$$p_1 = a_0b_1c_0d_0 + d_1a_0b_1c_0 + d_0a_1b_1c_0 + 2d_0c_1a_1b_1 + a_0b_0c_1d_0 + d_1a_0b_0c_1 + d_0c_1a_0b_1$$
$$d_1a_0b_0c_0 + d_1c_1a_1b_1 + a_1b_0c_0d_0 + d_1a_1b_0c_0 + d_0c_1a_1b_0$$

B.2 For $\mathbb{GF}(8)$

For the product $p = ab$

$$p_0 = a_0b_0 + a_2b_1 + a_1b_2$$
$$p_1 = a_0b_2 + a_2b_0 + a_2b_2 + a_1b_1$$
$$p_2 = a_0b_1 + a_2b_1 + a_2b_2 + a_1b_0 + a_1b_2$$

For the product $p = abc$

$$p_0 = c_0a_0b_0 + c_0a_2b_1 + c_0a_1b_2 + c_1a_0b_2 + c_1a_2b_0 + c_1a_2b_2 + c_1a_1b_1 + c_2a_0b_1 + c_2a_2b_1 + c_2a_2b_2$$
$$+ c_2a_1b_0 + c_2a_1b_2$$

$$p_1 = c_1a_0b_0 + c_1a_2b_1 + c_1a_1b_2 + c_1a_0b_2 + a_0b_2c_2 + c_1a_2b_0 + a_2b_0c_2 + c_1a_2b_2 + c_1a_1b_1 + c_2a_1b_2$$
$$+ a_1b_1c_2 + a_0b_1c_0 + c_2a_0b_1 + c_0a_2b_1 + c_2a_2b_1 + a_2b_2c_0 + a_1b_0c_0 + c_2a_1b_0 + c_0a_1b_2$$

$$p_2 = c_2a_0b_0 + c_2a_2b_1 + c_2a_1b_2 + a_0b_2c_0 + a_0b_2c_2 + a_2b_0c_0 + a_2b_0c_2 + a_2b_2c_0 + c_2a_2b_2 + a_1b_1c_0$$
$$+ a_1b_1c_2 + c_1a_0b_1 + c_1a_2b_1 + c_1a_2b_2 + c_1a_1b_0 + c_1a_1b_2$$

For the product $p = abcd$

$$p_0 = d_0c_0a_0b_0 + d_0c_0a_1b_2 + d_0c_0a_2b_1 + d_0c_1a_0b_2 + d_0c_1a_1b_1 + d_0c_1a_2b_2 + d_0c_1a_2b_0 + d_0c_2a_2b_2$$
$$+ d_0c_2a_2b_1 + d_0c_2a_0b_1 + d_1c_2a_2b_1 + d_1c_2a_0b_0 + d_0c_2a_1b_2 + d_0c_2a_1b_0 + d_1a_0b_2c_2 + d_1a_0b_2c_0$$
$$+ d_1c_2a_1b_2 + d_2c_2a_0b_1 + d_2c_0a_2b_1 + d_2c_2a_2b_1 + d_2a_2b_2c_0 + d_2a_1b_0c_0 + d_2c_2a_1b_0 + d_2c_0a_1b_2$$
$$+ d_2c_2a_1b_2 + d_1a_1b_1c_0 + d_1c_2a_2b_2 + d_1a_2b_2c_0 + d_1a_2b_0c_2 + d_1a_2b_0c_0 + d_1c_1a_0b_1 + d_1a_1b_1c_2$$
$$+ d_1c_1a_1b_0 + d_1c_1a_2b_2 + d_1c_1a_2b_1 + d_1c_1a_1b_2 + d_2c_1a_0b_2 + d_2c_1a_1b_2 + d_2c_1a_2b_1 + d_2c_1a_2b_0$$
$$+ d_2a_0b_2c_2 + d_2a_2b_0c_2 + d_2c_1a_2b_2 + d_2c_1a_0b_0 + d_2c_1a_1b_1 + d_2a_1b_1c_2 + d_2a_0b_1c_0$$

$$p_1 = d_0c_0a_1b_2 + d_0c_0a_2b_1 + d_0c_1a_0b_2 + d_0c_1a_1b_1 + d_0c_1a_2b_2 + d_2a_0b_1c_0 + d_0c_1a_2b_0 + d_0c_2a_2b_1$$
$$+ d_0c_2a_0b_1 + d_1c_2a_0b_0 + d_0c_2a_1b_2 + d_0c_2a_1b_0 + d_1a_0b_2c_2 + d_1a_0b_2c_0 + d_1c_1a_0b_2 + d_2c_1a_0b_0$$
$$+ d_2c_2a_0b_1 + d_2c_0a_2b_1 + d_2a_1b_0c_0 + d_2c_2a_1b_0 + d_2c_0a_1b_2 + d_1a_1b_1c_0 + d_1c_0a_0b_0 + d_2c_1a_1b_1$$
$$+ d_1a_2b_2c_0 + d_1a_2b_0c_2 + d_1a_2b_0c_0 + d_1c_1a_0b_1 + d_1a_1b_1c_2 + d_1c_1a_1b_0 + d_1c_1a_2b_1 + d_1c_1a_2b_0$$
$$+ d_1c_1a_1b_2 + a_0b_2c_0d_2 + a_2b_0c_0d_2 + a_1b_1c_0d_2 + c_1a_0b_1d_2 + d_1c_1a_1b_1 + d_1c_2a_0b_1 + d_1c_2a_1b_0$$
$$+ c_2a_0b_0d_2 + c_1a_1b_0d_2 + c_1a_0b_0d_0 + c_1a_2b_1d_0 + c_1a_1b_2d_0 + a_0b_2c_2d_0 + a_2b_0c_2d_0 + a_1b_1c_2d_0$$
$$+ a_0b_1c_0d_0 + a_2b_2c_0d_0 + a_1b_0c_0d_0 + d_1c_0a_2b_1 + d_2c_1a_0b_2 + d_2c_1a_2b_0 + d_2c_2a_2b_2 + d_1c_0a_1b_2$$

$$p_2 = d_0c_1a_2b_2 + d_0c_2a_2b_2 + d_0c_2a_2b_1 + d_1c_2a_2b_1 + d_0c_2a_1b_2 + d_1a_0b_2c_2 + d_1c_2a_1b_2 + d_2c_2a_0b_1$$
$$+ d_2c_0a_2b_1 + d_2a_2b_2c_0 + d_2c_2a_1b_0 + d_2c_0a_1b_2 + d_1a_2b_2c_0 + d_2a_1b_1c_2 + d_1c_1a_0b_2 + d_2c_1a_1b_1$$
$$+ d_1a_2b_0c_2 + d_1a_1b_1c_2 + d_1c_1a_2b_2 + d_1c_1a_2b_1 + d_1c_1a_1b_2 + a_0b_2c_0d_2 + a_2b_0c_0d_2 + a_1b_1c_0d_2$$
$$+ c_1a_0b_1d_2 + d_1c_1a_1b_1 + d_1c_2a_0b_1 + d_1c_2a_1b_0 + c_2a_0b_0d_2 + c_1a_1b_0d_2 + c_1a_2b_1d_0 + c_1a_1b_2d_0$$
$$+ a_0b_2c_2d_0 + a_2b_0c_2d_0 + a_1b_1c_2d_0 + a_2b_2c_0d_0 + c_2a_0b_0d_0 + a_0b_2c_0d_0 + a_2b_0c_0d_0 + a_1b_1c_0d_0$$
$$+ c_1a_0b_1d_0 + c_1a_1b_0d_0 + d_1c_0a_2b_1 + d_2c_0a_0b_0 + d_2c_1a_0b_2 + d_2c_1a_1b_2 + d_2c_1a_2b_1 + d_2c_1a_2b_0$$
$$+ d_2a_0b_2c_2 + d_2a_2b_0c_2 + d_1c_0a_1b_2 + d_1c_1a_0b_0 + d_1a_0b_1c_0 + d_1a_1b_0c_0 + d_1c_1a_2b_0$$

B.3 For $\mathbb{GF}(16)$

See Section 15.5 on Page 284.

B.4 For $\mathbb{GF}(32)$

For the product $p = ab$

$$p_0 = a_0b_0 + a_4b_1 + a_3b_2 + a_2b_3 + b_4a_1 + a_4b_4$$
$$p_1 = a_1b_0 + a_0b_1 + a_4b_2 + a_3b_3 + a_2b_4$$
$$p_2 = a_0b_2 + a_4b_3 + a_3b_2 + a_3b_4 + a_2b_0 + a_2b_3 + a_1b_1 + b_4a_1 + a_4b_1 + a_4b_4$$
$$p_3 = a_0b_3 + a_4b_4 + a_3b_0 + a_3b_3 + a_2b_1 + a_2b_4 + a_1b_2 + a_4b_2$$
$$p_4 = a_0b_4 + a_4b_0 + a_3b_1 + a_3b_4 + a_2b_2 + a_1b_3 + a_4b_3$$

For the product $p = abc$

$$
\begin{aligned}
p_0 = {}& c_1a_4b_3 + c_0a_2b_3 + c_4a_4b_2 + c_4a_4b_0 + c_1a_4b_0 + c_3a_3b_2 + c_1a_2b_2 + c_3a_1b_1 + c_4a_0b_4 + c_2a_3b_0 \\
&+ c_3a_2b_0 + c_2b_2a_1 + c_3a_3b_4 + c_3a_4b_1 + c_2a_0b_3 + c_0a_4b_1 + c_1a_3b_4 + c_2a_4b_2 + c_3a_2b_3 + c_3b_4a_1 \\
&+ c_0a_4b_4 + c_4a_4b_3 + c_2a_2b_4 + c_4a_3b_3 + c_4a_1b_0 + c_2a_4b_4 + c_2a_2b_1 + c_1a_3b_1 + c_2a_3b_3 + c_3a_4b_4 \\
&+ c_4a_3b_4 + c_0a_0b_0 + c_4a_2b_4 + c_1a_0b_4 + c_3a_0b_2 + c_1b_3a_1 + c_4b_3a_1 + c_4a_2b_2 + c_0b_4a_1 + c_4a_0b_1 \\
&+ c_3a_4b_3 + c_4a_3b_1 + c_0a_3b_2
\end{aligned}
$$

$$
\begin{aligned}
p_1 = {}& c_0a_2b_4 + c_0a_0b_1 + c_0a_1b_0 + c_3a_0b_3 + c_1a_4b_1 + c_3a_1b_2 + c_1a_4b_4 + c_2a_1b_3 + c_0a_4b_2 + c_2a_0b_4 \\
&+ c_4a_4b_3 + c_1a_3b_2 + c_1a_0b_0 + c_3a_4b_2 + c_3a_2b_4 + c_2a_3b_1 + c_3a_4b_4 + c_4a_3b_4 + c_2a_4b_0 + c_3a_2b_1 \\
&+ c_3a_3b_3 + c_2a_3b_4 + c_1a_2b_3 + c_2a_4b_3 + c_2a_2b_2 + c_4a_1b_1 + c_4a_1b_4 + c_4a_0b_2 + c_4a_4b_1 + c_4a_4b_4 \\
&+ c_4a_3b_2 + c_4a_2b_0 + c_4a_2b_3 + c_3a_3b_0 + c_1a_1b_4 + c_0a_3b_3
\end{aligned}
$$

$$
\begin{aligned}
p_2 = {}& c_1a_4b_3 + c_0a_2b_3 + c_4a_4b_0 + c_1a_4b_0 + c_3a_3b_2 + c_2a_1b_4 + c_1a_2b_2 + c_3a_1b_1 + c_4a_0b_4 + a_1b_0c_1 \\
& \mid c_2a_3b_0 + c_3a_2b_0 + c_2b_2a_1 + c_3a_2b_2 + c_3a_4b_1 + c_2a_0b_3 + c_0a_4b_1 + c_1a_3b_4 + c_2a_4b_2 + a_2b_4c_1 \\
&+ c_3a_2b_3 + c_3b_4a_1 + c_3a_0b_4 + c_0a_4b_4 + c_4a_4b_3 + c_2a_2b_4 + c_4a_1b_0 + c_3a_1b_3 + a_3b_3c_1 + a_4b_2c_1 \\
&+ c_2a_2b_1 + c_1a_3b_1 + c_2a_3b_3 + c_2a_4b_1 + c_3a_4b_4 + a_3b_0c_4 + c_4a_3b_4 + a_2b_1c_4 + c_3a_4b_0 + a_0b_1c_1 \\
&+ c_1a_0b_4 + c_3a_0b_2 + c_1b_3a_1 + c_4b_3a_1 + c_4a_2b_2 + c_0b_4a_1 + c_4a_0b_1 + c_2a_3b_2 + 2c_3a_4b_3 + c_4a_3b_1 \\
&+ a_1b_2c_4 + c_4a_4b_4 + c_2a_2b_3 + c_0a_3b_2 + a_0b_3c_4 + c_2a_0b_0 + c_3a_3b_1 + a_1b_1c_0 + a_0b_2c_0 + a_4b_3c_0 \\
&+ a_3b_4c_0 + a_2b_0c_0
\end{aligned}
$$

$$
\begin{aligned}
p_3 = {}& c_1a_4b_3 + c_0a_2b_4 + c_3a_0b_3 + c_4a_4b_0 + c_1a_4b_1 + c_3a_1b_2 + c_3a_3b_2 + c_1a_4b_4 + c_4a_0b_4 + c_3a_4b_1 \\
&+ c_3a_0b_0 + c_2a_1b_3 + c_1a_3b_4 + c_2a_4b_2 + c_0a_4b_2 + c_3a_2b_3 + c_3b_4a_1 + c_2a_0b_4 + c_0a_4b_4 + c_1a_3b_2 \\
&+ c_2a_2b_4 + c_3a_4b_2 + c_3a_2b_4 + c_2a_3b_3 + c_2a_3b_1 + c_2a_4b_0 + c_3a_2b_1 + c_3a_3b_3 + c_1a_1b_4 + c_0a_3b_3 \\
&+ c_4b_3a_1 + c_4a_2b_2 + c_2a_3b_4 + c_4a_3b_1 + c_1a_2b_3 + c_2a_4b_3 + c_2a_0b_1 + c_2a_1b_0 + a_0b_3c_0 + a_3b_0c_0 \\
&+ a_2b_1c_0 + b_2a_1c_0 + a_0b_2c_1 + a_2b_0c_1 + a_1b_1c_1 + c_2a_2b_2 + c_4a_1b_1 + c_4a_1b_4 + c_4a_0b_2 + c_4a_4b_1 \\
&+ c_4a_4b_4 + c_4a_3b_2 + c_4a_2b_0 + c_4a_2b_3 + c_3a_3b_0
\end{aligned}
$$

$$
\begin{aligned}
p_4 = {}& c_4a_4b_2 + c_2a_1b_4 + c_1a_4b_4 + c_3a_2b_2 + c_3a_3b_4 + c_3a_0b_4 + c_4a_3b_3 + c_3a_1b_3 + c_2a_4b_4 + c_3a_4b_2 \\
&+ c_0b_3a_1 + c_3a_2b_4 + c_4a_0b_0 + c_2a_4b_1 + a_3b_0c_4 + a_2b_1c_4 + c_3a_4b_0 + c_3a_3b_3 + c_4a_2b_4 + c_2a_3b_4 \\
&+ c_2a_3b_2 + c_3a_4b_3 + c_2a_4b_3 + a_1b_2c_4 + c_4a_1b_4 + c_4a_4b_1 + c_4a_3b_2 + c_4a_2b_3 + c_2a_2b_3 + c_3a_1b_0 \\
&+ a_0b_3c_4 + c_3a_3b_1 + c_0a_0b_4 + c_0a_4b_0 + c_0a_3b_1 + a_4b_3c_0 + a_3b_4c_0 + a_4b_2c_1 + a_3b_3c_1 + a_2b_4c_1 \\
&+ c_0a_2b_2 + a_0b_3c_1 + a_3b_0c_1 + a_2b_1c_1 + b_2a_1c_1 + c_2a_0b_2 + c_2a_2b_0 + c_2a_1b_1 + c_3a_0b_1
\end{aligned}
$$

For the product $p = abcd$, the formulas were too large to be efficiently copied from MAPLE, unfortunately.

B.5 For $\mathbb{GF}(64)$

For the product $p = ab$

$$p_0 = a_0b_0 + a_5b_1 + a_4b_2 + a_3b_3 + a_2b_4 + a_1b_5$$
$$p_1 = a_0b_1 + a_5b_1 + a_5b_2 + a_4b_2 + a_4b_3 + a_3b_3 + a_3b_4 + a_2b_4 + a_2b_5 + a_1b_0 + a_1b_5$$
$$p_2 = a_0b_2 + a_5b_2 + a_5b_3 + a_4b_3 + a_4b_4 + a_3b_4 + a_3b_5 + a_2b_0 + a_2b_5 + a_1b_1$$
$$p_3 = a_0b_3 + a_5b_3 + a_5b_4 + a_4b_4 + a_4b_5 + a_3b_0 + a_3b_5 + a_2b_1 + a_1b_2$$
$$p_4 = a_0b_4 + a_5b_4 + a_5b_5 + a_4b_0 + a_4b_5 + a_3b_1 + a_2b_2 + a_1b_3$$
$$p_5 = a_0b_5 + a_5b_0 + a_5b_5 + a_4b_1 + a_3b_2 + a_2b_3 + a_1b_4$$

For the product $p = abc$

$$
\begin{aligned}
p_0 =\ & c_0a_0b_0 + c_0a_5b_1 + c_0a_4b_2 + c_0a_3b_3 + c_0a_2b_4 + c_0a_1b_5 + c_1a_0b_5 + c_1a_5b_0 + c_1a_5b_5 + c_1a_4b_1 \\
& + c_1a_3b_2 + c_1a_2b_3 + c_1a_1b_4 + c_2a_0b_4 + c_2a_5b_4 + c_2a_5b_5 + c_2a_4b_0 + c_2a_4b_5 + c_2a_3b_1 + c_2a_2b_2 \\
& + c_2a_1b_3 + c_3a_0b_3 + c_3a_5b_3 + c_3a_5b_4 + c_3a_4b_4 + c_3a_4b_5 + c_3a_3b_0 + c_3a_3b_5 + c_3a_2b_1 + c_3a_1b_2 \\
& + c_4a_0b_2 + c_4a_5b_2 + c_4a_5b_3 + c_4a_4b_3 + c_4a_4b_4 + c_4a_3b_4 + c_4a_3b_5 + c_4a_2b_0 + c_4a_2b_5 + c_4a_1b_1 \\
& + c_5a_0b_1 + c_5a_5b_1 + c_5a_5b_2 + c_5a_4b_2 + c_5a_4b_3 + c_5a_3b_3 + c_5a_3b_4 + c_5a_2b_4 + c_5a_2b_5 + c_5a_1b_0 \\
& + c_5a_1b_5
\end{aligned}
$$

$$
\begin{aligned}
p_1 =\ & c_5a_5b_1 + c_5a_4b_2 + c_5a_3b_3 + c_5a_2b_4 + c_5a_1b_5 + c_1a_5b_5 + c_2a_5b_4 + c_2a_4b_5 + c_3a_5b_3 + c_3a_4b_4 \\
& + c_3a_3b_5 + c_4a_5b_2 + a_4b_5c_4 + a_3b_0c_3 + a_3b_0c_4 + a_2b_1c_3 + a_2b_1c_4 + a_1b_2c_3 + a_0b_3c_3 + a_0b_3c_4 \\
& + a_1b_2c_4 + a_0b_2c_4 + a_0b_2c_5 + a_5b_3c_5 + a_4b_4c_5 + a_3b_5c_5 + a_2b_0c_4 + a_3b_1c_3 + a_2b_2c_2 + a_2b_2c_3 \\
& + a_2b_0c_5 + a_1b_1c_4 + a_1b_1c_5 + a_0b_1c_0 + a_0b_1c_5 + a_5b_1c_0 + a_5b_2c_0 + a_4b_2c_0 + a_4b_3c_0 + a_5b_4c_4 \\
& + a_3b_3c_0 + a_3b_4c_0 + a_2b_4c_0 + a_2b_5c_0 + a_1b_0c_0 + a_1b_0c_5 + a_1b_5c_0 + c_4a_4b_3 + c_4a_3b_4 + c_4a_2b_5 \\
& + c_1a_0b_0 + c_1a_5b_1 + c_1a_4b_2 + c_1a_3b_3 + c_1a_2b_4 + c_1a_1b_5 + a_0b_5c_1 + a_0b_5c_2 + a_5b_0c_1 + a_5b_0c_2 \\
& + a_4b_1c_1 + a_4b_1c_2 + a_3b_2c_1 + a_3b_2c_2 + a_2b_3c_1 + a_2b_3c_2 + a_1b_4c_1 + a_1b_4c_2 + a_0b_4c_2 + a_1b_3c_3 \\
& + a_0b_4c_3 + a_5b_5c_3 + a_4b_0c_2 + a_4b_0c_3 + a_3b_1c_2 + a_1b_3c_2
\end{aligned}
$$

$$
\begin{aligned}
p_2 =\ & c_1a_0b_1 + a_5b_0c_3 + a_0b_5c_3 + c_2a_1b_5 + c_2a_2b_4 + c_2a_3b_3 + c_2a_4b_2 + c_2a_5b_1 + c_2a_0b_0 + a_5b_4c_5 \\
& + a_0b_3c_5 + a_1b_3c_4 + a_2b_2c_4 + a_3b_1c_4 + a_4b_0c_4 + a_5b_5c_4 + a_0b_4c_4 + a_1b_4c_3 + a_2b_3c_3 + a_3b_2c_3 \\
& + a_4b_1c_3 + c_1a_1b_0 + c_1a_2b_5 + c_1a_3b_4 + c_1a_4b_3 + c_1a_5b_2 + a_1b_1c_0 + a_2b_0c_0 + a_3b_5c_0 + a_4b_4c_0 \\
& + a_5b_3c_0 + a_0b_2c_0 + a_1b_2c_5 + a_2b_1c_5 + a_3b_0c_5 + a_4b_5c_5 + a_4b_4c_4 + a_3b_0c_4 + a_3b_5c_4 + a_5b_3c_4 \\
& + a_2b_1c_4 + a_1b_2c_4 + a_0b_2c_5 + a_5b_2c_5 + a_4b_3c_5 + a_3b_4c_5 + a_2b_2c_3 + a_1b_3c_3 + a_0b_3c_4 + a_3b_1c_3 \\
& + a_2b_0c_5 + a_2b_5c_5 + a_1b_1c_5 + a_5b_2c_0 + a_4b_3c_0 + a_3b_4c_0 + a_2b_5c_0 + c_1a_5b_1 + c_1a_4b_2 + c_1a_3b_3 \\
& + c_1a_2b_4 + c_1a_1b_5 + a_0b_5c_2 + a_5b_0c_2 + a_5b_5c_2 + a_4b_1c_2 + a_3b_2c_2 + a_2b_3c_2 + a_1b_4c_2 + a_0b_4c_3 \\
& + a_5b_4c_3 + a_4b_0c_3 + a_4b_5c_3
\end{aligned}
$$

$$
\begin{aligned}
p_3 =\ & a_5b_5c_5 + a_5b_0c_3 + a_0b_5c_3 + c_2a_1b_5 + c_2a_2b_4 + c_2a_3b_3 + c_2a_4b_2 + c_2a_5b_1 + a_0b_3c_5 + a_0b_4c_5 \\
& + a_1b_3c_4 + a_2b_2c_4 + a_3b_1c_4 + a_4b_0c_4 + a_0b_4c_4 + a_1b_4c_3 + a_2b_3c_3 + a_3b_2c_3 + a_4b_1c_3 + a_1b_4c_4 \\
& + c_1a_2b_5 + c_1a_3b_4 + c_1a_4b_3 + c_1a_5b_2 + a_3b_5c_0 + a_4b_4c_0 + a_5b_3c_0 + a_1b_2c_5 + a_2b_1c_5 + a_3b_0c_5 \\
& + c_2a_0b_1 + c_2a_5b_2 + c_2a_4b_3 + c_2a_3b_4 + c_2a_2b_5 + c_2a_1b_0 + a_4b_5c_4 + a_5b_3c_5 + a_4b_4c_5 + a_3b_5c_5 \\
& + a_5b_5c_3 + a_5b_4c_4 + a_4b_0c_5 + a_3b_1c_5 + a_2b_2c_5 + a_1b_3c_5 + a_0b_3c_0 + a_5b_4c_0 + a_4b_5c_0 + a_3b_0c_0 \\
& + a_2b_1c_0 + a_1b_2c_0 + c_1a_0b_2 + c_1a_5b_3 + c_1a_4b_4 + c_1a_3b_5 + c_1a_2b_0 + c_1a_1b_1 + c_3a_0b_0 + c_3a_5b_1 \\
& + c_3a_4b_2 + c_3a_3b_3 + c_3a_2b_4 + c_3a_1b_5 + a_0b_5c_4 + a_5b_0c_4 + a_4b_1c_4 + a_3b_2c_4 + a_2b_3c_4
\end{aligned}
$$

$$\begin{aligned}
p_4 =\ & a_0b_5c_5 + a_5b_0c_5 + a_5b_5c_0 + a_4b_1c_5 + a_3b_2c_5 + a_2b_3c_5 + a_1b_4c_5 + c_1a_5b_4 + c_1a_4b_5 + a_0b_4c_5 \\
& + a_5b_4c_5 + a_5b_5c_4 + a_4b_5c_5 + c_2a_5b_2 + c_2a_4b_3 + c_2a_3b_4 + c_2a_2b_5 + c_2a_5b_3 + c_2a_4b_4 + c_2a_3b_5 \\
& + c_3a_5b_2 + c_3a_4b_3 + c_3a_3b_4 + c_3a_2b_5 + c_4a_5b_1 + c_4a_4b_2 + c_4a_0b_0 + a_0b_4c_0 + a_4b_0c_0 + a_3b_1c_0 \\
& + a_2b_2c_0 + a_1b_3c_0 + c_1a_0b_3 + c_1a_3b_0 + c_1a_2b_1 + c_1a_1b_2 + c_2a_0b_2 + c_2a_2b_0 + c_2a_1b_1 + c_3a_0b_1 \\
& + c_3a_1b_0 + c_4a_3b_3 + c_4a_2b_4 + c_4a_1b_5 + a_4b_0c_5 + a_3b_1c_5 + a_2b_2c_5 + a_1b_3c_5 + a_5b_4c_0 + a_4b_5c_0 \\
& + c_1a_5b_3 + c_1a_4b_4 + c_1a_3b_5 + c_3a_5b_1 + c_3a_4b_2 + c_3a_3b_3 + c_3a_2b_4 + c_3a_1b_5 + a_0b_5c_4 + a_5b_0c_4 \\
& + a_4b_1c_4 + a_3b_2c_4 + a_2b_3c_4 + a_1b_4c_4 \\[6pt]
p_5 =\ & a_0b_5c_5 + a_5b_0c_5 + a_5b_5c_0 + a_5b_5c_5 + a_4b_1c_5 + a_3b_2c_5 + a_2b_3c_5 + a_1b_4c_5 + c_1a_5b_4 + c_1a_4b_5 \\
& + c_5a_0b_0 + c_5a_5b_1 + c_5a_4b_2 + c_5a_3b_3 + c_5a_2b_4 + c_5a_1b_5 + a_5b_0c_0 + c_2a_5b_3 + c_2a_4b_4 + c_2a_3b_5 \\
& + c_3a_5b_2 + c_3a_4b_3 + c_3a_3b_4 + c_3a_2b_5 + c_4a_5b_1 + c_4a_4b_2 + c_4a_3b_3 + c_4a_2b_4 + c_4a_1b_5 + a_0b_5c_0 \\
& + a_4b_1c_0 + a_3b_2c_0 + a_2b_3c_0 + a_1b_4c_0 + c_1a_0b_4 + c_1a_5b_5 + c_1a_4b_0 + c_1a_3b_1 + c_1a_2b_2 + c_1a_1b_3 \\
& + c_2a_5b_4 + c_2a_4b_5 + c_2a_3b_0 + c_2a_2b_1 + c_2a_1b_2 + c_3a_0b_2 + c_3a_5b_3 + c_3a_4b_4 + c_3a_3b_5 + c_3a_2b_0 \\
& + c_3a_1b_1 + c_4a_0b_1 + c_2a_0b_3 + c_4a_5b_2 + c_4a_4b_3 + c_4a_3b_4 + c_4a_2b_5 + c_4a_1b_0
\end{aligned}$$

For the product $p = abcd$, the formulas were too large to be efficiently copied from MAPLE, unfortunately.

Appendix C
Polynomials and Graph Coloring, with Other Applications

While this book is intended to inform the reader how polynomial systems of equations can be used to perform algebraic cryptanalysis, other applications of polynomials over finite fields exist. In particular, the connection to graph coloring is given in this appendix. This is a natural association, as both problems are NP-Complete; both problems are related to real-world applications; and both problems are quite solvable in small or special cases. Finally, both problems are unsolvable in the large, most general case. Thus a tool that works well for one might work well for the others.

C.1 A Very Useful Lemma

Before we begin, the following lemma is very useful at this point.

Lemma 93. *Let F be a finite field of order q. Then $x \in F$ is a root of $x^{q-1} - 1$ if and only if $x \neq 0$.*

Proof. Consider the equation $x^q - x = 0$, which is also known as Fermat's Little Theorem. This polynomial is satisfied for every element of the field $\mathbb{GF}(q)$. Since there are q elements in $\mathbb{GF}(q)$ and $x^q - x$ is a polynomial of degree q and thus has at most q distinct roots, then we know each value of $\mathbb{GF}(q)$ is a root of multiplicity one.

One can rewrite $x^q - x = x(x^{q-1} - 1)$ by factoring. Clearly $x = 0$ satisfies x but not $x^{q-1} - 1$. Therefore, the other roots of $x^q - x$, namely all the non-zero elements of $\mathbb{GF}(q)$ are roots of $x^{q-1} - 1$. $\qquad\square$

Corollary 94. *Let $\mathbb{F}$ be a finite field of order q. Then $(y, z) \in \mathbb{F} \times \mathbb{F}$ is a root of $(y - z)^{q-1} - 1$ if and only if $y \neq z$.*

Proof. Obvious. $\qquad\square$

Therefore, we can think of $x^{q-1} - 1 = 0$ as an encoding of $x \neq 0$ as a polynomial system of equations, and also $(y - z)^{q-1} - 1 = 0$ as an encoding of $y \neq z$.

C.2 Graph Coloring

The author assumes the following has been known for a long time. Suppose one wants to color the graph $G = (V, E)$ with c colors, or alternatively, determine if such a coloring exists. First, assume that $c = q$ is the size of some field—we will deal with other values of c shortly. Name each of the c colors by the elements of the field $\mathbb{GF}(q)$. Each vertex v_i shall have the color c_i. The $|V|$ variables $c_1, \ldots, c_{|V|}$ will be the $|V|$ variables of our system of polynomial equations.

Recall, since our field is of prime power order q, then $x^{q-1} - 1 = 0$ if and only if $x \neq 0$. Therefore, for each pair of vertices v_i and v_j, if an edge connects them, write the equation $(c_i - c_j)^{q-1} - 1 = 0$. (See Corollary 94 on Page 315). This is equivalent to requiring $c_i \neq c_j$.

Clearly, any solution set $c_1, \ldots, c_{|V|}$ will meet the requirements of a coloring. Some methods of solving polynomial systems of equations only consider solutions in the base field. Others will consider solutions that are found in extension fields, and so adding the equations $c_i^q - c_i = 0$ will ensure that the value of each c_i is strictly inside the chosen field. These extra equations are sometimes called "the field equations."

C.2.1 The $c \neq p^n$ Case

Lastly we consider colorings with a number of colors that cannot be the size of a field. Suppose $q \geq c$ and q is a number such that there is a field of size q. (That is, q is either a prime or a power of a prime). Then add $q - c$ "dummy vertices" to the graph, and connect each of these dummy vertices to every vertex of the graph, including the other dummies. Clearly, each dummy must have a color of its own, which no other vertex in the graph uses, because it is adjacent to every other vertex in the graph. Thus, we dispose of $q - c$ colors, and the rest of the graph will be c colored by any satisfactory solution. Likewise, any c coloring of the original $|V|$ vertices will produce an equivalent q coloring of the new $|V| + (q - c)$ vertex graph, after coloring the dummy vertices in one of the obvious $(q - c)!$ ways.

C.2.2 Application to $\mathbb{GF}(2)$ Polynomials

The above shows that we could restrict ourselves to polynomial system of equations of characteristic two if needed, by adding dummy vertices. We would simply choose

$$q = 2^{\lceil \log_2 c \rceil}$$

and use the field $\mathbb{GF}(q)$. But suppose we wanted only to use $\mathbb{GF}(2)$, how would we proceed to find a 4-coloring?

Give each vertex v_x two variables, c_x and d_x. For all edges, for example between v_x and v_y, if we write $(c_x - c_y) - 1 = 0$ and $(d_x - d_y) - 1 = 0$ then we ensure that $c_x \neq c_y$ and also that $d_x \neq d_y$. In $\mathbb{GF}(2)$, this would normally be written as $c_x + c_y + 1 = 0$. Unfortunately, we desire an "or", because if $c_x \neq c_y$ and $d_x = d_y$, then these are distinct colors, and vice versa. The equations in this paragraph, if both added to the system, result in a logical-AND, not a logical-OR as desired. This is because any solution to a polynomial system of equations must satisfy each and every equation.

The following equation provides what we need.

$$(c_x \neq d_x) \text{ OR } (c_y \neq d_y) \iff (c_x + d_x = 1) \text{ OR } (c_y + d_y = 1)$$
$$\iff (c_x + d_x) + (c_y + d_y) + (c_x + d_x)(c_y + d_y) = 1$$
$$\iff c_x + d_x + c_y + d_y + c_x c_y + d_x c_y + c_x d_y + d_x d_y + 1 = 0$$

And thus solving this particular sparse polynomial system of equations over $\mathbb{GF}(2)$, with $2|V|$ variables, is equivalent to 4-coloring $G = (V, E)$.

C.3 Related Applications

The following applications follow directly from graph coloring.

C.3.1 Radio Channel Assignments

Graph coloring problems come up in theoretical radio channel assignments, where cities near each other can cause interference, if they have radio channels on the same frequencies. This is complicated by the fact that it is not merely a geographic proximity. Mountains, for example, will block signals, while the Midwestern Plains of the USA create a "clear channel" of millions of square miles.

A more challenging example is a conference with many rooms connected by open public spaces, and many users with laptops who wish to use a wireless service. Two laptops might interfere if in the same large room some significant distance apart but not if a small distance apart in distinct rooms. Thus one could make a vertex for each user and draw an edge between those who would interfere, and color the graph. Each color is a set of users, any one of which cannot interfere with any other of the same color (otherwise there would be an edge between them), and so they can use the same frequency.

In mobile radio networks, this is crucial, as the radios are moving around, and so one must re-assign frequencies quite often. Recent work to develop sensor networks by scattering very small simple devices randomly about a region of space would also require this problem to be solved. See [108].

C.3.2 Register Allocation

Suppose a microprocessor has five registers named A, B, C, D, and E. In a particular function in a compiler, there might be 5 or fewer variables. In this case, all of them can be allocated to registers. This is useful because reading and writing to memory is often far slower than reading and writing from the registers.

If there are more than 5 variables it may still be possible to assign the variables to registers in such a way that reading or writing to memory is not required. First, compile the code from the higher-level language into the microprocessor's instruction set, but with an infinite number of registers. That is to say, every variable is a register variable.

The first time a variable is used, it is said to be born at the time of that instruction. Likewise, it is said to die at the time of the last instruction which uses it. Construct a graph with a vertex for each variable. If the "life-spans" of two variables overlap, then draw an edge between their vertices. When the graph is constructed, attempt a 5-coloring.

If a 5-coloring is possible, then each variable with a given color can be assigned to the same register. If two variables are assigned the same vertex color, then there is no edge between their vertices. This means that their life-spans do not overlap, and so there can be no conflict between them.

If a 5-coloring is not possible, then it is clear that not all the variables can be shared among the 5 registers. By empirical methods, one is chosen to become a memory location, and the process is repeated. This approach is found in [57].

C.4 Interval Graphs

At first glance, it seems that the previous example of microprocessor register scheduling might be very useful for assigning lecture halls to classes. However, we will show that this is actually a special subproblem. The graphs created by scheduling lecture halls for classes will always produce a graph that is an interval graph (to be defined momentarily)—provided that there is some moment when no classes are scheduled (e.g. at 4 o'clock in the morning).

While coloring a graph in general is NP-hard, to color an interval graph is polynomial time compared to the number of vertices. The interval graph concept was used in 1965 by Fulkerson and Gross [118], but might have been known earlier. Here we will solve the underlying scheduling problem, and produce the coloring as a side-effect. Many more general results are known, which we omit.

Suppose we are given a series of tasks $t_1, \ldots, t_n$ which must be done in time intervals of the form $[a_i, b_i)$, by some resource (e.g. scheduled classes which must be given a lecture hall, or tasks being assigned to a worker). We must outlaw tasks of zero duration (i.e. $a_i = b_i$). We wish to use the fewest number of colors (lecture halls or workers) possible. One can make a graph with a vertex for each task t. Then one can draw an edge between task t_i and t_j if their[1] intervals $[a_i, b_i)$ and $[a_j, b_j)$ are of non-empty overlap. More precisely, if $a_i < b_j$ or if $a_j < b_i$. Surely there is no harm in two tasks getting the same color (lecture hall or worker) if the times of those tasks do not overlap.

The graph produced in the above manner can then be colored. Because two vertices with the same color cannot have an overlapping interval, these tasks can be assigned to the same lecture hall/worker. Thus we need as many lecture halls/workers as there are colors—very similar to what we saw before.

The definition of an interval graph is a bit backward.

Definition 95. A graph $G = (V, E)$ is an interval graph if there exist intervals $[a_i, b_i)$ for each $v_i \in V$ such that there is an edge between v_i and v_j if and only if $[a_i, b_i)$ and $[a_j, b_j)$ are of non-empty intersection.

Given the above definition, one would wonder if there is any method to determine, given the graph alone, if it is an interval graph. The answer is yes, and can be found in [52]. In general, there is a rich literature on the subject. Searching for papers which cite [118] is sufficient to find more than one can read in a reasonable length of time. An excellent combination of interval graphs and register allocation can be found in the article [233].

C.4.1 Scheduling an Interval Graph Scheduling Problem

The algorithm for solving this problem is as follows. Number the colors $1, 2, \ldots, c$ and sort the events by their starting time. That is if $a_j > a_i$ then $j > i$ in the final ordering. As it comes to pass, we will not even construct the graph at all during the coloring. The algorithm was given by Fulkerson and Gross, and so we call it the FG algorithm.

The FG algorithm proceeds quite simply. We start with no colors. When a task begins, we see if any colors are available. If so, then we use one. If not, then we create one, and apply it to the given task. The color used is placed in a "used list". When a task ends, we simply move that color from the "used list" to the "available list." The invariant condition is that we never create a new color unless all the current colors are currently in use. For this reason the sorting should place all STOPs before all STARTs that have the same moment of time assigned to them. The details are found in Algorithm 29, where the used list is denoted $\mathscr{C}_u$ and the available list is denoted $\mathscr{C}_a$.

Suppose the FG algorithm says that the graph can be colored with c colors, but some other coloring exists with $c - 1$ colors. Let t_x be the first task in the coloring generated by the FG algorithm, to be assigned the cth color. At that instant, namely a_x we only could possibly have chosen the cth color if colors $1, 2, 3, \ldots, c - 1$ were already occupied on tasks that have not yet ended, otherwise we would certainly

---

[1] Here we use the interval notation from real analysis that $x \in [a, b)$ means $a \leq x < b$.

1: $\mathscr{L} \leftarrow \{\}$
2: For each event $[a_i, b_i]$ do

- Insert $(a_i, START, i)$ in the list $\mathscr{L}$.
- Insert $(b_i, STOP, i)$ in the list $\mathscr{L}$.

3: Sort the list $\mathscr{L}$ by comparing the first element of each triple. If there is a tie, STOPs are to come before STARTs.
4: $c \leftarrow 0$
5: $\mathscr{C}_a, \mathscr{C}_u \leftarrow \{\}$
6: For each triple in $\mathscr{L}$ do

- If it is of the form $(a_i, START, i)$ then
  - If $\mathscr{C}_a$ is empty, then
    - 1: $c \leftarrow c + 1$
    - 2: Print "task i gets color c."
    - 3: Insert (c, i) into $\mathscr{C}_u$.
  - Else $\mathscr{C}_a$ has something in it,
    - 1: $\mathscr{C}_a$ contains a positive integer in $[1, c]$. Call it c_1.
    - 2: Remove c_1 from $\mathscr{C}_a$.
    - 3: Print "task i gets color c_1."
    - 4: Insert (c_1, i) into $\mathscr{C}_u$.
- if it is of the form $(b_i, STOP, i)$ then
  - 1: Fetch the item from $\mathscr{C}_u$ whose second entry is i. This will be (c_x, i).
  - 2: Remove (c_x, i) from $\mathscr{C}_u$.
  - 3: Insert c_x into the list $\mathscr{C}_a$.

7: Print "A total of c colors were used."

Algorithm 29: Coloring an Interval Graph [Fulkerson and Gross]

---

have chosen a lower color. Thus, at this very moment, there will be c simultaneous activities. And if there are c simultaneous activities, then there is no way to use only $c - 1$ colors, and so the $c - 1$ coloring does not actually exist.

And therefore the FG algorithm is optimal in the following sense. If it uses c colors, then no coloring with fewer colors would be possible. And by contrapositive, if a coloring with fewer than c colors is possible, then the algorithm will not use c colors, but instead will use fewer.

Moreover, the complexity of the algorithm appears to have the longest step being the sorting of the list of tasks, which will be $n \log n$ time for n tasks. This is very distinct from the general case of graph coloring, where (assuming $P \neq NP$), super-polynomial time would be required.

C.4.2 Comparison to Other Problems

If a program had straight-line execution, with no loops at all, it is easy to see that the register-allocation problem could be handled as a lecture-hall scheduling problem. However, the difference becomes crucial in the presence of loops. In a loop, it is possible that A can depend on B, which must depend on C and C depends on A. This is not possible in the interval scenario, because of the transitive and antisymmetric properties of $\leq$.

Hence the requirement that there exist a time when no classes are scheduled. Otherwise the school day could be thought of as a loop.

C.4.3 Moral of the Story

The graph coloring problem is NP-Complete [63, Ch. 34.5], and so is MP (see Theorem 11.5 on Page 199), and so it is only natural that we found a way to map the graph coloring problem into solving a polynomial system of equations. But, for a major class of applications, namely interval graphs, the original scheduling problem can be solved in time slightly worse than linear to the number of vertices $O(|V| \log |V|)$. Thus, a very slight tweak to the problem renders the computation trivial instead of intractable.

And thus, we must be very careful with problems that we proclaim to be difficult. We must be ever watchful that no small tweak to the problem (perhaps made without even realizing it) suddenly renders the problem much easier than we first contemplated.

Appendix D
Options for Very Sparse Matrices

Here we will discuss what can be done about finding the null space of sparse matrices over $\mathbb{GF}(2)$, or solving linear systems of equations $Ax = \mathbf{b}$, when the matrices are of enormous dimensions. If a dense matrix of the same size can be stored in the computer in question, one can use dense matrix techniques, but they are far slower. If storing a dense matrix of the same size is not feasible, then sparse matrix techniques are obligatory.

The following is a general survey

- "Solving Large Sparse Linear Systems over Finite Fields", by Brian A. LaMacchia and Andrew M. Odlyzko [156], published at CRYPTO in 1991.

D.1 Preliminary Points

Before we begin to discuss the options that are available, we should review a few fundamental concepts.

D.1.1 Accidental Cancellations

An important factor must be discussed first, namely that of accidental cancellations. Take as an example, working in $\mathbb{R}$. When two sparse rows are added, the set of entries that are non-zero in the sum is the union of the sets of non-zeros in the originals, except when two elements in the same position in each vector are additive inverses of each other. If you assume that the sets of non-zeros in the original are disjoint, then the weight of the sum is the sum of the weights of the originals. If not, due to accidental cancellation, then the weight is at least less than or equal to the sums of the weights of the original rows.

However, if one adds a 1 to a 1 in $\mathbb{GF}(2)$, then one obtains a 0. This is called an accidental cancellation. It is clear that accidental cancellations in the real numbers will be very rare. In $GF(2)$, it is also clear that they will be common. This makes the subject of sparse linear algebra over $GF(2)$ a bit different, from $\mathbb{R}$ or $\mathbb{C}$.

D.1.2 Solving Equations by Finding a Null Space

A review of null spaces can be found in Section 10.5.5 on Page 174 with notes in Section 6.4 on Page 85 also. As stated in Corollary 28 on Page 88, a linear system of equations $A\mathbf{x} = \mathbf{b}$ over a finite field $GF(q)$ has either q^{n-r} or 0 solutions, where A is an $m \times n$ matrix and r is the rank of the matrix (note $r \leq n$, $r \leq m$). There is one solution (to $A\mathbf{x} = \mathbf{b}$) for each vector in the null space. For this reason, one might be inclined to believe that finding a null space and solving a linear system are basically equivalent. And they are in the following sense.

Suppose one had a machine which could rapidly compute the null space of a matrix A. Then, one could simply create a dummy variable, x_{n+1}, and solve instead $A\mathbf{x} - \mathbf{b}x_{n+1} = 0$ or

$$\underbrace{\left[A \ -\mathbf{b} \right]}_{A'} \begin{bmatrix} \mathbf{x} \\ x_{n+1} \end{bmatrix} = \begin{bmatrix} \mathbf{0} \end{bmatrix}$$

which is just finding the null space of the matrix A'.

Because the matrix A' is $m \times n + 1$, we expect q times as many solutions, because $q^{n-r+1} = q q^{n-r}$. But, the basis of the null space is what null space finding algorithms produce, and so we would have merely one extra basis vector, which is not a big deal.

We simply search through the null space and discard all vectors that do not have $x_{n+1} = 1$. In conclusion, all tools for finding a null space and for solving a linear system of equations are good for the opposite problem as well.

D.1.3 Data Structures and Storage

Consider a sparse matrix over $GF(2)$. It is mostly zeros. Anything left is a 1. Therefore, if we know the coordinates of each 1, then we know everything. There are basically three ways to do this. The first, and obvious way, is to store a list of ordered pairs (i, j) such that $A_{ij} = 1$ for every pair in the list, and of course $A_{ij} = 0$ otherwise. This is very inefficient, and the author does not know a name for it.

Instead, each row could be represented that way. Then there would be an array of rows. Therefore you can jump to a row immediately, and then use normal linked-list operations on the rows. The linked list would simply store the column number of the non-zero entries. This is the row-major format, and is analogous to the row-major format in dense matrices. It is good because most operations that work internal to a matrix will iterate over rows. This is by far the most common data structure used.

Sometimes, one desires a column-major representation. This is the same thing, but there is an array of columns, and each column is a linked list.

Last but not least, in fields larger than $\mathbb{GF}(2)$, one can store the entry along with the column number (row-major format) or row number (column-major format).

D.1.3.1 An Interesting Variation

In the matrices defined by the Linear Sieve, Quadratic Sieve (QS) and Number Field Sieve (NFS) methods of factoring, the first few columns are very dense, and then half of the entries are in the first $\sqrt{M}$ columns, followed by the others scattered sparsely in the remaining columns. Therefore, to optimize storage, the following might work in this special case.

Have the densest 1024 columns (or for simplicity, the first 1024 columns) be represented by 32 words of 32-bits each, i.e. as a dense matrix data structure. The rest of the entries in each row would be stored in the normal sparse matrix format, namely linked-lists only containing the non-zero entry. The operations of row-addition, for example, are very fast for 32-bit words for dense matrices, and the sparse part would take advantage of the linked lists—since only non-zero entries are stored.

D.2 Naïve Sparse Gaussian Elimination

Consider the most ordinary form of Gaussian Elimination as taught in high school. At iteration i, the $i-1$ columns at the left have been processed. Now, in column i, one must find a 1 at position A_{ii} or swap one into place, using row swaps. Thus the pivoting strategy could be said to be to ensure a non-zero entry at A_{ii}.

Instead, in Numerical Analysis, we learn that we may scan the entire column, and take the row that has an entry in that column with largest absolute value. This procedure, called "partial pivoting", reduces rounding error (see Section 7.1.3 on Page 91). The rarer practice of "full pivoting" involves scanning the entire matrix and selecting the value with largest absolute value. This requires column-swaps, which makes the code more complex.

For sparse matrices over finite fields, the obvious approach is to take the lowest weight row that happens to have a non-zero element in column i. Recall, the weight of a row is the number of non-zero entries in it. We call this "Naïve Sparse Gaussian Elimination". Suppose that in column i there are 3 rows which have a non-zero entry in that column. Let their weights be 3, 5, and 10. Then if the heaviest row is taken, these 9 non-zero entries off of the active column will be added to entries in the other two rows. It might come to pass that these two other rows already have entries in those columns, but if the matrix is very sparse, this is not likely to be the case. Thus those two rows will now acquire 9 or slightly less new entries. If instead, the row with weight 3 is chosen, then the other two rows will now acquire at most 2 new entries. This process of turning zero entries into non-zero entries is called "fill-in", and an enormous amount of research over 52 years has been done on reducing fill-in in sparse matrices; the seminal paper appeared in 1957, [169].

In joint work with Robert Miller and Seena Vali [34], the author is investigating better pivoting choices. We have already discovered that if instead of taking the lowest weight row, if one scans that row and picks its lowest weight column then this is a considerable improvement in reducing fill-in [34]. To be clear, in stage i, scan column i and note each row that has a non-zero entry there. Let the lowest-weight row be r. Next, scan row r and see which columns c have low weight, but A_{rc} non-zero. Then pivot by swapping column i and column c as well as row r and row i. We denote this "Semi-Naïve Sparse Gaussian Elimination," and it has been known for a long time. We also have a far more complex algorithm that is still being developed in [34].

D.2.1 Sparse Matrices can have Dense Inverses

As it turns out, a very sparse matrix can have a dense inverse. Also, the LUP factorization can be dense. This means that the standard method of solving a system of linear equations—doing the Gaussian Elimination to REF instead of RREF (see Section 7.3 on Page 93) is extremely unwise, as the REF contains U, and so would be dense. This is how we can expect that the fill-in effect is hard to avoid without some insight, effort, or structure.

D.3 Markowitz's Algorithm

The Markowitz algorithm is only a small modification of the above. In full-pivoting, we consider all non-zero entries in the region $A_{ii}, \ldots, A_{mn}$, and choose the one of largest absolute value. In Markowitz's algorithm [169], we calculate the weight of each row, denoted w_r, and the weight of each column w_c, and find a non-zero entry A_{rc} that minimizes $(w_r - 1)(w_c - 1)$. Alternatively, Pomerance suggests $(w_r - 2)(w_c - 2) - 2$ [193]. This is a raw estimate on the amount of fill-in created by pivoting at A_{rc}, and so in the spirit of the greedy algorithm, we pick that pivot.

D.4 The Block Wiedemann Algorithm

The Block Wiedemann algorithm by Don Coppersmith [61] is an excellent method for sparse $\mathbb{GF}(2)$ linear systems. Section 5.2.3.1 on Page 72 for a discussion of the running time, or [232] for a discussion of the application to the XL algorithm. Currently, Block-Wiedemann seems to be the method of choice, at least in crypt-analysis.

This algorithm is discussed in the following papers, and many others.

- "Solving sparse linear equations over finite fields", by D. H. Wiedemann [230], published in the IEEE Transactions on Information Theory in 1986.
- "On Wiedemann's Method of Solving Sparse Linear Systems", by Erich Kaltofen and B. David Saunders [147], published in the proceedings of the 9th International Symposium on Applied Algebra, Algebraic Algorithms and Error-Correcting Codes in 1991.
- "Solving homogeneous linear equations over $\mathrm{GF}(2)$ via block Wiedemann algorithm", by Don Coppersmith [61], published in Mathematics of Computation in 1994.
- "Analysis of Coppersmith's block Wiedemann algorithm for the parallel solution of sparse linear systems", by Erich Kaltofen [146], published in Mathematics of Computation in 1995.
- "Further Analysis of Coppersmith's Block Wiedemann Algorithm Using Matrix Polynomials", by Gilles Villard [219], published in the proceedings of the 1997 international symposium on symbolic and algebraic computation.

D.5 The Block Lanczos Algorithm

Like the Block Wiedemann algorithm, the Block Lanczos algorithm is also a descendent of algorithms based on the Krylov subspaces approach. The author is not familiar with this alternative, but the following references are highly cited.

- "A block Lanczos algorithm for finding dependencies over $\mathrm{GF}(2)$", by Peter Montgomery [178], published in CRYPTO 1995.
- "Solving linear equations over $\mathrm{GF}(2)$: the block Lanczos algorithm," by Don Coppersmith [60], published in Linear Algebra and its Applications in 1993.
- "Computational Aspects of Discrete Logarithms", the PhD Dissertation of Robert Lambert [157], from 1996.

The last one presents a unified framework for understanding the Block Lanczos and and Block Wiedemann algorithms.

D.6 The Pomerance-Smith Algorithm

This algorithm is sometimes called "Structured Gaussian Elimination" or "The Created Catastrophes Algorithm", but neither of these names is very descriptive. We will call it the Pomerance-Smith algorithm after its creators [193].

The Pomerance-Smith Algorithm consists of seven operations which are guaranteed to never increase fill-in, and all of which reduce the size of the matrix. Therefore, a series of these operations, which are all very simple, is guaranteed to shrink the problem at hand and simultaneously will not increase the memory used. Thus, it is in its own way a sort of "Greedy Algorithm". The key to understanding this algorithm is to take the Operations and convince yourself that they are safe, and will not create fill-in. Only then will the total picture take shape. The operations mostly consist of deleting rows and columns that no longer matter.

D.6.1 Overview

Mainly, the algorithm is used during factoring methods like the Quadratic Sieve (see Section 21 on Page 182) and those matrices are over $GF(2)$. The algorithm is usually presented over $GF(2)$ but it can work over any field with trivial modification which we omit.

The matrices produced by the algorithm have the unusual property that half the entries are in the leftmost $\sqrt{n}$ out of n columns. The remainder of the matrix is extremely sparse. Thus, while the matrix may have roughly 100,000–300,000 rows, but only 10–30 entries per row. Essentially, each column represents a prime, and as lots of numbers are divisible by 2, 3, 5, and 7, but far fewer by 101, it is only natural that the columns representing the smaller primes are more dense, and the larger primes far less dense. This is a very smooth transition. Since the matrix is actually divided into this "dense part" and "sparse part", the algorithm actually divides the matrix into a dense region, called inactive, and a sparse region, called active. Because the transition is smooth, it is very hard to decide where to draw the line.

D.6.1.1 Objective

The objective is to produce roughly 10 vectors in the basis of the null space. This will produce 1023 non-trivial vectors (or 512 linear system solutions, if used in the style of Section D.1.2 on Page 324). In the case of the Quadratic Sieve, each null space vector has the potential to give the factorization of the number desired with probability one-half. Therefore, even have 32 vectors would result in the number not being factorable with probability $2^{-32} \approx 2.5 \times 10^{-10}$. The choice of 10 is thus "over kill". This parameter, which we call "margin," can be set to other values, as might be useful in other applications. The algorithm will trim the matrix to have a number of rows equal to at most the number of columns plus the margin. Thus, excess rows are deleted, and this destroys some information.

If the null space has 100 vectors in its basis, finding only 10 of those might be undesirable for applications other than the Quadratic Sieve. We partially address this in Section D.6.7 on Page 334.

D.6.1.2 The Method

The general process is as follows. One begins with the entire matrix in memory as if it were a sparse matrix, minus the heaviest 5% of the columns. Those 5% highest weight columns are copied into a dense matrix data structure (two-dimensional array). Then one executes a series of the seven operations, including erasing rows and columns. Also, some columns are declared dense, and copied into a dense matrix data structure. Other rows and columns are "ejected", and deleted entirely. Eventually, the sparse matrix is all zeros. In addition, one stores a history of the operations that have been done.

The operations done are then to be replayed on the dense part of the matrix which may have roughly 10% of the columns in it. This will be reduced into REF with ordinary dense linear algebra techniques. Finally, one must reconstruct the null-space.

D.6.2 Inactive and Active Columns

As mentioned above, the matrix will be divided between a dense part and a sparse part. In particular, some columns will be declared inactive. In this case, they will be removed from the sparse matrix data structure entirely. They will be copied into a dense matrix data structure, with some sort of tag to represent which column they were in the original matrix. The easiest way to do this is to tag each column initially with its "id-number", and then this information will be retained throughout the algorithm.

Thus "declaring a column inactive" means adding a column to the dense matrix data structure, copying over the column from the sparse matrix to the dense matrix. This is non-trivial, because rows will be deleted, and so one must match-up with some sort of row id-numbers. Finally, delete the column from the sparse matrix.

It is absolutely vital to point out that by "weight" of a column, we mean the number of non-zero entries in the column. By "weight" of a row, we explicitly only count those entries which are in active columns (in the sparse matrix data structure) and we explicitly do not count those entries which are in the inactive columns.

D.6.3 The Operations

The following seven operations will be repeated frequently during the algorithm. They are named "Step n" in [193] and we rename them "Operation n" here only because our more explicit rendering of the algorithm will include some intermediate steps (which would otherwise have to be denoted as fractional steps).

First we will describe what each operation does.

Operation 0:   This operation declares k columns inactive. One searches for the
k heaviest columns, and does the copy-and-delete operation specified in Section D.6.2.

Operation 1:   Any column of weight 0 is uninvolved with any null-space vector.
Therefore, we can simply delete the column (not make it inactive, but actually
delete it). In a sense, it is a variable which we thought might occur in the system
of equations, but which is in fact not found there. Of course, this might be because
the rows that used it were deleted earlier. We will justify why it is okay to simply
delete rows in the description of Operation 3.

Operation 2:   A column of weight 1 has only one row that it shares an element
with. Call this entry A_{rc}. Since there is no A_{sc} with $s \neq r$ such that $1 = A_{sc}$, then
there is no way to cancel out the 1 in the cth column of row r. This means that row
r is not used in any null-space vector. Therefore, we delete row r. Furthermore,
we column c as well, because it is now empty, and will get picked out for deletion
during the next cycle of Operation 1.

Operation 3:   We desire roughly 10 vectors in the null space. If a matrix is full
rank, and it has 10 more columns than rows, it will have 10 vectors in the null
space. If the matrix is not full rank, it will have more, and we are even happier.
From Table 9.3 on Page 145 we might expect that we will have nullity 1 or 2 by
the time the matrix becomes dense, but it is good to be conservative. Therefore,
we will seek to maintain a number of rows equal to the number of columns plus
10 (or plus the margin if it is not 10). Any extra rows are superfluous. And so we
may wish to delete those rows which have the highest weight. Therefore in this
step we will delete the k heaviest rows, where $k = r - c + 10$.

Note: You might suppose we could eliminate rows of weight 0. This would basically represent a redundant equation, that earlier was actually equal to some
linear combination of other equations in the system, and which has now been
shown to be redundant. However, there may be useful information in the inactive
part of the matrix (the columns which have been deactivated). Recall, the weight
of a row does not include entries in the inactive columns, because they have been
removed from the sparse matrix data structure. Therefore, we do not touch rows
of weight 0.

Operation 4:   Suppose there is a row of weight 1. Then this means that the associated variable with that one entry (call it A_{rc}) is equal to zero (i.e. $x_c = 0$) in
any null-space solution. You might imagine that this means we can forget that
this variable exists, and delete column c. But because of the previous note, we do
not wish to do that. The inactive part of the row might be quite dense or useful.
Therefore, we will simulate adding row r to every row which which has an entry
in column c. For each such row s, we will store on our task-list that we must
add rows r and s, in the dense part later (and store the answer in row s). Then
we could artificially zero-out column c, except for entry A_{rc}. Of course, now row
r and column c will play no further role in the algorithm. Furthermore, column
c has only weight 1 now. Therefore, we can immediately execute Operation 2,
which would delete row r and delete column c as well.

Operation 5:   Now suppose there is a row of weight 2. Then this means that the row (call it r) has two entries that are 1, call them c and c', and without loss of generality assume $w(c) \leq w(c')$. Then we can simulate adding row r to every row s that has an entry in column c. This means that we must store on our task-list that we must add row r and row s in the dense part of the matrix later (and store the answer in row s). Also, in the sparse part, the entry A_{sc} will be zeroed out, and $A_{sc'}$ will be toggled (flipped from its current value to the opposite). Note that this will leave column c with only the entry A_{rc}, and so we can immediately execute an Operation 2. There is a very rapid way to execute all this. Column c' should be changed into the logical-XOR of column c and column c'. Column c and row r can now be deleted.

Operation 5^T: This step is mentioned in the paper as a suggestion by Odlyzko, but is not explicitly numbered. Suppose there is a column of weight 2. Call the two entries in it A_{rc} and A_{sc}, with $w(r) < w(s)$. Then we could add rows r and s together (storing the sum in row r). After this, column c has only one entry, namely A_{sc}. This means we can immediately execute Operation 2, which would delete column c and row s.

The reader may want to take a moment and review each of the operations, to understand what they do and to ensure that indeed these are not going to affect our search for null-space vectors. In fact, the Pomerance-Smith Algorithm just executes these six operations in a loop, some repeating more often than others. After this is done, the dense matrix data structure has to be updated. We simply go through the task-list, and execute the row additions stated there.

Careful examination of each operation will yield that only Operation 3 destroys null-space vectors. However, it does so in such a way as to ensure that at least 10 will be found. That means we will know about 1023 of the vectors in the null space (counting all linear combinations of the 10, but of course one of them is the zero vector, which is boring). Thus, we know the algorithm is correct.

D.6.4 The Actual Algorithm

Later researchers have made it clear that one can do the operations in different orders, and one can even leave some out. However, Pomerance-Smith, in the original paper worked as follows. First, do Operation 0 to remove 5% of the columns.

Next do Operation 1 as many times as possible. Then do Operation 2 as many times as possible. If Operation 2 was able to do anything at all, then restart with Operation 1. If not, then do Operation 3 as many times as possible. If Operation 3 was able to do anything at all, restart with Operation 1, but otherwise do Operation 4 as many times as possible. Again, if Operation 4 accomplished anything, return to Operation 1. If not, then Operation 5 and Operation 5^T are treated as equals—do them as many times as possible. If something happened, then return to Operation 1. Otherwise do Operation 0 to the $\lceil n/1000 \rceil$ heaviest columns and continue with Operation 1.

This seems awfully heuristic, and it is. But it is still an algorithm, because it is guaranteed to terminate. Each of the operations, if successfully applied, will reduce the size of the matrix. Sometimes an operation cannot be applied, and in fact, this is the only way to go from Operation i to Operation $i+1$. On the other hand, each time Operation 0 is called, it will succeed in removing some number of columns. It cannot fail. And so, with at most 950 calls to Operation 0, since the first removes 5% and all later applications of it remove 0.1%, the algorithm will terminate.

D.6.5 Fill-in and Memory Management

Now we will look at the impact on the number of non-zero entries in the sparse part of the matrix, for each of these operations. Careful examination will reveal that only Operations 5 and 5^T can result in an entry which is 0 turning into an entry which is 1, in the sparse part of the matrix. Thus, the number of non-zero entries in the sparse part of the matrix (the active columns) will never increase, other than with Operation 5 and 5^T. In fact, Operations 0, 1, 2, 3, and 4 will frequently reduce the number of ones, and in the case of Operations 0 and 3, perhaps very significantly. It turns out with Operations 5 and 5^T, the number of non-zeros will actually decrease as well, but it takes a bit of work to prove it.

For Operation 5, the net number of new non-zero entries introduced into column c' will be $w(c) - 2w(c \cap c')$, because those in the intersection are -1 to the weight and those not in the intersection are +1 to the weight. But, this is offset by erasing column c. And so we have a total of $-2w(c \cap c')$. In other words, if columns c and c' are totally disjoint, then the ones in column c simply move into column c', except for the one in row r which does not survive anyway. If the intersection is non-empty, those not in the intersection move over (which creates 0) and those in the intersection annihilate a one in column c' for a net reduction of two each. Of course, the erasure of row r would remove two entries as well (we would not be doing this operation if row r did not have 2 entries in it), for a net decrease of $1 + 2w(c \cap c')$, being careful to not count the benefit of zeroing out A_{rc} twice—once in the removal of row r and once in the removal of column c. Thus, Operation 5 also strictly reduces the number of ones in the sparse part of the matrix.

For Operation 5^T, the net number of new non-zero entries introduced into row r will be $w(s) - 2w(r \cap s)$, for reasons the same as in the previous paragraph. But, when we erase row s, we get a net change of $-2w(r \cap s)$. Next, the erasure of column c will remove two elements, namely A_{rc} and A_{sc}, the latter of which we have already counted. So the net decrease is $1 + 2w(r \cap s)$. The relationship between this step and the previous is the reason for the notation Operation 5^T.

From the point of view of memory-management, if a sparse data structure is used, then the amount of memory used is directly proportional to the number of entries. Since this never increases, that means if the computer can hold the matrix initially, it can process it to the end. This is certainly not the case for whole categories of matrix techniques.

In the paper [193], Pomerance and Smith note that the algorithm proceeds very slowly, until the average weight of each row in the sparse part is about three. Then he conjectures that some sort of threshold is crossed, because suddenly there is an enormous number of erasures of rows and columns. In the 1960s, René Thom created a theory called Catastrophe Theory which is the study of thresholds where functions change discontinuously, as well as thresholds where functions change very rapidly but are nevertheless mathematically continuous. Since this is the effect we see here, when the weight of the rows is roughly 3.3 ± 0.3, then it is natural to deem this a "catastrophe" from that theory, but in a very positive sense. Pomerance and Smith suggest the world "miracle" might have been a better choice. The catastrophe/miracle itself is suddenly a long series of Operation 1–$5, 5^T$ which dramatically reduce the non-zero count of the matrix, until it is zero, with no uses of Operation 0.

D.6.6 Technicalities

These are some unimportant technicalities that come up in the implementation.

D.6.6.1 Why not Do Operation 0 only Once?

It might seem that since Operation 0 might move many columns into the dense storage area, in several stages throughout the algorithm, one could just simply pick a larger number than 5%, and move them over earlier. However, this is essentially the algorithm by Odlyzko [183] on which this algorithm was based [193].

In practice, in [193] it is claimed that one moves fewer columns with Pomerance-Smith than with Odlyzko's implementation of this idea. Furthermore, there was never a good estimate of how many columns should be moved (i.e. declared dense)—with consequences in the event of overbidding. In [193] not only are exactly the right number moved, with no over-shooting, but information in the middle of the algorithm is used to determine that as the matrix evolves.

Also, it is noteworthy to mention the idea of splitting the matrix into a dense part and spare-part, reducing the sparse part, and then re-acting those operations on the dense part, finishing with dense linear algebra, is due to Odlyzko [183].

D.6.6.2 Random Matrices

For test purposes, it might be convenient to generate many random matrices of the kind that would occur in the QS or NFS, without actually carrying out the QS. This model is suggested by [193]. One requires a parameter D, typically $2 > D > 3$ for the QS and slightly more for the NFS. The probability of an entry being 1 for column i is given by $\min(1/2, D/i)$.

The number of columns C that get declared dense by Operation 0 is very predictable. Doing a power-regression on the data in [193] one gets that the following model

$$C \approx 0.10767 D^{2.1176} n^{0.84621} - 1296.5$$

with average error 1.3% and worse-case error of 3.63%.

D.6.6.3 Only Getting Part of the Null Space

This algorithm will produce a number of null-space basis-vectors equal to "margin", which is recommended to be roughly 10. This is a basis for a 10-dimensional subspace of the null space, and so since we are over $\mathbb{GF}(2)$, there would be 1023 non-zero vectors.

What if there are more null space vectors to be found? Suppose one has an $m \times n$ matrix A, and one believes that the rank r is between $r_1 \leq r \leq r_2$. Then the nullity is at most $n - r_2$, and so this is a safe value for margin. If one does not have a lower bound on the rank (an upper bound on the nullity) then perhaps one should try a different algorithm. One could imagine it might be useful to attempt using all the operations except Operation 3. However, the number of rows might not reduce fast enough in that case and it would be likely that the matrix would have (eventually) many more rows than columns, and things would bog down.

The previous paragraph applies only to operations other than factoring. When factoring, having 1023 null space vectors is overkill. There is no need for more than that. Even so, it is somewhat distressing to realize that if there were two very high weight rows, but which differed in only 1–4 positions, the rows would be deleted because of their weight. But, their sum would be a row with weight 1–4, which is excellently useful in this algorithm or any other.

One could imagine a matrix where Pomerance-Smith fails, because not enough of Operations 1–5 occur and Operation 0 is called too many times leaving far too large of a matrix for dense operations. Yet it could have two or three rows which were dense and thus deleted, yet whose sum is very low weight. That low weight vector could be used to clear out a few columns, producing new low weight rows, *et cetera...*

D.6.7 Cremona's Implementation

John Cremona mentioned at SAGE Days #10 in Nancy, Lorraine, France that he uses this algorithm for solving large sparse linear systems of equations. He has implemented the algorithm and it is part of SAGE. Furthermore, it can work over any field. Looking over the list of operations, one can see that this is true provided that one trivially add a few scalar multiplications or negations in the right spots.

D.6.8 Further Reading

The author suggests the following further reading

- "Reduction of Huge, Sparse Matrices over Finite Fields via Created Catastrophes," by Carl Pomerance and J. Smith [193], published in *Experimental Mathematics*, in 1992.
- "An Approximate Probabilistic Model for Structured Gaussian Elimination", by Edward Bender and E. Rodney Canfield [41] published in *The Journal of Algorithms*, in 1999.

Appendix E
Inspirational Thoughts, Poetry and Philosophy

The following reflects what I have long believed, though I must confess I have lacked the expressiveness to state it. This quote is notable not only for what it communicates, but for its age as well. Written about 450 years ago, it still rings true (despite the odd spelling). I am confident many mathematicians alive today would agree with it.

> The ignoraunte multitude doeth, but as it was euer wonte, enuie that knoweledge, whiche thei can not attaine, and wishe all men ignoraunt, like unto themself... Yea, the pointe in Geometrie, and the unitie in Arithmetike, though bothe be undiuisible, doe make greater woorkes, & increase greater multitudes, then the brutishe bande of ignoraunce is hable to withstande...
> (Robert Recorde, 1557, quoted from [121, Ch. 3]).

> But yet one commoditie moare... I can not omitte. That is the filying, sharpenyng, and quickenyng of the witte, that by practice of Arithmetike doeth insue. It teacheth menne and accustometh them, so certainly to remember thynges paste: So circumspectly to consider thynges presente: And so prouidently to forsee thynges that followe: that it maie truelie bee called the *File of witte*.
> (Robert Recorde, 1557, quoted from [121, Ch. 17]).

Note also, a third quote can be found on page xv of this book. The Welsh mathematician Robert Recorde is mostly known for inventing the equal sign. The following quote describes that invention.

> And to avoide the tediouse repetition of these woordes : is equalle to : I will sette as I doe often in woorke use, a pair of paralleles, or Gemowe lines of one lengthe, thus: =====, bicause noe .2. thynges, can be moare equalle.
> (Robert Recorde, quoted from [100]).

Recorde lived 1510 to 1558, and was a faculty member at Oxford and at Cambridge. He wrote several books on algebra and geometry, as is credited with the first translation of Euclid's *Elements* into English, because he believed "it would be of use to Merchants." The four quotes given in this book are from *The Whetstone of Witte*, an algebra text, also intended for the merchant community. A whetstone is a stone used for sharpening, serving like a file but natural instead of man-made.

This spirit of trying to bring mathematics into the lives of those otherwise interested in business is one that I can admire, since the majority of my teaching obligations are for the freshman mathematics requirement of students attempting degrees in business, accounting, finance, marketing, and related fields.

The innovation of the equal sign, or the utility of the Euclidean Geometry to "file the wit" of future merchants have both stuck with us, as any high school student is aware. Luckily for us, not all of his innovations have stuck with us however. He invented the term "zenzizenzizenzike" for x^8. This follows naturally from using "zenzike" for x^2, and "zenzizenzike" for x^4. Imagine how complex our vocabulary would become to describe 2^{128}.

References

1. BOINC: Berkeley open infrastructure for networked computing. Available at `http://boinc.berkeley.edu/`
2. Magma. Software Package. Available at `http://magma.maths.usyd.edu.au/magma/`
3. Maple. Software Package. Available at `http://www.maplesoft.com/`
4. Mathematica. Software Package. Available at `http://www.wolfram.com/products/mathematica/index.html`
5. Matlab. Software Package. Available at `http://www.mathworks.com/`
6. MiniSAT. Software Package. Available at `http://www.cs.chalmers.se/Cs/Research/FormalMethods/MiniSat/` or `http://minisat.se/Papers.html`
7. Sage. Software Package. Available at `http://www.sagemath.org/`
8. SETI@home classic. See `http://setiathome.berkeley.edu/classic.php`
9. Singular. Software Package. Available at `http://www.singular.uni-kl.de/`
10. Specification for the Advanced Encryption Standard: Federal information processing standards publication 197 (FIPS-197). Tech. rep., National Institute of Standards and Technology (NIST) (2001)
11. Personal communications with Adi Shamir
12. Aggarwal, D., Maurer, U.: Breaking RSA generically is equivalent to factoring. In: A. Joux (ed.) Advances in Cryptology—Proc. of EUROCRYPT, *Lecture Notes in Computer Science*, vol. 5479, pp. 36–53. Springer-Verlag (2009)
13. Aho, A., Hopcroft, J., Ullman, J.: The Design and Analysis of Computer Algorithms, second edn. Addison-Wesley (1974)
14. Albrecht, M.: Algebraic attacks on the courtois toy cipher. Master's thesis, University of Bremen (Universität Bremen), Department of Computer Science (2006)
15. Albrecht, M., Bard, G., Hart, W.: Efficient multiplication of dense matrices over $GF(2)$. Submitted to Transactions on Mathematical Software (2008). Available at `http://www.math.umd.edu/~bardg/albrecht_bard_hart.pdf`
16. Albrecht, M., Cid, C.: Algebraic techniques in differential cryptanalysis. Cryptology ePrint Archive, Report 2008/177 (2008). Available at `http://eprint.iacr.org/2008/177`
17. Albright, B.: An introduction to simulated annealing. College Mathematics Journal **38**(1), 37–42 (2007)
18. Apéry, R.: Irrationalité de $\zeta(2)$ et $\zeta(3)$. Astérisque **61**, 11–13 (1979)
19. Arditti, D., Berbain, C., Billet, O., Gilbert, H.: Compact FPGA implementations of QUAD. In: F. Bao, S. Miller (eds.) Proceedings of the 2007 ACM Symposium on Information, Computer and Communications Security, (ASIACCS'07), pp. 347–349. ACM (2007)

20. Arditti, D., Berbain, C., Billet, O., Gilbert, H., Patarin, J.: QUAD: Overview and recent developments. In: E. Biham, H. Handschuh, S. Lucks, V. Rijmen (eds.) Symmetric Cryptography, *Dagstuhl Seminar Proceedings*, vol. 07021. Internationales Begegnungs- und Forschungszentrum fuer Informatik (IBFI), Schloss Dagstuhl, Germany (2007)

21. Arlazarov, V., Dinic, E., Kronrod, M., Faradzev, I.: On economical construction of the transitive closure of a directed graph. Dokl. Akad. Nauk. SSSR **194**(11) (1970). (in Russian), English Translation in Soviet Math Dokl

22. Armknecht, F.: A linearization attack on the Bluetooth key stream generator. Cryptology ePrint Archive, Report 2002/191 (2002). Available at http://eprint.iacr.org/2002/191

23. Armknecht, F.: Improving fast algebraic attacks. In: B.K. Roy, W. Meier (eds.) Proc. of Fast Software Encryption (FSE'04), *Lecture Notes in Computer Science*, vol. 3017, pp. 65–82. Springer-Verlag (2004)

24. Armknecht, F., Ars, G.: Introducing a new variant of fast algebraic attacks and minimizing their successive data complexity. In: E. Dawson, S. Vaudenay (eds.) Proc. of Mycrypt, *Lecture Notes in Computer Science*, vol. 3715, pp. 16–32. Springer-Verlag (2005)

25. Armknecht, F., Krause, M.: Algebraic attacks on combiners with memory. In: D. Boneh (ed.) Advances in Cryptology—Proc. of CRYPTO, *Lecture Notes in Computer Science*, vol. 2729, pp. 162–175. Springer-Verlag (2003)

26. Ars, G., Faugère, J.C., Imai, H., Kawazoe, M., Sugita, M.: Comparison between XL and Gröbner Basis algorithms. In: P.J. Lee (ed.) Advances in Cryptology—Proc. of ASIACRYPT, *Lecture Notes in Computer Science*, vol. 3329, pp. 338–353. Springer-Verlag (2004)

27. Atkins, D., Graff, M., Lenstra, A.K., Leyland, P.C.: The magic words are squeamish ossifrage. In: J. Pieprzyk, R. Safavi-Naini (eds.) Advances in Cryptology—Proc. of ASIACRYPT, *Lecture Notes in Computer Science*, vol. 917, pp. 263–277. Springer-Verlag (1994)

28. Atkinson, M., Santoro, N.: A practical algorithm for boolean matrix multiplication. Information Processing Letters (1988)

29. Bard, G.: Algorithms for fast matrix operations. Tech. rep., University of Maryland, Applied Mathematics and Scientific Computation Program (2005). Scholarly Paper for M.Sc. in Applied Math, available on the author's web-page

30. Bard, G.: Achieving a log(n) speed up for boolean matrix operations and calculating the complexity of the dense linear algebra step of algebraic stream cipher attacks and of integer factorization methods. Cryptology ePrint Archive, Report 2006/163 (2006). Available at http://eprint.iacr.org/2006/163

31. Bard, G.: Algorithms for the solution of linear and polynomial systems of equations over finite fields, with applications to cryptanalysis. Ph.D. thesis, Department of Applied Mathematics and Scientific Computation, University of Maryland at College Park (2007). Available at http://www.math.umd.edu/~bardg/bard_thesis.pdf

32. Bard, G.: Extending SAT-Solvers to low-degree extension fields of $GF(2)$. Presented at the Central European Conference on Cryptography (2008). Available at http://www.math.umd.edu/~bardg/extension_fields.pdf

33. Bard, G., Courtois, N., Jefferson, C.: Efficient methods for conversion and solution of sparse systems of low-degree multivariate polynomials over $GF(2)$ via SAT-Solvers. Cryptology ePrint Archive, Report 2007/024 (2006). Available at http://eprint.iacr.org/2007/024.pdf

34. Bard, G., Miller, R., Vali, S.: Lowering fill-in for Gaussian Elimination on sparse matrices over finite fields. In preparation (2009)

35. Bard, G.V.: A challenging but feasible blockwise-adaptive chosen-plaintext attack on SSL. In: M. Malek, E. Fernández-Medina, J. Hernando (eds.) Proceedings of the International Conference on Security and Cryptography (SECRYPT'06), pp. 99–109. INSTICC Press (2006). Available at http://eprint.iacr.org/2006/136/

36. Barkan, E., Biham, E., Keller, N.: Instant ciphertext-only cryptanalysis of GSM Encrypted communication. In: D. Boneh (ed.) Advances in Cryptology—Proc. of CRYPTO, *Lecture Notes in Computer Science*, vol. 2729, pp. 600–616. Springer-Verlag (2003)

37. Barkee, B., Can, D.C., Ecks, J., Moriarty, T., Ree, R.F.: Why you cannot even hope to use Gröbner bases in public-key cryptography—an open letter to a scientist who failed and a challenge to those who have not yet failed. Journal of Symbolic Computations **18**(6), 497–501 (1994)
38. Bartee, T.C.: Digital Computer Fundamentals, sixth edn. McGraw Hill (1985)
39. Baur, W., Strassen, V.: The complexity of partial derivatives. Theoretical Computer Science (1983)
40. Bellare, M., Desai, A., Jokipii, E., Rogaway, P.: A concrete security treatment of symmetric encryption: Analysis of the DES modes of operation. In: Proc. 38th Symposium on Foundations of Computer Science. IEEE (1997)
41. Bender, E., Canfield, E.R.: An approximate probabilistic model for structured Gaussian Elimination. The Journal of Algorithms (1999)
42. Berbain, C., Gilbert, H., Patarin, J.: QUAD: A practical stream cipher with provable security. In: S. Vaudenay (ed.) Advances in Cryptology—Proc. of EUROCRYPT, *Lecture Notes in Computer Science*, vol. 4004, pp. 109–128. Springer-Verlag (2006)
43. Bernstein, D.: Matrix inversion made difficult. Unpublished Manuscript (1995). Available on http://cr.yp.to/papers/mimd.ps
44. Bernstein, D.: Response to slid pairs in Salsa20 and Trivium. Technical Report: Available at: http://cr.yp.to/snuffle/reslid-20080925.pdf (2008)
45. Bernstein, D., Chen, T.R., Cheng, C.M., Lange, T., Yang, B.Y.: ECM on graphics cards. In: A. Joux (ed.) Advances in Cryptology—Proc. of EUROCRYPT, *Lecture Notes in Computer Science*, vol. 5479, pp. 483–501. Springer-Verlag (2009)
46. Berre, D.L., Simon, L.: Special volume on the SAT 2005 competitions and evaluations. Journal of Satisfiability (2006)
47. Biham, E., Dunkelman, O., Indesteege, S., Keller, N., Preneel, B.: How to steal cars—a practical attack on Keeloq (2008)
48. Biryukov, A., Shamir, A., Wagner, D.: Real time cryptanalysis of A5/1 on a PC. In: B. Schneier (ed.) Fast Software Encryption (FSE'00), *Lecture Notes in Computer Science*, vol. 1978, pp. 1–18. Springer-Verlag (2000)
49. Bogdanov, A.: Attacks on the Keeloq block cipher and authentication systems. In: 3rd Conference on RFID Security 2007 (RFIDSec'07) (2007)
50. Bogdanov, A.: Cryptanalysis of the Keeloq block cipher. Cryptology cPrint Archive, Report 2007/055 (2007). Available at http://eprint.iacr.org/2007/055/
51. Bogdanov, A.: Linear slide attacks on the Keeloq block cipher. In: D. Pei, M. Yung, D. Lin, C. Wu (eds.) INSCRYPT'07, *Lecture Notes in Computer Science*, vol. 4990, pp. 66–80. Springer-Verlag (2007)
52. Booth, K.S., Lueker, G.S.: Testing for the consecutive ones property, interval graphs, and graph planarity using PQ-tree algorithms. J. Comput. Systems Sci. **13**, 335–379 (1976)
53. Boothby, T.J., Bradshaw, R.W.: Bitslicing and the method of four russians over larger finite fields. Submitted to a journal (2009). Available at http://arxiv.org/abs/0901.1413
54. Bressourd, D.M.: Factorization and Primality Testing. Undergraduate Texts in Mathematics. Springer-Verlag (1989)
55. Bunch, J., Hopcroft, J.: Triangular factorization and inversion by fast matrix multiplication. Math Comp. **28**(125) (1974)
56. Cannière, C.D.: Trivium: A stream cipher construction inspired by block cipher design principles. In: S.K. Katsikas, J. Lopez, M. Backes, S. Gritzalis, B. Preneel (eds.) Proceedings of the 9th International Conference on Information Security, ISC'06, *Lecture Notes in Computer Science*, vol. 4176, pp. 171–186. Springer-Verlag (2006)
57. Chaitin, G.J., Auslander, M.A., Chandra, A.K., Cocke, J., Hopkins, M.E., Markstein, P.W.: Register allocation via coloring. Computer Languages **6**, 47–57 (1981)
58. Cid, C., Albrecht, M., Augot, D., Canteaut, A., Weinmann, R.P.: Algebraic cryptanalysis of symmetric primitives. Tech. Rep. D.STVL.7, ECRYPT: The European Union Network of Excellence in Cryptography (2008)

59. Cohen, H.: A course in Computational Algebraic Number Theory. Springer-Verlag (1993)
60. Coppersmith, D.: Solving linear equations over $\mathbb{GF}(2)$: the Block Lanczos algorithm. Linear Algebra and its Applications **192**, 33–60 (1993)
61. Coppersmith, D.: Solving homogeneous linear equations over $\mathbb{GF}(2)$ via Block Wiedemann algorithm. Mathematics of Computation **62**(205), 333–350 (1994)
62. Coppersmith, D., Winograd, S.: Matrix multiplication via arithmetic progressions. J. of Symbolic Computation **9** (1990)
63. Cormen, T., Leiserson, C., Rivest, R., Stein, C.: Introduction to Algorithms, second edn. MIT Press, McGraw-Hill Book Company (2001)
64. Courtois, N.: The security of cryptographic primitives based on multivariate algebraic problems: MQ, MinRank, IP, HFE. Ph.D. thesis, Paris VI (2001). Available at http://www.nicolascourtois.net/phd.pdf
65. Courtois, N.: The security of Hidden Field Equations (HFE). In: D. Naccache (ed.) Cryptographers' Track, RSA Conference, *Lecture Notes in Computer Science*, vol. 2020, pp. 266–281. Springer-Verlag (2001)
66. Courtois, N.: Higher order correlation attacks, XL algorithm and cryptanalysis of Toyocrypt. In: P.J. Lee, C.H. Lim (eds.) Proc. of ICISC, *Lecture Notes in Computer Science*, vol. 2587, pp. 182–199. Springer-Verlag (2002)
67. Courtois, N.: Fast algebraic attacks on stream ciphers with linear feedback. In: D. Boneh (ed.) Advances in Cryptology—Proc. of CRYPTO, *Lecture Notes in Computer Science*, vol. 2729, pp. 176–194. Springer-Verlag (2003)
68. Courtois, N.: Generic attacks and the security of Quartz. In: Y. Desmedt (ed.) Public Key Cryptography (PKC'03), *Lecture Notes in Computer Science*, vol. 2567, pp. 351–364. Springer-Verlag (2003)
69. Courtois, N.: Algebraic attacks on combiners with memory and several outputs. In: C. Park, S. Chee (eds.) Proc. of ICISC, *Lecture Notes in Computer Science*, vol. 3506, pp. 3–20. Springer-Verlag (2004)
70. Courtois, N.: Algebraic attacks over $\mathbb{GF}(2^k)$, application to HFE Challenge 2 and Sflash-v2. In: F. Bao, R.H. Deng, J. Zhou (eds.) Public Key Cryptography (PKC'04), *Lecture Notes in Computer Science*, vol. 2947, pp. 201–217. Springer-Verlag (2004)
71. Courtois, N.: General principles of algebraic attacks and new design criteria for components of symmetric ciphers. In: H. Dobbertin, V. Rijmen, A. Sowa (eds.) Proc. AES 4 Conference, *Lecture Notes in Computer Science*, vol. 3373, pp. 67–83. Springer-Verlag (2004)
72. Courtois, N.: Short signatures, provable security, generic attacks, and computational security of multivariate polynomial schemes such as HFE, Quartz and Sflash. Cryptology ePrint Archive, Report 2004/143 (2004). Available at http://eprint.iacr.org/2004/143
73. Courtois, N.: How fast can be algebraic attacks on block ciphers? In: E. Biham, H. Handschuh, S. Lucks, V. Rijmen (eds.) Symmetric Cryptography, *Dagstuhl Seminar Proceedings*, vol. 07021. Internationales Begegnungs- und Forschungszentrum fuer Informatik (IBFI), Schloss Dagstuhl, Germany (2007)
74. Courtois, N., Bard, G.: Algebraic and slide attacks on Keeloq. Cryptology ePrint Archive, Report 2007/062 (2007). Available at http://eprint.iacr.org/2007/062
75. Courtois, N., Bard, G., Ault, S.: Statistics of random permutations and the cryptanalysis of periodic block ciphers. Cryptology ePrint Archive, Report 2009/186 (2009). Available at http://eprint.iacr.org/2009/186
76. Courtois, N., Bard, G.V.: Algebraic cryptanalysis of the data encryption standard. In: S.D. Galbraith (ed.) IMA International Conference on Cryptography and Coding Theory, *Lecture Notes in Computer Science*, vol. 4887, pp. 152–169. Springer-Verlag (2007). Available at http://eprint.iacr.org/2006/402
77. Courtois, N., Daum, M., Felke, P.: On the security of HFE, HFEv- and Quartz. In: Y. Desmedt (ed.) Public Key Cryptography (PKC'03), *Lecture Notes in Computer Science*, vol. 2567, pp. 337–350. Springer-Verlag (2003)

78. Courtois, N., Goubin, L., Patarin, J.: SFLASHv3, a fast asymmetric signature scheme. Cryptology ePrint Archive, Report 2003/211 (2003). Available at http://eprint.iacr.org/2003/211

79. Courtois, N., Meier, W.: Algebraic attacks on stream ciphers with linear feedback. In: E. Biham (ed.) Advances in Cryptology—Proc. of EUROCRYPT, *Lecture Notes in Computer Science*, vol. 2656, pp. 345–359. Springer-Verlag (2003)

80. Courtois, N., Patarin, J.: About the XL algorithm over $\mathbb{GF}(2)$. In: M. Joye (ed.) Cryptographers' Track, RSA Conference, *Lecture Notes in Computer Science*, vol. 2612, pp. 141–157. Springer-Verlag (2003)

81. Courtois, N., Pieprzyk, J.: Cryptanalysis of block ciphers with overdefined systems of equations. In: Y. Zheng (ed.) Advances in Cryptology—Proc. of ASIACRYPT, *Lecture Notes in Computer Science*, vol. 2501, pp. 267–287. Springer-Verlag (2002). Available at http://eprint.iacr.org/2002/044/

82. Courtois, N., Shamir, A., Patarin, J., Klimov, A.: Efficient algorithms for solving overdefined systems of multivariate polynomial equations. In: B. Preneel (ed.) Advances in Cryptology—Proc. of EUROCRYPT, *Lecture Notes in Computer Science*, vol. 1807, pp. 392–407. Springer-Verlag (2000)

83. Courtois, N.T., Bard, G.V., Bogdanov, A.: Periodic ciphers with small blocks and cryptanalysis of Keeloq. Tatra Mountains Mathematical Publications, Slovak Academy of Sciences **41**, 167–188 (2008)

84. Courtois, N.T., Bard, G.V., Wagner, D.: Algebraic and slide attacks on Keeloq. In: K. Nyberg (ed.) Proc. of Fast Software Encryption (FSE'08), *Lecture Notes in Computer Science*, vol. 5086, pp. 97–115. Springer-Verlag (2008)

85. Courtois, N.T., Nohl, K., O'Neil, S.: Algebraic attacks on the Crypto-1 stream cipher in MiFare Classic and Oyster Cards. Cryptology ePrint Archive, Report 2008/166 (2008). Available at http://eprint.iacr.org/2008/166

86. Cox, D., Little, J., O'Shea, D.: Ideals, Varieties, and Algorithms: An Introduction to Computational Algebraic Geometry and Commutative Algebra, second edn. Undergraduate Texts in Mathematics. Springer Verlag (2006)

87. Creignou, N., Daude, H.: Satisfiability threshold for random XOR-CNF Formulas. Discrete Applied Mathematics (1999)

88. Curtin, M.: Brute Force: Cracking the Data Encryption Standard. Springer-Verlag (2005)

89. Daemen, J., Rijmen, V.: Rijndael. AES Proposal (1999). Available at http://csrc.nist.gov/CryptoToolkit/aes/rijndael/Rijndael-ammended.pdf

90. Danzig, G.: Maximization of a linear function of variables subject to linear inequalities. In: T.J.C. Koopmans (ed.) Activity Analysis of Production and Allocation, pp. 339–347. Wiley (1951)

91. Davis, M., Logemann, G., Loveland, D.: A machine program for theorem proving. Communications of the ACM **5**(7), 394–397 (1962)

92. Davis, M., Putnam, H.: A computing procedure for quantification theory. Journal of the Association of Computing Machinery **7**(3), 201–215 (1960)

93. Davis, T.: Direct Methods for Sparse Linear Systems. Society for Industrial and Applied Mathematics (2006)

94. Dawson, S.: Code hopping decoder using a PIC16C56. Tech. Rep. Technical Report AN642, Microchip Corporation. Available at http://www.keeloq.boom.ru/decryption.pdf

95. Ding, J., Gower, J., Schmidt, D.: Zhuang-Zi: A new algorithm for solving multivariate polynomial equations over a finite field. Tech. rep., University of Cincinnati (2006)

96. Ding, J., Gower, J., Schmidt, D.: Zhuang-Zi: A new algorithm for solving multivariate polynomial equations over a finite field. Cryptology ePrint Archive, Report 2006/038 (2006). Available at http://eprint.iacr.org/2006/038, and presented at the IMA Annual Workshop on Algorithms in Algebraic Geometry

97. Ding, J., Gower, J.E., Schmidt, D.S.: Multivariate Public Key Cryptosystems. Springer Verlag (2006)

98. Dinur, I., Shamir, A.: Cube attacks on tweakable black box polynomials. In: A. Joux (ed.)
     Advances in Cryptology—Proc. of EUROCRYPT, *Lecture Notes in Computer Science*, vol.
     5479, pp. 278–299. Springer-Verlag (2009)
99. Dixon, J.: Asymptotically fast factorization of integers. Mathematics of Computation **36**,
     255–260 (1981)
100. Dominus, M.J.: The universe of discourse, Friday 07 April 2006. Blog. See http://
     blog.plover.com/math/recorde.html
101. Dummit, D., Foote, R.: Abstract Algebra, third edn. Wiley (2003)
102. Een, N., Biere, A.: Effective preprocessing in SAT through variable and clause elimination.
     In: F. Bacchus, T. Walsh (eds.) Theory and Applications of Satisfiability Testing (SAT'05),
     Lecture Notes in Computer Science, vol. 3569, pp. 61–75. Springer-Verlag (2005)
103. Eén, N., Sörensson, N.: An extensible SAT-solver. In: E. Giunchiglia, A. Tacchella (eds.)
     Theory and Applications of Satisfiability Testing (SAT'03), *Lecture Notes in Computer Sci-
     ence*, vol. 2919, pp. 333–336. Springer-Verlag (2003)
104. Eén, N., Sörensson, N.: Minisat — a SAT solver with conflict-clause minimization. In:
     F. Bacchus, T. Walsh (eds.) Proc. Theory and Applications of Satisfiability Testing (SAT'05),
     Lecture Notes in Computer Science, vol. 3569, pp. 61–75. Springer-Verlag (2005)
105. Eibach, T., Pilz, E., Völkel, G.: Attacking Bivium using SAT solvers. In: H.K. Büning,
     X. Zhao (eds.) Theory and Applications of Satisfiability Testing (SAT '08), *Lecture Notes in
     Computer Science*, vol. 4996, pp. 63–76. Springer-Verlag (2008)
106. Eisenbarth, T., Kasper, T., Moradi, A., Paar, C., Salmasizadeh, M., Shalmani, M.T.M.: On the
     power of power analysis in the real world: A complete break of the KeeLoqCode Hopping
     Scheme. In: D. Wagner (ed.) CRYPTO, *Lecture Notes in Computer Science*, vol. 5157, pp.
     203–220. Springer-Verlag (2008)
107. Erickson, J.: Algebraic cryptanalysis of SMS4. Tech. rep., Taylor University (2008)
108. Eschenauer, L., Gligor, V.: A key-management scheme for distributed sensor networks. In:
     V. Atluri (ed.) In Proceedings of the 9th ACM Conference on Computer and Communication
     Security (CCS'02), pp. 41–47. ACM (2002)
109. Faugère, J.C.: A new efficient algorithm for computing Gröbner Bases (F_4). Journal of Pure
     and Applied Algebra **139**, 61–88 (1999)
110. Faugère, J.C.: A new efficient algorithm for computing Gröbner bases without reduction to
     zero (F_5). In: Workshop on Applications of Commutative Algebra. ACM Press, Catania,
     Italy (2002)
111. Faugère, J.C., Joux, A.: Algebraic cryptanalysis of Hidden Field Equation (HFE) cryptosys-
     tems using Gröbner Bases. In: D. Boneh (ed.) Advances in Cryptology—Proc. of CRYPTO,
     Lecture Notes in Computer Science, vol. 2729, pp. 44–60. Springer-Verlag (2003)
112. Fedin, S.S., Kulikov, A.S.: Automated proofs of upper bounds on the running time of splitting
     algorithms [English translation, original is in Russian]. Journal of Mathematical Sciences
     134(5), 2383–2391 (2006)
113. Ferguson, N., Schroeppel, R., Whiting, D.: A simple algebraic representation of Rijndael. In:
     S. Vaudenay, A.M. Youssef (eds.) Proc. Selected Areas in Cryptography (SAC01), *Lecture
     Notes in Computer Science*, vol. 2259, pp. 103–111. Springer-Verlag (2001)
114. Fiorini, C., Martinelli, E., Massacci, F.: How to fake an RSA signature by encoding modular
     root finding as a SAT problem. Discrete Applied Mathematics **130**(2), 101–127 (2003)
115. Fischer, S.: Analysis of lightweight stream ciphers. Ph.D. thesis, École Polytechnique
     Fédéral de Lausanne (2008)
116. Flajolet, P., Sedgewick, R.: Analytic Combinatorics. Cambridge University Press (2008).
     http://algo.inria.fr/flajolet/Publications/book.pdf
117. Fleischmann, P., Michler, G., Roelse, P., Rosenboom, J., Staszewski, R., Wagner, C., Weller,
     M.: Linear algebra over small finite fields on parallel machines. Tech. rep., Universität Essen,
     Fachbereich Mathematik, Essen (1995)
118. Fulkerson, D.R., Gross, O.A.: Incidence matrices and interval graphs. Pacific Journal of
     Mathematics **15**, 835–855 (1965)
119. Gallian, J.A.: Contemporary Abstract Algebra. Heath (1986)

120. Garfinkel, S., Spafford, G.: Practical Unix & Internet Security, 2nd edn. O'Reilly (1996)
121. von zur Gathen, J., Gerhard, J.: Modern Computer Algebra, second edn. Cambridge University Press (2003)
122. Gerver, J.: Factoring large numbers with a quadratic sieve. Mathematics of Computation **41**, 287–294 (1983)
123. Giusti, M.: Some effectivity problems in polynomial ideal theory. In: J. Fitch (ed.) Proc. of EUROSAM 84, *Lecture Notes in Computer Science*, vol. 174, pp. 159–171. Springer-Verlag (1984)
124. Goldstein, R.: Incompleteness: The Proof and Paradox of Kurt Gödel. Great Discoveries. W. W. Norton & Company (2005)
125. Golomb, S.W.: Shift Register Sequences. Agean Park Press (1981)
126. Golub, G., Loan, C.V.: Matrix Computations, third edn. Johns Hopkins University Press (1996)
127. Goubin, L., Courtois, N.: Cryptanalysis of the TTM Cryptosystem. In: T. Okamoto (ed.) Advances in Cryptology—Proc. of ASIACRYPT, *Lecture Notes in Computer Science*, vol. 1976, pp. 44–57. Springer-Verlag (2000)
128. Granboulan, L., Pornin, T.: Perfect block ciphers with small blocks. In: A. Biryukov (ed.) Proc. of Fast Software Encryption (FSE'07), *Lecture Notes in Computer Science*, vol. 4593, pp. 452–463. Springer-Verlag (2007)
129. Gray, F.: Pulse code communication (1953). USA Patent 2,632,058
130. Greenberg, H.J.: Klee-Minty Polytope shows exponential time complexity of simplex method. Tech. rep., University of Colorado at Denver (1997). Available at http://glossary.computing.society.informs.org/notes/Klee-Minty.pdf
131. Greub, W.: Linear Algebra, fourth edn. Graduate Texts in Mathematics. Springer-Verlag (1981)
132. Hårvard, Semaev, I.: New technique for solving sparse equation systems. Cryptology ePrint Archive, Report 2006/475 (2006). Available at http://eprint.iacr.org/2006/475
133. Havas, G., Wagner, C.: Some performance studies in exact linear algebra, english summary. In: G. Cooperman, E. Jessen, G.O. Michler (eds.) Worksop on Wide Area Networks and High Performance Computing, *Lecture Notes in Control and Information Science*, vol. 249, pp. 161–170. Springer-Verlag, Essen (1998)
134. Hawkes, P., Rose, G.: Rewriting variables: The complexity of fast algebraic attacks on stream ciphers. In: Advances in Cryptology—Proc. of CRYPTO, *Lecture Notes in Computer Science*, vol. 3152, pp. 390–406. Springer-Verlag (2004)
135. Hietalahti, M., Massacci, F., Niemelä, I.: DES: A challenge problem for nonmonotonic reasoning systems. In: Proc. 8th International Workshop on Non-Monotonic Reasoning (2000)
136. Higham, N.: Accuracy and Stability of Numerical Algorithms, second edn. Society for Industrial and Applied Mathematics (2002)
137. Hirsch, E.A.: New worst-case upper bounds for SAT. Journal of Automated Reasoning **24**(4), 397–420 (2000)
138. Hofmeister, T., Schöning, U., Schuler, R., Watanabe, O.: A probabilistic 3-SAT algorithm further improved. In: H. Alt, A. Ferreira (eds.) 19th Annual Symposium on Theoretical Aspects of Computer Sciences (STACS'02), *Lecture Notes in Computer Science*, vol. 2285, pp. 192–202. Springer-Verlag (2002)
139. Hojsík, M., Rudolf, B.: Differential fault analysis of Trivium. In: K. Nyberg (ed.) Fast Software Encryption (FSE'08), *Lecture Notes in Computer Science*, vol. 5086, pp. 158–172. Springer-Verlag (2008)
140. Hojsík, M., Rudolf, B.: Floating fault analysis of Trivium. In: D.R. Chowdhury, V. Rijmen, A. Das (eds.) Progress in Cryptology (INDOCRYPT'08), *Lecture Notes in Computer Science*, vol. 5365, pp. 239–250. Springer-Verlag (2008)
141. Ibara, O.H., Moran, S., Hui, R.: A generalization of the fast LUP matrix decomposition algorithm and applications. Journal of Algorithms **1**(3), 45–56 (1982)

142. Indesteege, S., Keller, N., Dunkelman, O., Biham, E., Preneel, B.: A practical attack on Keeloq. In: N.P. Smart (ed.) Advances in Cryptology—Proc. of EUROCRYPT, *Lecture Notes in Computer Science*, vol. 4965, pp. 1–18. Springer-Verlag (2008)

143. Jovanovic, D., Janicic, P.: Logical analysis of hash functions. In: B. Gramlich (ed.) Proceedings of the Frontiers of Combining Systems, *Lecture Notes in Artificial Intelligence*, vol. 3717, pp. 200–215. Springer-verlag (2005)

144. Joyner, D., Kreminski, R., Turisco, J.: Applied Abstract Algebra. Free Internet Textbook (2002). Available at http://www.usna.edu/~wdj/book/book.html

145. Kahn, D.: The Codebreakers, The Comprehensive History of Secret Communication from Ancient Times to the Internet, 2nd edn. Scribner (1996). First published in 1967

146. Kaltofen, E.: Analysis of Coppersmith's Block Wiedemann algorithm for the parallel solution of sparse linear systems. Mathematics of Computation **64**(210), 777–806 (1995)

147. Kaltofen, E., Saunders, B.D.: On Wiedemann's method of solving sparse linear systems. In: H.F. Mattson, T. Mora, T.R.N. Rao (eds.) Proceedings of the 9th International Symposium on Applied Algebra, Algebraic Algorithms and Error-Correcting Codes, *Lecture Notes in Computer Science*, vol. 539, pp. 29–38. Springer-Verlag (1991)

148. Karp, R.: Reducibility among combinatorial problems. In: Proc. of Symposium on Complexity of Computer Computations, pp. 85–103. IBM Thomas J. Watson Res. Center, Plenum, Yorktown Heights, New York (1972)

149. Knuth, D.E.: The Art of Computer Programming, Volume 4A: Enumeration and Backtracking. Addison-Wesley (2004)

150. Krantz, S.G.: A Primer of Mathematical Writing: Being a Disquisition on Having Your Ideas Recorded, Typeset, Published, Read & Appreciated. American Mathematical Society (1997)

151. Krantz, S.G.: How to Teach Mathematics, second edn. American Mathematical Society (1999)

152. Krantz, S.G.: A Mathematician's Survival Guide: Graduate School and Early Career Development. American Mathematical Society (2003)

153. Krantz, S.G.: Mathematical Publishing: A Guidebook. American Mathematical Society (2005)

154. Krishnamurthy, E.V.: Error-Free Polynomial Matrix Computations. Springer-Verlag (1985)

155. Kutz, M.: The complexity of boolean matrix root computation. In: T. Warnow, B. Zhu (eds.) Proc. of Computing and Combinatorics, *Lecture Notes in Computer Science*, vol. 2697, pp. 212–221. Springer-Verlag (2003)

156. LaMacchia, B., Odlyzko, A.: Solving large sparse linear systems over finite fields. In: A. Menezes, S.A. Vanstone (eds.) Advances in Cryptology—Proc. of CRYPTO, *Lecture Notes in Computer Science*, vol. 537, pp. 109–133. Springer-Verlag (1990)

157. Lambert, R.J.: Computational aspects of discrete logarithms. Ph.D. thesis, University of Waterloo (1996)

158. Lazard, D.: Gröbner-bases, Gaussian Elimination and resolution of systems of algebraic equations. In: J.A. van Hulzen (ed.) EUROCAL 1983, *Lecture Notes in Computer Science*, vol. 162, pp. 146–156. Springer-Verlag (1983)

159. Lenstra, A.K., Jr., H.W.L.: The Development of the Number Field Sieve. Lecture Notes in Mathematics. Springer-Verlag (1993)

160. Lewis, R.: The Dixon Resultant following Kapur-Saxena-Yung. Tech. rep., Fordham University (2002). Available at http://fordham.academia.edu/RobertLewis/Papers

161. Lewis, R.: Heuristics to accelerate the dixon resultant. Mathematics and Computers in Simulation **77**(4), 400–407 (2008)

162. Lewis, R., Wester, M.: Comparison of polynomial-oriented computer algebra systems. SIGSAM Bulletin **33**(4), 5–13 (1999)

163. Lewis, R.H., Bridgett, S.: Conic tangency equations and Apollonius problems in biochemistry and pharmacology. Mathematics and Computers in Simulation **61**(2), 101–114 (2003)

164. Lidl, R., Niederreiter, H.: Introduction to Finite Fields and Their Applications, revised edn. Cambridge University Press (1994)

165. Lidl, R., Niederreiter, H.: Finite Fields. Encyclopedia of Mathematics and its Applications. Cambridge University Press (2008)
166. van Lint, J.H., Wilson, R.M.: A Course in Combinatorics, second edn. Cambridge University Press (2001)
167. Liu, F.H., Lu, C.J., Yang, B.Y.: Secure PRNGs from specialized polynomial maps over any F_q. In: J. Buchmann, J. Ding (eds.) Post-Quantum Cryptography (PQC'08), *Lecture Notes in Computer Science*, vol. 5299, pp. 181–202. Springer-Verlag (2008)
168. Lovász, L.: Normal hypergraphs and the perfect graph conjecture. Discrete Mathematics **2**, 253–267 (1972)
169. Markowitz, H.M.: The elimination form of the inverse and its application to linear programming. Management Science **3**(3), 255–269 (1957)
170. Marques-Silva, J.P., Sakallah, K.A.: GRASP: a search algorithm for propositional satisfiability. IEEE Transactions on Computers **45**(5), 506–521 (1999)
171. Marraro, L., Massacci, F.: Towards the formal verification of ciphers: Logical cryptanalysis of DES. In: Proc. Third LICS Workshop on Formal Methods and Security Protocols, Federated Logic Conferences (FLOC-99) (1999)
172. Massacci, F.: Using Walk-SAT and Rel-SAT for cryptographic key search. In: T. Dean (ed.) Proc. 16th International Joint Conference on Artificial Intelligence, pp. 290–295. Morgan Kaufmann Publishing (1999)
173. Massacci, F., Marraro, L.: Logical cryptanalysis as a SAT-problem: Encoding and analysis of the US data encryption standard. Journal of Automated Reasoning **24** (2000)
174. Personal communications with Mate Soos
175. Maximov, A., Biryukov, A.: Two trivial attacks on Trivium. In: C.M. Adams, A. Miri, M.J. Wiener (eds.) Proc. Selected Areas in Cryptography (SAC07), *Lecture Notes in Computer Science*, vol. 4876, pp. 36–55. Springer-Verlag (2007). Available from http://eprint.iacr.org/2007/021
176. McDonald, C., Charnes, C., Pieprzyk, J.: An algebraic analysis of Trivium ciphers based on the boolean satisfiability problem. Cryptology ePrint Archive, Report 2007/129 (2007). Available at http://eprint.iacr.org/2007/129, and presented at the International Conference on Boolean Functions: Cryptography and Applications (BFCA'2008)
177. Mironov, I., Zhang, L.: Applications of SAT solvers to cryptanalysis of hash functions. In: A. Biere, C.P. Gomes (eds.) Proc. Theory and Applications of Satisfiability Testing (SAT'06), *Lecture Notes in Computer Science*, vol. 4121, pp. 102–115. Springer-Verlag (2006). Also available as IACR E-print 2006/254
178. Montgomery, P.L.: A Block Lanczos algorithm for finding dependencies over $\mathbb{GF}(2)$. In: L.C. Guillou, J.J. Quisquater (eds.) Advances in Cryptology—Proc. of EUROCRYPT, *Lecture Notes in Computer Science*, vol. 921, pp. 106–120. Springer-Verlag (1995)
179. Moore, E.: On the reciprocal of the general algebraic matrix. Bulletin of the American Mathematical Society **26** (1920)
180. Morgenstern, J.: How to compute fast a function and all its derivatives: a variation on the theorem of Baur-Strassen. SIGACT News **16**(4), 60–62 (1985). DOI http://doi.acm.org/10.1145/382242.382836
181. Morrison, M.A., Brillhart, J.: A method of factorization and the factorization of F_7. Mathematics of Computation **29**, 183–205 (1975)
182. Moskewicz, M.W., Madigan, C.F., Zhao, Y., Zhang, L., Malik, S.: Chaff: Engineering an efficient SAT Solver. In: Proc. of 28th Design Automation Conference (DAC'01), pp. 530–535. ACM (2001)
183. Odlyzko, A.M.: Discrete logarithms in finite fields and their cryptographic significance. In: N. Cot (ed.) Advances in Cryptology—Proc. of EUROCRYPT, *Lecture Notes in Computer Science*, vol. 209, pp. 224–316. Springer-Verlag (1984)
184. Paar, C.: Remote keyless entry system for cars and buildings is hacked; rub security experts discover major vulnerability; access from a distance of 300 feet without traces. Tech. rep., University of Bochum, Germany (2008)
185. Pan, V.: How to Multiply Matrices Faster. No. 179 in Lecture Notes in Computer Science. Springer-Verlag (1984)

186. Papadimitriou, C.H.: On selecting a satisfying truth assignment. In: Proceedings of the Conference on the Foundations of Computer Science (FOCS'91), pp. 163–169. IEEE (1991)

187. Patarin, J.: Hidden Field Equations (HFE) and Isomorphisms of Polynomials (IP): two new families of asymmetric algorithms. In: N. Koblitz (ed.) Advances in Cryptology—Proc. of EUROCRYPT, *Lecture Notes in Computer Science*, vol. 1070, pp. 33–48. Springer-Verlag (1996)

188. Paturi, R., Pudlák, P., Saks, M.E., Zane, F.: An improved exponential-time algorithm for k-SAT. The Journal of the Association of Computing Machinery **52**(3), 337–364 (2005)

189. Penrose, R.: A generalized inverse for matrices. Proc. of the Cambridge Phil. Soc. **51** (1955)

190. Pernet, C.: Implementation of Winograds algorithm over finite fields using ATLAS Level 3 BLAS. Tech. rep., ID-Laboratory (2001)

191. Pilz, E.: Boolsche gleichungssysteme, SAT Solver und stromchiffren. Master's thesis, Universität Ulm, Institut für Theoretische Informatik (2008)

192. Pomerance, C.: A tale of two sieves. Notices of the American Mathematical Society **43**(12), 1473–1485 (1996)

193. Pomerance, C., Smith, J.W.: Reduction of huge, sparse matrices over finite fields via created catastrophes. Experimental Mathematics **1**(2), 89–94 (1992)

194. Preneel, B., Biryukov, A., Cannière, C.D., Ors, S.B., Oswald, E., van Rompay, B., Granboulan, L., Dottax, E., Martinet, G., Murphy, S., Dent, A., Shipsey, R., Swart, C., White, J., Dichtl, M., Pyka, S., Schafheutle, M., Serf, P., Biham, E., Barkan, E., Braziler, Y., Dunkelman, O., Furman, V., Kenigsberg, D., Stolin, J., Quisquater, J.J., Ciet, M., Sica, F.: NESSIE: Final report of european project number IST-1999-12324, named new european schemes for signatures, integrity, and encryption. Final Technical Report: Available at: https://www.cosic.esat.kuleuven.be/nessie/Bookv015.pdf (2004)

195. Priemuth-Schmid, D., Biryukov, A.: Slid pairs in Salsa20 and Trivium. In: D.R. Chowdhury, V. Rijmen, A. Das (eds.) Progress in Cryptology—INDOCRYPT'08, *Lecture Notes in Computer Science*, vol. 5365, pp. 1–14. Springer-Verlag (2008)

196. Raddum, H.: Cryptanalytic results on Trivium. eStream Report: 2006/039" (2006). Available at www.ecrypt.eu.org/stream/papersdir/2006/039.ps

197. Raddum, H., Semaev, I.: Solving multiple right hand sides linear equations. Des. Codes Cryptography **49**(1-3), 147–160 (2008)

198. Raz, R., Shpilka, A.: Lower bounds for matrix product, in bounded depth circuits with arbitrary gates. In: STOC, pp. 409–418 (2001)

199. Raz, R., Shpilka, A.: Lower bounds for matrix product in bounded depth circuits with arbitrary gates. SIAM J. Comput. **32**(2), 488–513 (2003)

200. Riedel, M.R.: Random permutation statistics. Paper available on the Internet (2006). http://www.geocities.com/markoriedelde/papers/randperms.pdf

201. Rijmen, V., Oswald, E.: Representations and Rijndael descriptions. In: H. Dobbertin, V. Rijmen, A. Sowa (eds.) 4th International Conference on the Advanced Encryption Standard (AES'04), *Lecture Notes in Computer Science*, vol. 3373, pp. 148–158. Springer-Verlag (2004)

202. Ryser, H.J.: Combinatorial properties of matrices of zeros and ones. Canadian Journal of Mathematics **9**, 371–377 (1957)

203. Saad, Y.: Iterative Methods for Sparse Linear Systems. PWS Publishing Company (1996)

204. Santoro, N.: Extending the Four-Russians bound to general matrix multiplication. Information Processing Letters (1979)

205. Santoro, N., Urrutia, J.: An improved algorithm for boolean matrix multiplication. Computing **36** (1986)

206. Schönhage, A.: Partial and total matrix multiplication. Journal of Computing **10**(3) (1981)

207. Selman, B., Kautz, H., Cohen, B.: Local search strategies for satisfiability testing. In: D.S. Johnson, M.A. Trick (eds.) Cliques, Coloring, and Satisfiability: Second DIMACS Implementation Challenge (DIMACS'93), vol. 26. AMS (1996)

208. Selman, B., Levesque, H.J., Mitchell, D.G.: A new method for solving hard satisfiability problems. In: 10th National Conference on Artificial Intelligence (AAAI'92), pp. 440–446 (1992)

209. Semaev, I.: On solving sparse algebraic equations over finite fields II. Cryptology ePrint Archive, Report 2007/280 (2007). Available at http://eprint.iacr.org/2007/280, and presented at the Eleventh International Workshop on Algebraic and Combinatorial Coding Theory (ACCT'2008)

210. Semaev, I.: On solving sparse algebraic equations over finite fields. Designs Codes and Cryptography **49**(1-3), 47–60 (2008)

211. Stamp, M., Low, R.M.: Applied Cryptanalysis: Breaking Ciphers in the Real World. Wiley-IEEE Press (2007)

212. Strassen, V.: Gaussian Elimination is not optimal. Numerische Mathematik **13**(3) (1969)

213. Strassen, V.: Relative bilinear complexity and matrix multiplication. J. Reine Angew. Math. **375–376** (1987). This article is so long that it is split among two volumes.

214. Swenson, C.: Modern Cryptanalysis: Techniques for Advanced Code Breaking. Wiley (2008)

215. Tang, X., Feng, Y.: A new efficient algorithm for solving systems of multivariate polynomial equations. Cryptology ePrint Archive, Report 2005/312 (2005). Available at http://eprint.iacr.org/2005/312

216. Trappe, W., Washington, L.C.: Introduction to Cryptography with Coding Theory, second edn. Pearson Prentice-Hall (2006)

217. Trefethen, L., III, D.B.: Numerical Linear Algebra. Society for Industrial and Applied Mathematics (1997)

218. Vielhaber, M.: Breaking One.Fivium by AIDA an algebraic IV differential attack. Cryptology ePrint Archive, Report 2007/413 (2007). Available at http://eprint.iacr.org/2007/413

219. Villard, G.: Further analysis of Coppersmith's Block Wiedemann algorithm using matrix polynomials. In: Proceedings of the 1997 international symposium on Symbolic and algebraic computation (ISSAC'97), pp. 32–39 (1997)

220. Warner, M., Benson, R.L.: Venona and beyond: Thoughts on work undone. Intelligence and National Security **12**(3), 1–13 (1997)

221. Warren, H.S.: Hacker's Delight. Addison-Wesley Longman Publishing Co., Inc., Boston, MA, USA (2002)

222. Watkins, D.: Fundamentals of Matrix Computations, second edn. Wiley (2002)

223. Weisstein, E.W.: Apéry's constant. From MathWorld—A Wolfram Web Resource. Available at http://mathworld.wolfram.com/AperysConstant.html

224. Weisstein, E.W.: Apollonius problem. From MathWorld—A Wolfram Web Resource. Available at http://mathworld.wolfram.com/PrimeNumberTheorem.html

225. Weisstein, E.W.: Berlekamp-Massey algorithm. From MathWorld—A Wolfram Web Resource. Available at http://mathworld.wolfram.com/PrimeNumberTheorem.html

226. Weisstein, E.W.: Prime number theorem. From MathWorld—A Wolfram Web Resource. Available at http://mathworld.wolfram.com/PrimeNumberTheorem.html

227. Weisstein, E.W.: Relatively prime. From MathWorld—A Wolfram Web Resource. Available at http://mathworld.wolfram.com/RelativelyPrime.html

228. Weisstein, E.W.: Sylvester matrix. From MathWorld—A Wolfram Web Resource. Available at http://mathworld.wolfram.com/SylvesterMatrix.html

229. Whaley, R.C., Petitet, A., Dongarra, J.: Automated empirical optimization of software and the ATLAS Project. Parallel Computing **27**(1–2), 3–35 (2001)

230. Wiedemann, D.H.: Solving sparse linear equations over finite fields. IEEE Transactions on Information Theory **32**(1), 54–62 (1986)

231. Wong, K., Bard, G., Lewis, R.: Partitioning multivariate polynomial equations via vertex cuts for algebraic cryptanalysis and other applications. Submitted to a journal (2008). Available at http://www.math.umd.edu/~bardg/wong_bard_lewis.pdf

232. Yang, B.Y., Chen, O.C.H., Bernstein, D.J., Chen, J.M.: Analysis of QUAD. In: A. Biryukov (ed.) Proc. of Fast Software Encryption (FSE'07), *Lecture Notes in Computer Science*, vol. 4593, pp. 290–308. Springer-Verlag (2007)

233. Zeitlhofer, T., Wess, B.: List-coloring of interval graphs with application to register assignment for heterogeneous register-set architectures. Signal Processing **83**(7), 1411–1425 (2003)

234. Zetter, K.: Researchers crack Keeloq code for car keys. Wired Magazine (2007)

Index